Therapeutic Oligonucleotides

RSC Biomolecular Sciences

Editorial Board:

Professor Stephen Neidle (Chairman), *The School of Pharmacy, University of London, UK*
Dr Simon F. Campbell CBE, FRS
Dr Marius Clore, *National Institutes of Health, USA*
Professor David M. J. Lilley FRS, *University of Dundee, UK*

This Series is devoted to coverage of the interface between the chemical and biological sciences, especially structural biology, chemical biology, bio- and chemo-informatics, drug discovery and development, chemical enzymology and biophysical chemistry. Ideal as reference and state-of-the-art guides at the graduate and post-graduate level.

Titles in the Series:

Biophysical and Structural Aspects of Bioenergetics
Edited by Mårten Wikström, *University of Helsinki, Finland*
Computational and Structural Approaches to Drug Discovery: Ligand-Protein Interactions
Edited by Robert M. Stroud and Janet Finer-Moore, *University of California in San Francisco, San Francisco, CA, USA.*
Exploiting Chemical Diversity for Drug Discovery
Edited by Paul A. Bartlett, *Department of Chemistry, University of California, Berkeley* and Michael Entzeroth, *S*Bio Pte Ltd, Singapore*
Metabolomics, Metabonomics and Metabolite Profiling
Edited by William J. Griffiths, *University of London, The School of Pharmacy, University of London, London, UK*
Protein–Carbohydrate Interactions in Infectious Disease
Edited by Carole A. Bewley, *National Institutes of Health, Bethesda, Maryland, USA*
Protein-Nucleic Acid Interactions: Structural Biology
Edited by Phoebe A. Rice, *Department of Biochemistry & Molecular Biology, The University of Chicago, Chicago IL, USA* and Carl C. Correll, *Dept of Biochemistry and Molecular Biology, Rosalind Franklin University, North Chicago, IL, USA*
Quadruplex Nucleic Acids
Edited by Stephen Neidle, *The School of Pharmacy, University of London, London, UK* and Shankar Balasubramanian, *Department of Chemistry, University of Cambridge, Cambridge, UK*
Ribozymes and RNA Catalysis
Edited by David M. J. Lilley FRS, *University of Dundee, Dundee, UK* and Fritz Eckstein, *Max-Planck-Institut for Experimental Medicine, Goettingen, Germany*
Sequence-specific DNA Binding Agents
Edited by Michael Waring, *Department of Pharmacology, University of Cambridge, Cambridge, UK*
Structural Biology of Membrane Proteins
Edited by Reinhard Grisshammer and Susan K. Buchanan, *Laboratory of Molecular Biology, National Institutes of Health, Bethesda, Maryland, USA*
Structure-based Drug Discovery: An Overview
Edited by Roderick E. Hubbard, *University of York, UK and Vernalis (R&D) Ltd, Cambridge, UK*
Therapeutic Oligonucleotides
Edited by Jens Kurreck, *Institute of Industrial Genetics, University of Stuttgart, Stuttgart, Germany*

Visit our website on www.rsc.org/biomolecularsciences

For further information please contact:
Sales and Customer Care, Royal Society of Chemistry, Thomas Graham House, Science Park, Milton Road, Cambridge, CB4 0WF, UK
Telephone: +44 (0)1223 432360, Fax: +44 (0)1223 426017, Email: sales@rsc.org

Therapeutic Oligonucleotides

Edited by

Jens Kurreck
Institute of Industrial Genetics, University of Stuttgart, Stuttgart, Germany

RSCPublishing

The cover picture shows the structural model for a 19-nucleotide siRNA guide strand and its target RNA bound to the Piwi protein from *Archaeoglobus fulgidus* (J. S. Parker, S. M. Roe and D. Barford, *Nature,* 2005, **434**, 663; the coordinates have been taken from the Protein Data Bank with accession number 2bgg).

ISBN: 978-0-85404-116-9

A catalogue record for this book is available from the British Library

Published by The Royal Society of Chemistry,
Thomas Graham House, Science Park, Milton Road,
Cambridge CB4 0WF, UK

Registered Charity Number 207890

For further information see our web site at www.rsc.org

Preface

Oligonucleotides are a class of biomacromolecule with great potential for research and therapeutic applications. By definition, oligonucleotides consist of approximately ten to 100 DNA or RNA monomers or chemically modified analogs thereof. Although being rather homogeneous in composition, their mechanisms of action are extremely diverse. It is the aim of this volume to illuminate these different aspects of oligonucleotides. The story of their use is a rollercoaster of enthusiasm and depression. Recent progress made in clinical testing, as well as the hope that the most recently discovered RNA interference technology will help to overcome efficacy problems of the previous approaches, are the basis for this collection of contributions from leading experts in the different fields of application for oligonucleotides.

Three decades ago, in 1978, Paul Zamecnik and Mary Stephenson initiated the era of antisense research when they reported on the inhibition of virus replication in cell culture by a DNA oligonucleotide complementary to the target RNA. This finding launched the expectation that oligonucleotides might be employed to treat any disease that results from deleterious gene expression. We now know that this view was too optimistic, since only one antisense drug, Vitravene (to treat cytomegalovirus-induced retinitis) has been approved for the market to date. In Chapter 2 of this volume, Cy Stein and colleagues give a critical review about another advanced candidate, Genasense, which is an antisense oligonucleotide targeted against Bcl-2 and which has been tested in Phase III clinical trials to treat patients with cancer.

Most of the first-generation antisense oligonucleotides consist of phosphorothioate linkages. It soon became clear, however, that modified nucleotides with improved properties are desirable for many types of application of oligonucleotides, as outlined in Chapter 1. Several examples of antisense oligonucleotides consisting of modified building blocks are discussed in the first part of the book. Hong and Jon Moulton describe in Chapter 3 the use of phosphorodiamidate morpholino oligomers (PMOs) that have already been tested in clinical trials for different indications. Michael Gait and colleagues introduce

RSC Biomolecular Sciences
Therapeutic Oligonucleotides
Edited by Jens Kurreck
© Royal Society of Chemistry 2008

peptide nucleic acids (PNAs) in Chapter 4. Both PMOs and PNAs act as steric blocks of RNA cellular processes and, in each case, they can be coupled to peptides to improve their cellular uptake. In recent years, locked nucleic acids (LNAs) have been considered as very promising analogs. Troels Koch *et al.* describe in Chapter 5 their properties as well as the first results from clinical applications. This chapter also addresses another point with major relevance. For clinical applications, large amounts of oligonucleotides must be produced under current good manufacturing practices (cGMP) conditions at reasonable costs. Another recently developed application of LNAs that is dealt with by Koch and colleagues is the inhibition of endogenously expressed microRNAs.

Single-stranded oligodeoxynucleotides, however, do not only act as inhibitors of gene expression. CpG sequence motifs activate the innate immune response in mammals. This unwanted side effect of antisense oligonucleotides, which had been apparent for many years, is now exploited to deliberately stimulate the human immune system against viral infections and cancer, as outlined by Eugen Uhlmann in Chapter 6.

Another breakthrough in the field of oligonucleotide research was the finding that certain RNA molecules possess the ability to catalyse hydrolysis of phosphodiester bonds. For this unexpected discovery of *ribo*nucleotides with *enzym*atic activity, now known as ribozymes, Thomas Cech and Sydney Altman were awarded the Nobel Prize in Chemistry for 1989. In recent years, a number of new types of ribozymes were found in nature or identified by *in vitro* selection approaches. Several ribozymes have been tested in clinical trials to treat viral infections and cancer. The outcomes of these approaches, however, did not meet expectations. Thus, ribozymes can still be considered to be a fascinating area of research, but their relevance for therapeutic applications has decreased somewhat. One current approach is to combine a ribozyme with other types of RNA-based drugs to treat patients infected with human immunodeficiency virus-1 (HIV-1) and is described by John Rossi in Chapter 14 (see also below).

While antisense oligonucleotides and ribozymes usually act as single-stranded molecules, decoy oligonucleotides represent a class of double-stranded DNA molecules that can be employed to modulate gene expression. As described by Marcus Hecker and colleagues in Chapter 7, decoy oligonucleotides mimic promoter-binding regions to bind specifically and inactivate transcription factors. Several decoy oligonucleotides have been tested in clinical trials primarily for the treatment of inflammatory diseases.

In 1990, the groups of Larry Gold and Jack Szostak simultaneously came up with the concept of *in vitro* selection of oligonucleotides with a high affinity to a target molecule. This class of oligonucleotides, referred to as aptamers, can be obtained by an iterative process of selection and amplification called SELEX (systematic evolution of ligands by exponential enrichment). It was a great breakthrough for the field, when the aptamer Macugen was approved for the treatment of age-related macular degeneration in 2004. Nigel Courtenay-Luck and Donald Miller describe the clinical development of an anticancer aptamer targeting nucleolin (Chapter 8). To obtain highly stable molecules, a method was developed that results in aptamers in the unnatural L-form of nucleotides.

This sophisticated technology is described by Sven Klussmann and colleagues in Chapter 9.

In the late 1990s double-stranded RNA (dsRNA) molecules were found to efficiently inhibit expression of homologue genes in *Caenorhabditis elegans*, a phenomenon nowadays known as RNA interference (RNAi). For this discovery, Craig Mello and Andrew Fire were awarded the Nobel Prize in Physiology or Medicine for 2006. Interest in oligonucleotide-based approaches dramatically increased when Thomas Tuschl's group was able to specifically silence genes in mammalian cells with short dsRNA molecules, dubbed small or short interfering RNAs (siRNAs). This new technology not only revolutionized life-science research, but also opened the road for new therapeutic strategies.

Since siRNAs (as well as antisense oligonucleotides) act inside cells, a major hurdle for their successful application remains delivery of the highly charged molecules into cells. Two chapters therefore address this problem. Christian Reinsch *et al.* give a general introduction into strategies for systemic delivery of antisense oligonucleotides and siRNAs in Chapter 10. In the following chapter, Ian MacLachlan focuses on one of the most advanced systems for the efficient delivery of siRNAs, stable nucleic acid lipid particles (SNALPs). An alternative to the use of chemically pre-synthesized siRNA is the endogenous expression of short hairpin RNAs (shRNAs). This methodology, which is described in detail in Chapter 12, allows long-term inhibition of target gene expression, regulation of the silencing, and delivery by viral vectors.

In 2004, the first clinical trials based on RNAi were initiated. Two companies are developing siRNAs targeting the vascular endothelial growth factor (VEGF) and its receptor, respectively, to treat age-related macular degeneration, a major ocular disease. In another trial, the safety of siRNAs against respiratory syncytial virus is being evaluated. While these approaches involve chemically synthesized siRNAs that are delivered locally, the last two chapters of this volume deal with vector-based approaches to deliver shRNAs against HIV-1. In Chapter 13, Olivier ter Brake and Ben Berkhout describe their strategy to use vectors, which express multiple shRNAs against different sites in the HIV genome simultaneously, to prevent the development of escape mutants. John Rossi and colleagues (Chapter 14) follow a slightly different strategy. They combine three different oligonucleotide-based principles to cope with the problem of viral escape. In their approach, a lentiviral vector is employed to express simultaneously an shRNA targeting the viral *rev* and *tat* mRNAs, a nucleolar-localizing transactivation response element (TAR) RNA decoy and a ribozyme to inhibit the expression of the CCR5 receptor in the host cell. As outlined in this chapter, a clinical trial to use the triple vector for *ex vivo* delivery to haematopoietic progenitor cells of HIV-1-infected patients has just commenced.

I thank all authors for their contributions to this book. Furthermore, I am grateful to Annie Jacob, Janet Freshwater and Katrina Harding of the Royal Society of Chemistry for their continuous support during the production process of this book and I thank Michael Gait and David Lilley for their advice. I am very thankful to Diana Rothe and Denise Werk for their help with the cover figure and all members of my laboratory for their contributions. I am furthermore grateful to Volker Erdmann for his support and scientific advice. Finally, I dedicate this book to my father, Harry Kurreck, for everything he did for me and for passing on his enthusiasm for science.

Jens Kurreck
Berlin/Stuttgart

Contents

RSC Biomolecular Sciences
Therapeutic Oligonucleotides
Edited by Jens Kurreck
© Royal Society of Chemistry 2008

Chapter 3   Antisense Morpholino Oligomers and Their Peptide Conjugates
Hong M. Moulton and Jon D. Moulton

Chapter 6   Immune Stimulatory Oligonucleotides
Eugen Uhlmann

Chapter 9   Spiegelmer NOX-E36 for Renal Diseases
Dirk Eulberg, Werner Purschke, Hans-Joachim Anders,
Norma Selve and Sven Klussmann

Chapter 12   Vector-Mediated and Viral Delivery of Short Hairpin RNAs
Henry Fechner and Jens Kurreck

**Chapter 13   Development of an RNAi-Based Gene Therapy
               against HIV-1**
Olivier ter Brake and Ben Berkhout

Chapter 14   RNA Based Therapies for Treatment of HIV Infection
Lisa Scherer, Marc S. Weinberg and John J. Rossi

CHAPTER 1

The Role of Backbone Modifications in Oligonucleotide-Based Strategies

JENS KURRECK

Institute for Chemistry and Biochemistry, Free University Berlin, Thielallee 63, 14195 Berlin, Germany, and Institute of Industrial Genetics, University of Stuttgart, Allmandring 31, 70569 Stuttgart, Germany

1.1 Introduction

Inside cells, long deoxyribonucleic acid (DNA) or ribonucleic acid (RNA) molecules are enzymatically generated from monomeric nucleotides by DNA and RNA polymerases, respectively. The development of a method for solid-phase synthesis in the 1970s allowed the artificial generation of oligonucleotides (ONs) of up to ∼100 nucleotides in length. Moreover, this technology enabled the incorporation of modified building blocks into the growing chain composed of nucleotides. The chemical synthesis and (partial) modification of ONs opened the road for new research applications and novel therapeutic strategies.

Various classes of ONs have been developed in the meantime: antisense oligonucleotides (AS ONs), ribozymes and small interfering RNAs (siRNAs) specifically inhibit gene expression by Watson–Crick base pairing to a complementary messenger RNA (mRNA), but in contrast, decoy ONs and aptamers bind to their target by structural recognition. ON-based applications have been used widely for research purposes and some approaches have proceeded to the status of clinical investigations, which will be the focus of this review. Approximately 30 clinical trials of various phases with AS ONs are currently

RSC Biomolecular Sciences
Therapeutic Oligonucleotides
Edited by Jens Kurreck
© Royal Society of Chemistry 2008

underway[1] and an AS drug (Vitravene) to treat cytomegalovirus-induced retinitis was the first ON ever to be approved by the US Food and Drug Administration (FDA). Several chemically pre-synthesized ribozymes that target mRNAs of oncogenes or viral RNAs have been tested in early clinical phases. In 2004, only three-and-a-half years after the first demonstration that siRNAs can be used to specifically silence a target gene in mammalian cells, the first clinical trials based on RNA interference (RNAi) have been initiated. In the same year, the approval of Macugen, an aptamer to treat age-related macular degeneration, was another breakthrough in the field of ON therapeutics.

With only a few exceptions, the above-mentioned ONs are composed of modified building blocks. Biological fluids, like blood serum or intracellular liquids, contain highly active nucleases to destroy deleterious nucleic acids. As can be seen in Figure 1.1 (upper row), an unmodified DNA ON is completely degraded within only a few hours in a solution that contains 10% fetal calf serum. ONs composed of RNA are even more susceptible to nucleolytic degradation. The usability of unmodified ONs in animals or human patients is thus limited, since the ONs will be mainly degraded before they even reach their destination.

To overcome this problem, modified nucleotides have been developed that possess higher resistance against enzymatic degradation, since they are not recognized as substrates by nucleases. Early attempts in this direction focussed on the phosphodiester linkage that connects two nucleotides. Since then, the $2'$-position of the ribose has been used widely as a site for the introduction of functional groups that enhance the stability of ONs. Figure 1.1 (rows 2 and 3) shows that the incorporation of modified building blocks into an ON increases

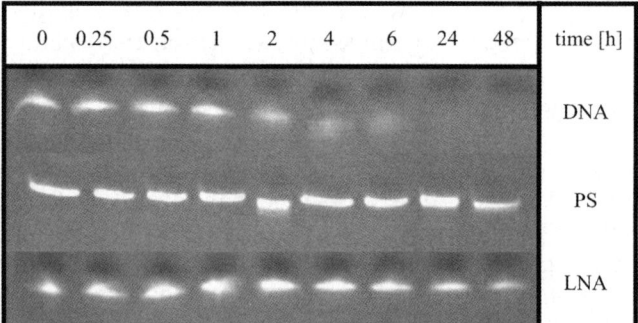

Figure 1.1   Stability of oligonucleotides in cell culture medium containing 10% fetal calf serum. 19-mer ONs (CCTATTGTACTTTAATGTC) were incubated at 37 °C at a final concentration of 10 μM. 10 μl aliquots were taken at the time points indicated and analyzed on a denaturing 20% polyacrylamide gel that was stained with ethidium bromide. Stability experiments are shown for an unmodified DNA ON, an all-phosphorothioate (PS) and a gapmer that contained five locked nucleic acids monomers at each end, while the centre consists of unmodified DNA.

its resistance against nucleolytic degradation. Even after two days of incubation at 37 °C in a medium that contains fetal calf serum the intact full-length ON can still be detected.

The introduction of modified building blocks, however, not only increases nuclease resistance of an ON, but also changes further pharmacokinetic parameters that are highly relevant for *in vivo* applications. Important features are the circulation time in the blood stream and the biodistribution (including cellular uptake). Furthermore, the use of non-natural nucleotides can lead to toxic side effects either of the complete ON or of breakdown products. It is also important to maintain the biological function that depends on the mode of action of the ON. For example, alterations of the binding site of an aptamer can result in decreased target affinity, ribozymes tend to lose their catalytic activity when modified nucleotides are introduced into the catalytic centre and siRNAs do not tolerate the addition of functional groups in certain positions while they tolerate this addition in other positions. It is therefore a great challenge to optimize ONs with respect to nuclease stability, functional activity, pharmacological properties and toxic side effects. In the following sections basic principles to fulfil this task are described for the different types of ONs.

1.2 Antisense Oligonucleotides

ONs, being 15 to 20 nucleotides in length, can be employed to inhibit gene expression specifically. This was originally discovered in the late 1970s, when Zamecnik and Stephenson demonstrated that an antisense agent can be used to inhibit virus replication in cell culture.[2] Two major mechanisms of action have been identified to mediate post-transcriptional gene silencing by AS ONs: first, most AS ONs are designed to activate ribonuclease H (RNase H), which is primarily located in the nucleus. RNase H recognizes hybrids composed of DNA and RNA and cleaves the RNA moiety of this heteroduplex. Second, AS ONs that do not induce target RNA cleavage by RNase H can be designed to inhibit translation by a steric blockade of the ribosome. For this approach to be highly efficient it is advisable to direct the AS ONs against the 5′-end or the AUG initiating codon region of the target RNA to prevent binding and assembly of the ribosome. Furthermore, AS ONs can be used to correct aberrant splicing.

Although the initial antisense experiments were carried out with unmodified DNA it soon became clear that ONs have to be protected against nucleolytic degradation for prolonged silencing. In the meantime, several hundred analogues of naturally occurring nucleic acids have been described in the literature. Space restraints in this review mean that only those building blocks that have reached the status of testing in clinical trials will be discussed here: phosphorothioates (PS), 2′-*O*-methoxyethyl RNA (MOE), locked nucleic acids (LNAs), phosphorodiamidate morpholino oligomers (PMOs) and N3′→P5′ phosphoramidates (NPs; Figure 1.2). Peptide nucleic acids (PNAs) are another class of widely used ONs, in which the ribose phosphate backbone is replaced

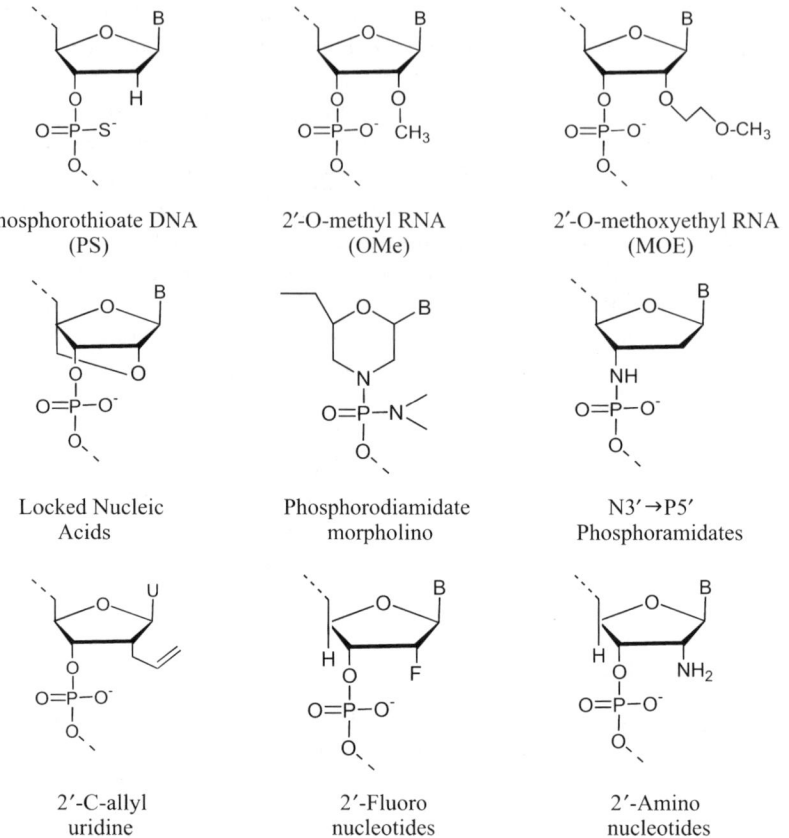

Figure 1.2   Selection of modified nucleic acid building blocks discussed herein. B denotes one of the bases.

by polyamide linkages. Chapter 4 describes the development of peptide conjugates of PNAs for enhanced cellular uptake and intracellular activity by a steric block mechanism.

1.2.1   Phosphorothioates

One of the first and still widely used modifications to stabilize AS ONs is the introduction of phosphorothioates (for a review, see Eckstein[3]). In this class of ONs, one of the non-bridging oxygen atoms is replaced by sulfur. Phosphorothioates were first synthesized by solid-phase chemistry in the 1960s.[4] They are easy to synthesize, highly water soluble and resistant against nucleolytic degradation (Figure 1.1). Just like unmodified DNA ONs, phosphorothioates bind to complementary RNAs by Watson–Crick base pairing and activate target RNA cleavage by RNase H.

As a result of their favourable properties, phosphorothioates have the longest history in clinical testing and most of the advanced studies are based on this class of AS ONs. The only AS ON approved by the FDA to date is the 21-mer phosphorothioate Vitravene (fomivirsen) which targets the immediate early mRNA of the human cytomegalovirus (CMV).[5] The ON is injected intravitreally and is used to treat CMV-induced retinitis in immunodeficient patients with acquired immunodeficiency syndrome (AIDS). The complicated mode of administration and the existence of efficient alternative drugs, however, hinder its broad application. Two additional phosphorothioates that have been tested in advanced stages of clinical investigations to treat cancer are Genasense (see Chapter 2) and Affinitak, which target Bcl-2 and PKC-α, respectively. The results of clinical trials with these ONs, however, did not meet the expectations. The primary mode of action of Genasense still remains somewhat ambiguous and may not even be antisense inhibition of the targeted gene.[6] But, despite this uncertainty, it might be a general problem for AS therapeutics for cancer that the single-target approach might be too narrow. Incomplete knockdown of the target gene might be insufficient to stop tumour growth and the loss of function may be compensated for by other pathways in the cancer cell.

Several disadvantageous properties of phosphorothioates further limit their broad applicability. First of all, they display a reduced affinity towards complementary RNA molecules in comparison to their isosequential unmodified DNA counterpart. Even more important is the tendency of phosphorothioates to bind to certain proteins. This feature has some positive effects for the pharmacokinetic profile because binding to plasma proteins protects them from rapid filtration from the blood stream, but it may also cause cellular toxicity.[7]

1.2.2   2′-*O*-Methyl and 2′-*O*-Methoxyethyl Ribonucleotides

To overcome these limitations, building blocks with modifications at the 2′-position of the ribose have been introduced into AS ONs. RNA derivatives with a methyl or a methoxyethyl group at the 2′-position of the ribose (Figure 1.2) have been used to obtain AS ONs with improved properties. These modifications confer high nuclease resistance with reduced toxicity as compared to phosphorothioate ONs. A major disadvantage of this second generation of modified nucleic acids, however, is their inability to induce RNase H cleavage of the targeted RNA. This problem can be circumvented by the use of so-called gapmers (Figure 1.3): blocks of nucleotides with a modified ribose at both termini protect the ON against dominant exonucleases and increase RNA-binding strength, while the stretch of deoxyribonucleotides in the centre

Figure 1.3   Design of gapmers. Gapmers usually consist of phosphodiester or phosphorothioate linkages throughout the length and 2′-modified nucleotides (*e.g.* 2′-*O*-methyl or 2′-*O*-methoxyethyl RNA or LNA) at both ends. U, nucleotides with unmodified ribose; M, nucleotides with modified ribose.

of the ON is sufficient to activate RNase H.[8] Interestingly, replacement of the phosphodiester bond by phosphorothioates throughout the gapmer allows further gain in nuclease stability without increasing the toxicity of the ON – a phenomenon that is not yet fully understood. Gapmer ONs that consist of MOE and phosphorothioate DNA monomers are currently in clinical development against a broad range of diseases, including diabetes, high cholesterol level, multiple sclerosis, psoriasis and cancer (Table 1.1).

However, even ONs that do not recruit RNase H have been shown to be potent antisense agents. In one of the first attempts in this direction, a fully modified 2′-*O*-methoxyethyl RNA ON that targeted the 5′-end of the mRNA of intercellular adhesion molecule 1 (ICAM-1) efficiently inhibited translation, most likely through interference with the assembly of the ribosome.[9]

Table 1.1 Examples of antisense oligonucleotides in clinical trials. Information based on Pan and Clawson,[1] Corey[82] and corporate websites. Phase I trials are normally conducted with a small group of (healthy) volunteers to assess safety and pharmacokinetics of a new drug in development. Phase II trials are performed on larger groups and are designed to assess clinical efficacy of the therapy and to optimize the dosing. Phase III studies are randomized controlled trials on large patient groups and are aimed at the definitive assessment of the efficacy of a new drug. PS, phosphorothioate; LNA, locked nucleic acids; MOE, methoxyethyl; PMO, phosphorodiamidate morpholino oligomer; NPS, N3′→P5′ thiophosphoramidate.

Company	Product	Disease	Modification	Status
First generation				
ISIS	Vitravene	CMV Retinitis	PS	Approved
Genta	Genasense	Cancer	PS	Phase III
ISIS	Alicaforsen	Ulcerative Colitis	PS	Phase II
Methylgene	MG98	Cancer	PS	Phase II
Antisense Pharma	AP 12009	Cancer	PS	Phase II
EpiGenesis	Epi-2010	Asthma	PS	Phase II
Second generation				
ISIS	ISIS 113715	Diabetes	MOE	Phase II
ISIS	ISIS 301012	High cholesterol	MOE	Phase II
ISIS/ATL	ATL-1102	Multiple sclerosis	MOE	Phase II
ISIS/OncoGeneX	OGX-011	Prostate cancer	MOE	Phase II
ISIS/ATL	ATL-1101	Psoriasis	MOE	Phase I
ISIS/Lilly	Ly2181308	Cancer	MOE	Phase I
ISIS/Lilly	Ly2275796	Cancer	MOE	Phase I
Third generation				
Santaris Pharma	SPC2996	B-cell lymphoma	LNA	Phase I/II
Avi BioPharma	AVI4126	Cardiac restenosis	PMO	Phase II
Avi BioPharma	AVI4065	Hepatitis C virus	PMO	Phase I/II
Avi BioPharma	AVI4557	Drug metabolism	PMO	Phase I
Geron	GRN163L	Cancer	NPS	Phase I

Furthermore, the seemingly undesirable property of 2'-*O*-methyl RNA not to activate RNase H is an indispensable prerequisite for the attempt to correct an mRNA rather than to destroy it. Roughly 60% of human genes are alternatively spliced, and close to 50% of genetic disorders are considered to result from mutations that cause defects in pre-mRNA splicing. Chemically modified AS ONs have successfully been employed to correct splicing by blocking aberrant splice sites.[10]

1.2.3 Locked Nucleic Acids, Phosphorodiamidate Morpholino Oligomers and N3' → P5' Phosphoramidates

In recent years, antisense strategies have received increasing attention because of the advances made by the development of new types of modifications. A large number of DNA or RNA analogues have been tested for their potential to improve antisense agents. Here, only LNAs, PMOs and NPs are discussed in more detail, since these building blocks have already made their way into clinical trials. Readers interested in further modifications are referred to previous review articles, which have exhaustively dealt with modern nucleic acids chemistry used in ONs for biological and therapeutic applications.[1,11–13]

LNAs were initially synthesized in the laboratories of Imanishi[14] and Wengel[15] in 1998. They are conformationally restricted in a 3'-*endo*/N-type sugar conformation by a methylene bridge that connects the 2'-oxygen atom of the ribose with the 4'-carbon atom (Figure 1.2). LNAs combine a number of desirable properties, including nuclease resistance (Figure 1.1) and an unprecedented hybridization affinity towards complementary ONs (for reviews, see Jepsen *et al.*[16] and Karkare and Bhatnagar[17]). Just like most of the nucleotides with modifications at the 2'-position of the ribose, LNAs do not activate RNase H. Gapmers that consist of five LNA monomers at both ends and a central stretch of eight DNA nucleotides in the centre, however, were shown to be potent inducers of RNase H cleavage.[18] Furthermore, LNA gapmers were found to be significantly more efficient inhibitors of gene expression than phosphorothioates or 2'-*O*-methyl RNA gapmers.[19] LNAs do not only confer high target affinity, but also enhance cellular as well as nuclear uptake of ONs after transfection with cationic lipids.[20] This property further accounts for their good antisense potency.

The high efficiency of LNA ONs as antisense agents was also confirmed *in vivo*. Chimeric AS LNA/DNA ONs that target the delta opioid receptor mRNA were found to efficiently reduce the antinociceptive effect of the agonist deltorphin II.[21] In this study, the LNA ONs did not elicit any histologically detectable toxicity when injected into rat brains. Furthermore, gapmers directed against H-ras with standard β-D-LNA or its diastereomer, α-L-LNA, at the termini inhibited tumour growth at very low dosages and did not show toxic side effects.[22] These promising findings prompted the development of an LNA ON against Bcl-2 for the treatment of B-cell lymphoma in patients with chronic lymphocytic leukaemia (CLL), as is outlined in Chapter 5. However, signs of hepatotoxicity after intraperitoneal (i.p.) injection of LNA gapmers were reported recently.[23]

In PMOs, the five-membered ribose ring is replaced by a six-membered morpholino moiety and a dimethylaminophosphoroamidate (phosphorodiamidate) intersubunit linkage is used instead of the phosphodiester bond (Figure 1.2). PMOs are resistant to nucleases, but, like most of the third-generation modifications, they do not activate RNase H. They are thus usually targeted against the 5′-untranslated region (UTR) or the first bases downstream of the AUG start codon to inhibit translation by preventing ribosomes from binding. PMOs are a widely used knockdown tool in developmental biology because of their efficient cytosolic delivery into embryos by microinjection (for a review, see Karkare and Bhatnagar[17]). Furthermore, Avi BioPharma is developing PMO AS ONs in clinical trials (see Table 1.1 and Chapter 3). In addition to antisense applications, PMO ONs have been employed to correct aberrant splicing of β-globin precursor mRNA in blood cells from patients with β-thalassemia. *Ex vivo* treatment of erythroid progenitor cells with a PMO ON restored correct splicing and synthesis of haemoglobin A.[24]

NPs are DNA analogs, in which the 3′-hydroxyl group of the 2′-deoxyribose ring is replaced by a 3′-amino group (Figure 1.2). NPs exhibit high affinity towards a complementary RNA strand and good nuclease resistance.[25] Since NPs do not activate RNase H, they have been employed for strategies that do not depend on this classical mode of antisense inhibition. Human telomerase is a reverse transcriptase that maintains telomers in rapidly dividing cells. The enzyme is inactive in most somatic cells, but more than 90% of all cancer cells display robust activation of telomerase, thus making it a suitable target for anticancer drugs. Telomerase consists of an RNA component and a proteinaceous catalytic subunit. A N3′→P5′ thiophosphoroamidate with a palmitoyl moiety conjugated to the 5′-end was directed against the template region of telomerase RNA and was found to inhibit telomerase activity in a human lung cancer cell line and to prevent lung metastases *in vivo* in xenograft animal models.[26] A more recent study suggests, however, that the antimetastatic potential of this AS ON, named GRN163L, might rather be by antiadhesive effects conferred via specific structural determinants than by its inhibition of telomerase.[27] According to the website of Geron Corporation, clinical Phase I studies with GRN163L for patients with CLL as well as for patients with solid tumours have been initiated.

1.3 Ribozymes

In the early 1980s, the groups of Thomas Cech and Sydney Altman discovered catalytically active ONs and coined the term ribozyme for these *ribo*nucleic acids with en*zym*atic activity. In the meantime, several classes of ribozymes have been discovered in nature, most of which catalyse intramolecular splicing or cleavage reactions (for reviews, see Schubert and Kurreck[28] and Fedor and Williamson[29]). The present review focuses exclusively on the hammerhead ribozyme, which has the greatest relevance for therapeutic applications.

The hammerhead ribozyme was initially found in plant pathogens, in which it processes linear concatamers that are generated during rolling circle replication. The use of hammerhead ribozymes for practical applications was made possible with the development of shortened variants that cleave a separate substrate RNA molecule *in trans*.[30,31] These ribozymes identify their target RNA by complementary base pairing and subsequently destroy it by a transesterification reaction.

Just as for the other types of intracellular use of ON, the efficient delivery of ribozymes into the cells is one of the most challenging aspects of their application. Two principle strategies can be distinguished: ribozymes can either be expressed inside cells from vectors that encode their sequence under control of a suitable promoter or they can be delivered exogenously as pre-synthesized molecules. While the first approach guarantees continuous generation of ribozymes for a certain period of time, the latter strategy again requires protection of the RNA molecules, being highly susceptible against nucleolytic degradation. This task, however, is even more challenging than protection of AS ONs, since the modified nucleotides may interfere with the three-dimensional structure, thereby reducing catalytic activity.

Extensive work was thus necessary to stabilize hammerhead ribozymes without deleterious consequences on their function. In a systematic study, selective modification of the hammerhead ribozyme with various DNA and RNA analogues resulted in the nuclease-resistant variant depicted in Figure 1.4.[32,33] This ribozyme consists primarily of 2'-O-methyl RNA monomers (lower case in Figure 1.4) and the termini are protected by an inverted 3'-3' deoxyabasic sugar (iB) and four phosphorothioate linkages (s), respectively. A 2'-C-allyl uridine (Figure 1.2) was used to further stabilize the core, while five unmodified ribonucleotides (upper case in Figure 1.4) were required to maintain catalytic activity. The protected variant had a half-life of more than 10 days in serum compared to a half-life of less than 1 minute for the unmodified RNA ribozyme.

Stabilized hammerhead ribozymes with the above-mentioned modification pattern were tested in clinical trials to treat cancer and hepatitis C virus (HCV)

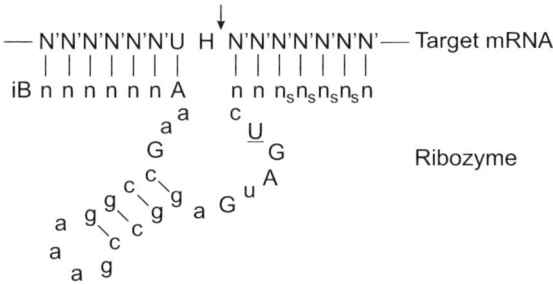

Figure 1.4 Design of a hammerhead ribozyme with high catalytic activity and nuclease resistance. Lower case, 2'-O-methyl RNA; upper case, RNA; s, phosphorothioate linkages; iB, inverted 3'-3' deoxyabasic sugar; underlined U, 2'-C-allyluridine; H, A, C or U. Schematic representation after Ref. 33.

infections. The first ribozyme named 'ANGIOZYME' was designed to inhibit the expression of the receptor for the vascular endothelial growth factor (VEGF) thereby blocking angiogenesis. Clinical tests with cancer patients demonstrated that the drug was well-tolerated,[33] but did not result in convincing therapeutic benefit. The second ribozyme to target the 5'-UTR of HCV (HEPTAZYME) was tested in a Phase II trial, either alone or in combination with interferon α to evaluate safety and efficiency in treating chronic hepatitis C. Although a slight reduction of the viral RNA level in the serum of patients treated with the ribozyme was observed, these results, as well as the outcome of the toxicology studies, did not justify further development of HEPTAZYME.[34]

As an alternative to the modification of a naturally occurring ribozyme, stable nucleic acid catalysts can be generated by *in vitro* selection. With this method, a 36-mer ribozyme was obtained that consisted of 2'-*O*-methyl RNA and 2'-NH_2 moieties, as well as an inverted abasic sugar and phosphorothioate linkages at the 3'- and 5'-ends, respectively.[35] The ribozyme displayed reasonable catalytic activity under physiological conditions and had a half-life of >100 hours in human serum. A variant of the selected molecule that targeted the mRNA of the human epidermal growth factor receptor-2 (HER-2), which is overexpressed in aggressive breast cancer, was found to be well-tolerated in a clinical Phase I trial.

In vitro selection strategies were also used to obtain catalytically active ONs that are composed entirely of deoxyribonucleotides. The most prominent representative of this class of molecular scissors is the 10–23 DNA enzyme.[36] Even though DNA has an intrinsically higher resistance against nucleolytic degradation than RNA, the DNA enzyme has to be further stabilized for biological applications. To protect it against the dominant 3'-exonucleases, an inverted thymidine has been added to its 3'-terminus.[37] In other studies, the binding arms have been modified with 2'-*O*-methyl RNA,[38] LNA,[39] and phosphoramidate building blocks,[40] respectively. We performed a systematic substitution study to introduce 2'-*O*-methyl RNA and LNA monomers into the substrate binding arms as well as into the catalytic core and obtained a modified variant with an approximately tenfold higher catalytic activity and substantially increased resistance against exo- and endonucleases as compared to the unmodified DNA enzyme.[41] In addition, the introduction of modified nucleotides with a high target affinity into the substrate binding arms was found to enhance the activity of DNA enzymes against seemingly uncleavable target sites, most likely because of their ability to compete with stable secondary structures of the target RNA.[42]

Nuclease resistance, however, is not the only problem associated with the application of (deoxy)ribozymes. Catalytically active ONs require comparably high concentrations of divalent cations as co-factors. For *in vitro* experiments, magnesium ions are usually added to the reaction mixture at a concentration of 10 mM (or even higher) to obtain good cleavage rates, while the intracellular concentration of magnesium ions is less than 1 mM. Interestingly, it was demonstrated that a loop outside the catalytic core of hammerhead ribozymes is required for high enzymatic activity under physiological conditions.[43,44]

The attempts to minimize the hammerhead motif for practical applications have obviously gone too far in that domains required for the efficient catalysis at low magnesium ion concentrations were removed. This failure might explain why ribozymes did not meet the expectations that arose with their development as a new class of therapeutics. Nowadays, structures and catalytic mechanisms of ribozymes are still important topics in research, but their relevance as tools for therapeutic interventions have decreased with the advent of the more efficient RNAi technology. Still, lessons learned from the application of ribozymes are valuable for the development of RNAi approaches, and it is no surprise that the leading company in the ribozyme field, Ribozyme Pharmaceuticals, belongs to the first companies that initiated clinical trials with siRNAs under the name Sirna Therapeutics.

1.4 Small Interfering RNAs

RNAi is an evolutionary conserved mechanism of post-transcriptional gene silencing mediated by double-stranded RNA molecules. It was first described in the nematode *Caenorhabditis elegans* by Andrew Fire and Craig Mello in 1998.[45] Since long double-stranded RNA molecules activate an interferon response in mammalian cells, the application of RNAi was initially restricted to lower eukaryotes that served as model organisms. Elucidation of the RNAi pathway helped to overcome this drawback: The long double-stranded RNA molecules are initially chopped by an RNase known as Dicer into smaller duplexes of 21–23 nucleotides in length with two nucleotide overhangs at both termini. The resulting so-called small or short interfering RNAs (siRNAs) are then incorporated into the RNA-induced silencing complex (RISC). The antisense strand of the siRNA guides RISC to the target RNA, which becomes cleaved by the Argonaute 2 protein, a major component of RISC (for a review on the RNAi mechanism, see Rana[46]). The demonstration that siRNAs are suitable to silence genes in mammalian cells without the induction of unspecific effects[47] opened the door for the application of RNAi as a tool for functional studies in higher eukaryotes as well as for therapeutic purposes in humans (for a review on the development of RNAi against human diseases, see Kim and Rossi[48]).

Just like ribozymes, siRNAs can either be expressed from vector systems as outlined in Chapter 12 or they can be delivered exogenously as pre-synthesized molecules. Although siRNAs were found to have a surprisingly high half-life in serum, most likely because proteins shield them from nucleases, it is advisable to modify siRNAs for *in vivo* applications. The introduction of RNA analogues and non-nucleosidic functional groups, however, is not only intended to increase the stability of the siRNA, but also it can help to reduce off-target effects caused by the siRNAs and to facilitate cellular uptake of the duplex.

Improvement of siRNAs by the introduction of modified nucleotides is a rather challenging task since the changes must not interfere with the silencing activity. We have to keep in mind that the two strands of the siRNA duplex have different functions. The antisense strand is of major functional relevance since it will guide RISC to the target RNA, while the sense strand – also referred to as the passenger strand – will be discarded during or after the loading of RISC. It is therefore likely to be more tolerant to the introduction of modifications. But even for the antisense (or guide) strand, RNA analogues can be used, depending on the position of the nucleotide to be substituted and on the type (and size) of the chemical group to be introduced. Many of the modified nucleotides that have been used previously for AS ONs and ribozymes have also been tested to improve siRNAs. Only a few selected examples can be discussed here to demonstrate the principle challenges of this approach. For a more comprehensive summary, an overview is already published.[49]

1.4.1   Introduction of Modified Nucleotides

The introduction of phosphorothioate linkages was among the first steps to protect siRNAs against ribonucleases. This modification was found to be very compatible with the silencing function of siRNAs, but a high content (50%) of phosphorothioates resulted in cytotoxic properties and reduced cell growth and viability.[50] Further studies have been carried out with modifications of the 2′-OH group of the ribose. Unlike alternative substitutions, the 2′-fluoro modification (Figure 1.2) does not add steric bulk. It is therefore not surprising that this analogue is well-tolerated in siRNA applications.[50–52] In contrast, the bulky CH_3 group of 2′-O-methyl-modified nucleotides completely abolished RNAi when incorporated throughout the siRNA.[52] It was thus necessary to find a modification pattern that improves stability of the siRNA in serum without compromising RNAi activity. Blunt-ended siRNAs with alternating 2′-O-methyl nucleotides on both strands, whereby unmodified nucleotides face modified ones on the opposite strand (Figure 1.5A), were found to confer high nuclease resistance without losing gene-silencing capacity.[53] In a subsequent study, this type of modification was used for the systemic application of RNAi

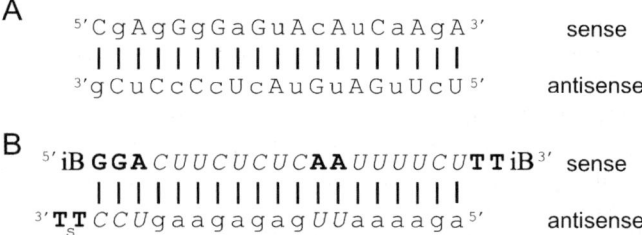

Figure 1.5   Modification patterns for siRNAs. (A) Design of blunt-ended siRNAs against Akt1 with alternating 2′-O-methyl nucleotides adapted from.[53] Modified nucleotides on one strand always face unmodified ones on the other strand. (B) Fully modified siRNA targeting hepatitis B virus adapted from.[60] iB, inverted deoxyabasic sugar; bold, deoxyribonucleotides; lower case, 2′-O-methyl RNA; upper case: RNA; s, phosphorothioate linkages; italics, 2′-fluoro nucleotides.

in vivo.[54] Lipid-mediated delivery of the modified siRNAs resulted in significant uptake of fluorescently labelled siRNAs into mouse vascular endothelium and downregulation of the target mRNA and protein.

As mentioned above, not only can 2′-*O*-methyl modifications be used to increase nuclease resistance, but also they can diminish unspecific effects induced by siRNAs. In contrast to early expectations, gene silencing by RNAi is not exquisitely specific and off-target gene regulations have been observed on the mRNA and protein level.[55] Recent investigations revealed that it is not the overall degree of homology that accounts for the probability that a particular gene is regulated by a given siRNA. Instead, off-target regulation was found to be associated with perfect matches of the 3′-UTR with the seed region (positions 2–7 or 2–8) of the antisense strand of the siRNA.[56] This finding helped to reduce unintended silencing by a substitution strategy. A single 2′-*O*-methyl nucleotide at the second position of the antisense strand was found to reduce silencing of off-target transcripts by 66% on average.[57] Importantly, this decrease of unspecific effects elicited by siRNAs was found to be sequence-independent.

LNAs have already been described as a modification that leads to a significant improvement of AS ONs. When using LNAs for RNAi approaches, we have to keep in mind that LNAs have a severe impact on the structure of a duplex and exert an extremely high target affinity, and thereby interfere with unwinding of the double-stranded RNA. It does not, therefore, come as a surprise that extensive substitutions of RNA with LNA are detrimental to the silencing function of an siRNA, but a few LNAs can be incorporated into the duplex without compromising the efficiency of RNAi.[51] Positions that tolerate the introduction of LNA monomers were identified in a systematic study in which each single RNA monomer of the antisense strand was substituted by an LNA.[58] The introduction of LNAs into the siRNA not only improves serum stability, but also it can help to reduce off-target effects of the siRNA and to enhance the efficiency of mediocre siRNAs by improving RISC loading with the antisense strand relative to the sense strand. Furthermore, we used LNA modifications of an siRNA for a functional study to investigate the mode of RNAi-mediated silencing of picornaviruses, a prominent class of human pathogens. Picornaviruses contain a plus-strand RNA genome and synthesize a template RNA in a minus-strand orientation during replication. To solve the question as to whether targeting of the genomic plus-strand or of the intermediate minus-strand is more efficient, we used an siRNA that is active against the viral RNA in both orientations. By selectively inactivating either of the two siRNA strands, we could clearly demonstrate that only siRNAs that targeted the genomic plus-strand exert antiviral activity.[59]

1.4.2 Combination of Different Modifications

The studies described above that employ a single type of modification showed that different types of RNA analogues have varying impact at certain positions of the siRNA. It is thus reasonable to optimize the modification pattern by combining different derivatives and take advantage of the respective local compatibility. In one of these attempts, researchers from the company Sirna Therapeutics (a MERCK subsidiary) stabilized an siRNA by substituting all the OH groups (Figure 1.5B).[60] In the sense strand, 2'-fluoro monomers were used for all pyrimidine and deoxyriboses for all purine positions. The ends were protected with 5'- and 3'-inverted abasic caps. The antisense strand contained 2'-fluoro substitutents in all pyrimidine positions as well, but 2'-O-methyl building blocks in all the purine positions and a single phosphorothioate linkage at the 3'-terminus. While the unmodified siRNA was found to have a half-life of only a few minutes in human serum, the modification pattern described increased stability by approximately 900-fold to a half-life of several days. In a vector-based *in vivo* model for hepatitis B virus (HBV) infections, the modified siRNA was significantly more efficient than the unmodified counterpart. In a follow-up study, the modified siRNA was encapsulated in lipid nanoparticles to form stable nucleic acid–lipid particles (SNALPS; see Chapter 11).[61] These formulations inhibited HBV replication highly efficiently and marked a further important improvement in the attempts to develop siRNA as a new therapeutic option. Consequently, Sirna was the first company to initiate clinical trials with chemically modified siRNAs that targeted the VEGF receptor 1 to treat patients with age-related macular degeneration. In a mouse model, intravitreous injection of this siRNA reduced choroidal neovascularization lesion size by 66% compared to a saline control.[62]

1.4.3   Terminal Modifications

Various non-nucleosidic modifications have been attached to the termini of siRNAs to introduce new functional properties. Fluorescent chromophores such as fluorescein and Cy-3 can be used to monitor cellular uptake of siRNAs (Figure 1.6).[50,63,64] Terminal modifications are usually well tolerated in the sense strand, while the attachment of groups to the guide strand may compromise silencing activity.

Furthermore, attachment of lipophilic moieties had been considered to increase cellular uptake of the negatively charged siRNA. A series of derivatives of cholesterol, lithocholic acid or lauric acid, covalently linked to the 5'-end of either of the two siRNA strands, was tested with respect to uptake into liver cells in the absence of transfection agents and to knockdown of a target gene.[65] siRNAs with a lipid conjugated to the sense strand were found to downregulate gene expression to a higher extent than siRNAs with the lipophilic moiety attached to the antisense strand or to both strands. Two modifications, a cholesterol derivative and a derivative of 12-hydroxy-lauric acid, were found to exert the best silencing effect. Consequently, a cholesterol-modified siRNA was used to deliver siRNAs into the liver of mice after injection into a tail vein at normal pressure.[66] The siRNAs contained additional terminal 2'-O-methyl-modified nucleotides and phosphorothioate linkages. The cholesterol-coupled siRNA was found to be taken up by cells in the liver and jejunum and RNAi-mediated knockdown of the target gene (apolipoprotein B) was

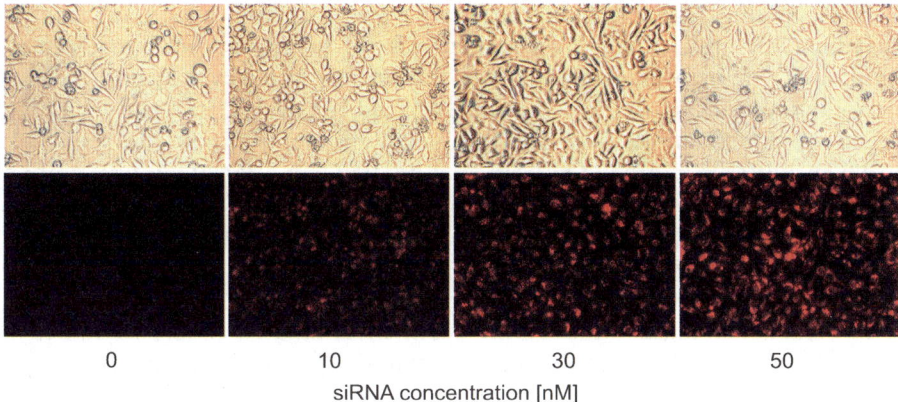

0   10   30   50

siRNA concentration [nM]

Figure 1.6 Fluorescence of Cy-3 labelled siRNA. HeLa cells were transfected with siRNAs carrying a Cy-3 label at the 5′ end of the sense strand with Lipofectamine 2000 as a transfection agent. 24 hours after transfection, images were taken with an Olympus IX 50 fluorescence microscope. Phase contrast and fluorescence images are shown for transfections with the indicated concentrations of the siRNA. The punctuate pattern reflects the endosomal uptake of the siRNA.

confirmed. In addition, cell-penetrating peptides have been covalently conjugated to an siRNA, to achieve cellular uptake in the absence of liposomal transfection agents.[67] Further attempts were made to achieve not only efficient, but also cell-type specific uptake of siRNAs. These approaches include non-covalent coupling of the siRNA to an antibody fragment,[68] as well as linkage of an siRNA to an aptamer.[69]

1.5 Aptamers and Decoy Oligonucleotides

Aptamers are nucleic acids with the capacity to bind to target molecules with high affinity and specificity. They are typically single-stranded ONs consisting of 30–70 monomers that form elaborate three-dimensional structures (for further details, see Famulok and G. Mayer,[70] Rimmele,[71] and Nimjee *et al.*[72]). The most common way to obtain nucleic acids with high target affinity is by an *in vitro* selection procedure dubbed SELEX (systematic evolution of ligands by exponential enrichment). This technology comprises an iterative process of selection and amplification of target-binding ONs, while unbound molecules are discarded. The procedure starts with a large library consisting of 10^{14} to 10^{15} different DNA or RNA species that fold into distinct three-dimensional structures, depending on their particular sequence. The library is then incubated with the target molecule to separate the DNA or RNA molecules that bind to the ligand from those that do not. The retained ONs are amplified and undergo further rounds of enrichment. The selection process is repeated until ONs with high affinity to the target molecule can be isolated, cloned, sequenced and further optimized.

Since aptamers are selected by a pure *in vitro* procedure, some shortcomings that limit the applicability of a related class of target-binding biomolecules, antibodies, can be overcome and aptamers can be selected even against toxic or non-immunogenic substances. Aptamers have been generated against a broad spectrum of target molecules, including metal ions, organic molecules, peptides, proteins and even complete viruses. The binding conditions (*e.g.* temperature and salt concentration) can be freely chosen and the specificity of aptamers can be enhanced by a procedure called counter-SELEX, during which binders to a structurally related molecule are withdrawn from the library prior to incubation with the actual target molecule.

A great advantage of the use of aptamers compared to the anti-mRNA strategies described above is the opportunity to direct aptamers against extracellular targets like signalling peptides or extracellular domains of proteins, and thus prevent the need to transfer the ONs into cells. Still, the high nuclease activity in the blood serum requires extensive chemical modification of the aptamer to prevent immediate degradation. Conceptually, two strategies can be differentiated: modified nucleotides can be introduced during the SELEX procedure by their incorporation into the initial library or post-SELEX by site-specific engineering of the selected aptamer.

For the selection of nuclease-resistant aptamers the substituted nucleotides must be acceptable substrates for the enzymes that are required for the SELEX process, *i.e.* the T7 RNA polymerase and the reverse transcriptase. Nucleotides containing a fluorine (F) or amino (NH_2) group at the 2′-position of the ribose of pyrimidines (Figure 1.2) fulfil this requirement and can therefore be included in the library.

Pegaptanib sodium (also known as Macugen) was the first aptamer to be approved by the FDA for the treatment of age-related macular degeneration.[72] This disease is caused by ocular neovascularization promoted by VEGF. In an attempt to develop anti-VEGF therapies, an aptamer against the growth factor was initially isolated from a library composed of 2′-ribo purines and 2′-fluoro pyrimidines. After the selection procedure the sequence motive was optimized and the aptamer was further stabilized by the introduction of 2′-*O*-methyl nucleotides. Except for two positions, all purines could be substituted without significant loss of binding activity. The predicted secondary structure and the modification pattern of the resulting aptamer are depicted in Figure 1.7. Human clinical studies with this highly stabilized ON suggest a half-life of 10 days.[12] In a randomized trial, patients treated with pegaptanib for two years demonstrated a 45% relative benefit in mean change in vision compared with those who received the usual care.[73]

To facilitate the selection of modified aptamers, mutated versions of the T7 RNA polymerase were evolved that accept 2′-*O*-methyl-modified nucleotides. These attempts resulted in the discovery of variants with high processivity that allow the introduction of 2′-*O*-methyl A, C and U (but not G) during *in vitro* transcription.[74] Under optimized conditions, a mutated version of the T7 polymerase even accepts all four 2′-*O*-methyl-substituted nucleotides.[75] A fully

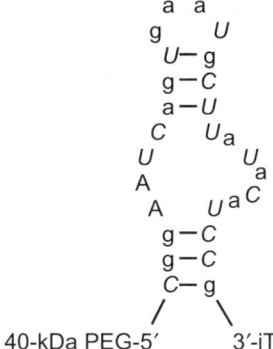

Figure 1.7 Predicted secondary structure and modification pattern of the aptamer pegaptanib. 2′-*O*-methylated purines are shown in lower case, 2′-fluorine-modified pyrimidines are in italics and two unmodified ribonucleotides are shown as uppercase A.

2′-*O*-methyl-modified aptamer against VEGF was selected with this method, which was highly stable and displayed good pharmacokinetic properties.

Another elegant approach to improve the stability of aptamers was introduced with the 'Spiegelmer' concept, which relies on the high enantioselectivity of nucleases. Spiegelmers consist of the unnatural L-form of nucleotides that are not recognized by nucleases and are therefore highly stable in human serum.[76] This advantage, however, has to be paid for with a highly sophisticated SELEX procedure, during which a regular aptamer is selected against the enantiomer of the biological target that consists of D-amino acids. The isolated aptamer is then converted into the mirror-image molecule built out of L-form nucleotides that recognize the natural target. The pharmacological potential of Spiegelmers has been demonstrated in animal models[77,78] and a Spiegelmer binding to the monocyte chemoattractant protein-1 (MCP-1) is currently in pre-clinical development as outlined in Chapter 9.

We recently described an alternative strategy for post-SELEX optimization of aptamers.[79] LNA monomers were incorporated into various regions of an aptamer that targeted tenascin-C, an extracellular tumour marker. Introduction of LNA nucleotides into the supposed binding region significantly enhanced the nuclease resistance of the aptamer, but resulted in a complete loss of affinity to the target molecule. In contrast, modification of a terminal stem, which is not involved in ligand binding, improved the thermodynamic properties and plasma stability of the aptamer without compromising its binding properties. The optimized aptamer displayed favourable pharmacokinetic properties and enhanced tumour uptake in nude mice. We consider the approach to improve the properties of an aptamer by post-SELEX incorporation of modified nucleotides to be generally applicable.

Decoy ONs are conceptually related to aptamers (see Chapter 7). They usually mimic promoter elements and compete with their cellular counterparts for nucleic acid binding proteins, like transcription factors, thereby altering the expression of the target gene. Introduction of non-natural nucleotides into decoy ONs is advisable to improve their properties (like nuclease resistance), but must be balanced against a potential loss of binding affinity. For example, decoy ONs that bind to the transcription nuclear factor kappa beta (NF-κB) have been stabilized successfully by the introduction of LNAs. Gapmer-type decoy ONs that contain LNA monomers at the termini of the double-stranded DNA molecule were protected against degradation by exonucleases.[80] Further improvement of stability was achieved when internal positions were modified as well, but these alterations decreased binding affinity of the transcription factor for the decoy ON. Replacement of conventionally used β-D-LNA by its diastereomeric form, α-L-LNA, leaves the DNA duplex in its native B-type form and thus permitted almost complete reversal of the adverse effect of LNA on NF-κB binding.[81]

1.6   Concluding Remarks

After the discovery of RNAi as an efficient mechanism for sequence-specific gene silencing and the approval of an aptamer to treat an important ocular disease, ON-based strategies have gained increasing attention again in recent years. The different approaches in this field have in common that *in vivo* applications necessitate the use of modified monomers to protect the ONs against nucleolytic degradation. Significant advancements were made by bioorganic chemists, who developed RNA analogues with high nuclease resistance, good binding properties and low toxicity. These novel modifications even gave new hope for the successful development of traditional antisense ONs. Lessons learnt from antisense and ribozyme approaches also allow improvement of the properties of siRNAs at high speed. Since it is now possible to design highly stable molecules with improved functionality, the major challenge remains delivery of ONs for intracellular applications, for which, however, promising approaches are in progress.

Acknowledgement

I thank Volker A. Erdmann for his continuous support, and all the current and previous laboratory members for sharing my enthusiasm for ON-based strategies. Furthermore, I thank Mariola Dutkiewicz, Michael Gait, Harry Kurreck and Diana Rothe for critical reading of the manuscript and I apologize to those whose work I could not refer to because of space restraints. Financial support by the DFG (Ku-1436; SFB/TR 19), the RNA network/BMBF and the Fonds der Chemischen Industrie is gratefully acknowledged.

References

1. W. H. Pan and G. A. Clawson, *J. Cell. Biochem.*, 2006, **98**, 14.
2. P. C. Zamecnik and M. L. Stephenson, *Proc. Natl. Acad. Sci. U.S.A.*, 1978, **75**, 280.

3. F. Eckstein, *Antisense Nucleic Acid Drug Dev.*, 2000, **10**, 117.
4. E. De Clercq, E. Eckstein and T. C. Merigan, *Science*, 1969, **165**, 1137.
5. C. Marwick, *JAMA*, 1998, **280**, 871.
6. J. C. Lai, W. Tan, L. Benimetskaya, P. Miller, M. Colombini and C. A. Stein, *Proc. Natl. Acad. Sci. U.S.A.*, 2006, **103**, 7494.
7. A. A. Levin, *Biochim. Biophys. Acta*, 1999, **1489**, 69.
8. B. P. Monia, E. A. Lesnik, C. Gonzalez, W. F. Lima, D. McGee, C. J. Guinosso, A. M. Kawasaki, P. D. Cook and S. M. Freier, *J. Biol. Chem.*, 1993, **268**, 14514.
9. B. F. Baker, S. S. Lot, T. P. Condon, S. Cheng-Flournoy, E. A. Lesnik, H. M. Sasmor and C. F. Bennett, *J. Biol. Chem.*, 1997, **272**, 11994.
10. P. Sazani and R. Kole, *J. Clin. Invest.*, 2003, **112**, 481.
11. J. Kurreck, *Eur. J. Biochem.*, 2003, **270**, 1628.
12. C. Wilson and A. D. Keefe, *Curr. Opin. Chem. Biol.*, 2006, **10**, 607.
13. M. M. Fabani, J. J. Turner and M. J. Gait, *Curr. Opin. Mol. Ther.*, 2006, **8**, 108.
14. S. Obika, D. Nanbu, Y. Hary, J.-I. Andoh, K.-I. Morio, T. Doi and T. Imanishi, *Tetrahedron Lett.*, 1998, **39**, 5401.
15. A. A. Koshkin, S. K. Singh, P. Nielsen, V. K. Rajwanshi, R. Kumar, M. Meldgaard, C. E. Olsen and J. Wengel, *Tetrahedron*, 1998, **54**, 3607.
16. J. S. Jepsen, M. D. Sorensen and J. Wengel, *Oligonucleotides*, 2004, **14**, 130.
17. S. Karkare and D. Bhatnagar, *Appl. Microbiol. Biotechnol.*, 2006, **71**, 575.
18. J. Kurreck, E. Wyszko, C. Gillen and V. A. Erdmann, *Nucleic Acids Res.*, 2002, **30**, 1911.
19. A. Grunweller, E. Wyszko, B. Bieber, R. Jahnel, V. A. Erdmann and J. Kurreck, *Nucleic Acids Res.*, 2003, **31**, 3185.
20. A. Arzumanov, D. A. Stetsenko, A. D. Malakhov, S. Reichelt, M. D. Sorensen, B. R. Babu, J. Wengel and M. J. Gait, *Oligonucleotides*, 2003, **13**, 435.
21. C. Wahlestedt, P. Salmi, L. Good, J. Kela, T. Johnsson, T. Hokfelt, C. Broberger, F. Porreca, J. Lai, K. Ren, M. Ossipov, A. Koshkin, N. Jakobsen, J. Skouv, H. Oerum, M. H. Jacobsen and J. Wengel, *Proc. Natl. Acad. Sci. U.S.A.*, 2000, **97**, 5633.
22. K. Fluiter, M. Frieden, J. Vreijling, T. Koch and F. Baas, *Oligonucleotides*, 2005, **15**, 246.
23. E. E. Swayze, A. M. Siwkowski, E. V. Wancewicz, M. T. Migawa, T. K. Wyrzykiewicz, G. Hung, B. P. Monia and C. F. Bennett, *Nucleic Acids Res.*, 2007, **35**, 687.
24. G. Lacerra, H. Sierakowska, C. Carestia, S. Fucharoen, J. Summerton, D. Weller and R. Kole, *Proc. Natl. Acad. Sci. U.S.A.*, 2000, **97**, 9591.
25. S. Gryaznov and J.-K. Chen, *J. Am. Chem. Soc.*, 1994, **116**, 3143.
26. Z. G. Dikmen, G. C. Gellert, S. Jackson, S. Gryaznov, R. Tressler, P. Dogan, W. E. Wright and J. W. Shay, *Cancer Res.*, 2005, **65**, 7866.

27. S. R. Jackson, C. H. Zhu, V. Paulson, L. Watkins, Z. G. Dikmen, S. M. Gryaznov, W. E. Wright and J. W. Shay, *Cancer Res.*, 2007, **67**, 1121.
28. S. Schubert and J. Kurreck, *Curr. Drug Targets*, 2004, **5**, 667.
29. M. J. Fedor and J. R. Williamson, *Nat. Rev. Mol. Cell Biol.*, 2005, **6**, 399.
30. O. C. Uhlenbeck, *Nature*, 1987, **328**, 596.
31. J. Haseloff and W. L. Gerlach, *Nature*, 1988, **334**, 585.
32. L. Beigelman, J. A. McSwiggen, K. G. Draper, C. Gonzalez, K. Jensen, A. M. Karpeisky, A. S. Modak, J. Matulic-Adamic, A. B. DiRenzo, P. Haeberli, D. Sweedler, D. Tracz, S. Grimm, F. E. Wincott, V. G. Thackaray and N. Usman, *J. Biol. Chem.*, 1995, **270**, 25702.
33. N. Usman and L. M. Blatt, *J. Clin. Invest.*, 2000, **106**, 1197.
34. A. Peracchi, *Rev. Med. Virol.*, 2004, **14**, 47.
35. S. P. Zinnen, K. Domenico, M. Wilson, B. A. Dickinson, A. Beaudry, V. Mokler, A. T. Daniher, A. Burgin and L. Beigelman, *RNA*, 2002, **8**, 214.
36. S. W. Santoro and G. F. Joyce, *Proc. Natl. Acad. Sci. U.S.A.*, 1997, **94**, 4262.
37. P. O. Iversen, P. D. Emanuel and M. Sioud, *Blood*, 2002, **99**, 4147.
38. M. Sioud and M. Leirdal, *J. Mol. Biol.*, 2000, **296**, 937.
39. B. Vester, L. B. Lundberg, M. D. Sorensen, B. R. Babu, S. Douthwaite and J. Wengel, *J. Am. Chem. Soc.*, 2002, **124**, 13682.
40. H. Takahashi, H. Hamazaki, Y. Habu, M. Hayashi, T. Abe, N. Miyano-Kurosaki and H. Takaku, *FEBS Lett.*, 2004, **560**, 69.
41. S. Schubert, D. C. Gul, H. P. Grunert, H. Zeichhardt, V. A. Erdmann and J. Kurreck, *Nucleic Acids Res.*, 2003, **31**, 5982.
42. S. Schubert, H. P. Grunert, H. Zeichhardt, D. Werk, V. A. Erdmann and J. Kurreck, *J. Mol. Biol.*, 2005, **346**, 457.
43. A. Khvorova, A. Lescoute, E. Westhof and S. D. Jayasena, *Nat. Struct. Biol.*, 2003, **10**, 708.
44. M. De la Pena, S. Gago and R. Flores, *EMBO J.*, 2003, **22**, 5561.
45. A. Fire, S. Xu, M. K. Montgomery, S. A. Kostas, S. E. Driver and C. C. Mello, *Nature*, 1998, **391**, 806.
46. T. M. Rana, *Nat. Rev. Mol. Cell Biol.*, 2007, **8**, 23.
47. S. M. Elbashir, J. Harborth, W. Lendeckel, A. Yalcin, K. Weber and T. Tuschl, *Nature*, 2001, **411**, 494.
48. D. H. Kim and J. J. Rossi, *Nat. Rev. Genet.*, 2007, **8**, 173.
49. H. Y. Zhang, Q. Du, C. Wahlestedt and Z. Liang, *Curr. Top. Med. Chem.*, 2006, **6**, 893.
50. J. Harborth, S. M. Elbashir, K. Vandenburgh, H. Manninga, S. A. Scaringe, K. Weber and T. Tuschl, *Antisense Nucleic Acid Drug Dev.*, 2003, **13**, 83.
51. D. A. Braasch, S. Jensen, Y. Liu, K. Kaur, K. Arar, M. A. White and D. R. Corey, *Biochemistry*, 2003, **42**, 7967.
52. Y. L. Chiu and T. M. Rana, *RNA*, 2003, **9**, 1034.
53. F. Czauderna, M. Fechtner, S. Dames, H. Aygun, A. Klippel, G. J. Pronk, K. Giese and J. Kaufmann, *Nucleic Acids Res.*, 2003, **31**, 2705.
54. A. Santel, M. Aleku, O. Keil, J. Endruschat, V. Esche, G. Fisch, S. Dames, K. Loffler, M. Fechtner, W. Arnold, K. Giese, A. Klippel and J. Kaufmann, *Gene Ther.*, 2006, **13**, 1222.

55. A. L. Jackson and P. S. Linsley, *Trends Genet.*, 2004, **20**, 521.
56. A. Birmingham, E. M. Anderson, A. Reynolds, D. Ilsley-Tyree, D. Leake, Y. Fedorov, S. Baskerville, E. Maksimova, K. Robinson, J. Karpilow, W. S. Marshall and A. Khvorova, *Nat. Methods*, 2006, **3**, 199.
57. A. L. Jackson, J. Burchard, D. Leake, A. Reynolds, J. Schelter, J. Guo, J. M. Johnson, L. Lim, J. Karpilow, K. Nichols, W. Marshall, A. Khvorova and P. S. Linsley, *RNA*, 2006, **12**, 1197.
58. J. Elmen, H. Thonberg, K. Ljungberg, M. Frieden, M. Westergaard, Y. Xu, B. Wahren, Z. Liang, H. Orum, T. Koch and C. Wahlestedt, *Nucleic Acids Res.*, 2005, **33**, 439.
59. S. Schubert, D. Rothe, D. Werk, H. P. Grunert, H. Zeichhardt, V. A. Erdmann and J. Kurreck, *Antiviral Res.*, 2007, **73**, 197.
60. D. V. Morrissey, K. Blanchard, L. Shaw, K. Jensen, J. A. Lockridge, B. Dickinson, J. A. McSwiggen, C. Vargeese, K. Bowman, C. S. Shaffer, B. A. Polisky and S. Zinnen, *Hepatology*, 2005, **41**, 1349.
61. D. V. Morrissey, J. A. Lockridge, L. Shaw, K. Blanchard, K. Jensen, W. Breen, K. Hartsough, L. Machemer, S. Radka, V. Jadhav, N. Vaish, S. Zinnen, C. Vargeese, K. Bowman, C. S. Shaffer, L. B. Jeffs, A. Judge, I. MacLachlan and B. Polisky, *Nat. Biotechnol.*, 2005, **23**, 1002.
62. J. Shen, R. Samul, R. L. Silva, H. Akiyama, H. Liu, Y. Saishin, S. F. Hackett, S. Zinnen, K. Kossen, K. Fosnaugh, C. Vargeese, A. Gomez, K. Bouhana, R. Aitchison, P. Pavco and P. A. Campochiaro, *Gene Ther.*, 2006, **13**, 225.
63. A. Grunweller, C. Gillen, V. A. Erdmann and J. Kurreck, *Oligonucleotides*, 2003, **13**, 345.
64. T. Holen, M. Amarzguioui, M. T. Wiiger, E. Babaie and H. Prydz, *Nucleic Acids Res.*, 2002, **30**, 1757.
65. C. Lorenz, P. Hadwiger, M. John, H. P. Vornlocher and C. Unverzagt, *Bioorg. Med. Chem. Lett.*, 2004, **14**, 4975.
66. J. Soutschek, A. Akinc, B. Bramlage, K. Charisse, R. Constien, M. Donoghue, S. Elbashir, A. Geick, P. Hadwiger, J. Harborth, M. John, V. Kesavan, G. Lavine, R. K. Pandey, T. Racie, K. G. Rajeev, I. Rohl, I. Toudjarska, G. Wang, S. Wuschko, D. Bumcrot, V. Koteliansky, S. Limmer, M. Manoharan and H. P. Vornlocher, *Nature*, 2004, **432**, 173.
67. J. J. Turner, S. Jones, M. M. Fabani, G. Ivanova, A. A. Arzumanov and M. J. Gait, *Blood Cells Mol. Dis.*, 2007, **38**, 1.
68. E. Song, P. Zhu, S. K. Lee, D. Chowdhury, S. Kussman, D. M. Dykxhoorn, Y. Feng, D. Palliser, D. B. Weiner, P. Shankar, W. A. Marasco and J. Lieberman, *Nat. Biotechnol.*, 2005, **23**, 709.
69. J. O. McNamara, 2nd, E. R. Andrachek, Y. Wang, K. D. Viles, R. E. Rempel, E. Gilboa, B. A. Sullenger and P. H. Giangrande, *Nat. Biotechnol.*, 2006, **24**, 1005.
70. M. Famulok and G. Mayer, *Chembiochem*, 2005, **6**, 19.
71. M. Rimmele, *Chembiochem*, 2003, **4**, 963.
72. S. M. Nimjee, C. P. Rusconi and B. A. Sullenger, *Annu. Rev. Med.*, 2005, **56**, 555.

73. E. W. Ng, D. T. Shima, P. Calias, E. T. Cunningham Jr., D. R. Guyer and A. P. Adamis, *Nat. Rev. Drug Discov.*, 2006, **5**, 123.
74. J. Chelliserrykattil and A. D. Ellington, *Nat. Biotechnol.*, 2004, **22**, 1155.
75. P. E. Burmeister, S. D. Lewis, R. F. Silva, J. R. Preiss, L. R. Horwitz, P. S. Pendergrast, T. G. McCauley, J. C. Kurz, D. M. Epstein, C. Wilson and A. D. Keefe, *Chem. Biol.*, 2005, **12**, 25.
76. S. Klussmann, A. Nolte, R. Bald, V. A. Erdmann and J. P. Furste, *Nat. Biotechnol.*, 1996, **14**, 1112.
77. B. Wlotzka, S. Leva, B. Eschgfaller, J. Burmeister, F. Kleinjung, C. Kaduk, P. Muhn, H. Hess-Stumpp and S. Klussmann, *Proc. Natl. Acad. Sci. U.S.A.*, 2002, **99**, 8898.
78. W. G. Purschke, D. Eulberg, K. Buchner, S. Vonhoff and S. Klussmann, *Proc. Natl. Acad. Sci. U.S.A.*, 2006, **103**, 5173.
79. K. S. Schmidt, S. Borkowski, J. Kurreck, A. W. Stephens, R. Bald, M. Hecht, M. Friebe, L. Dinkelborg and V. A. Erdmann, *Nucleic Acids Res.*, 2004, **32**, 5757.
80. R. Crinelli, M. Bianchi, L. Gentilini and M. Magnani, *Nucleic Acids Res.*, 2002, **30**, 2435.
81. R. Crinelli, M. Bianchi, L. Gentilini, L. Palma, M. D. Sorensen, T. Bryld, R. B. Babu, K. Arar, J. Wengel and M. Magnani, *Nucleic Acids Res.*, 2004, **32**, 1874.
82. D. R. Corey, *Nat. Chem. Biol.*, 2007, **3**, 8.

CHAPTER 2

Genasense (G3139): An Antisense Bcl-2 Oligodeoxyribonucleotide with Substantial Clinical Activity and a Complex Mechanism of Action

CY A. STEIN,* NOAH KORNBLUM, JOHNATHAN LAI AND LUBA BENIMETSKAYA

Albert Einstein-Montefiore Cancer Center, Department of Oncology, Montefiore Medical Center, 111 E. 210 St., Bronx, NY 10467, USA

2.1 Early Studies on G3139

Genasense (G3139 or oblimersen) is an 18-mer phosphorothioate oligodeoxyribonucleotide (oligo) that is complementary to codons 1–6 of the Bcl-2 messenger RNA (mRNA).[1] The compound was ultimately derived from a 20-mer phosphorothioate oligo called T1-AS that straddled the Bcl-2 initiation codon region,[2] and which thus had significant partial homology to G3139. Pre-B ALL cells (line 697), which lack the t14:18 translocation, were treated under serum-free conditions with this oligo, which, in the time before lipofection protocols had been developed, was employed at concentrations from 25 to 150 μM. A diminution in the expression of Bcl-2 protein was seen and also, as a harbinger of future events, extensive cellular apoptosis. The sequence currently known as G3139 was initially described by Kitada et al.,[3] who treated t(14:18) translocation-bearing SU-DHL-4 lymphoma cells with a phosphodiester congener of G3139. Inhibition

RSC Biomolecular Sciences
Therapeutic Oligonucleotides
Edited by Jens Kurreck
© Royal Society of Chemistry 2008

of Bcl-2 translation in a cell-free system required a remarkable 10 μM concentration of oligonucleotide (ON), increased to 200 μM (*sic*) in the SU-DHL-4 cell line. Significant apoptosis occurred after five days, but since the experiment was terminated at that time, and since apoptosis was simultaneously occurring, albeit to a lesser extent, in the sense control, the extent of the contribution of Bcl-2 downregulation to the apoptotic process is unclear. Subsequently, Kitada *et al.*[4] treated several hematopoietic cell lines with phosphodiester G3139 and a variant in which the 3′-internucleotide linkage was phosphorothioate. The ONs were complexed with Lipofectin, which decreased the optimal transfection concentration to 75–300 nM. Putatively specific reductions in Bcl-2 protein expression were quite variable, but the presence of the 3′-phosphorothioate did not increase silencing. This is not particularly surprising, because 50% of these 3′-phosphorothioate linkages would consist of the Rp diastereomer, which is as sensitive to nuclease degradation as a phosphodiester.[5] Nevertheless, despite sub-optimal Bcl-2 downregulation, significant pro-apoptotic synergy was observed in SU-DHL-4 cells with methotrexate, cytosine arabinoside and dexamethasone. Several other reports[6–8] have tended to confirm the anti-Bcl-2 activity of G3139 in hematopoietic cells and the increased apoptosis in these cells after treatment with chemotherapeutic agents. In a rather striking example of the *in vitro* and *in vivo* activity of G3139, Cotter *et al.*[9] cultured t(14:18)-bearing cells obtained from a patient with a B-cell lymphoma and treated them with G3139. These treated cells could not grow as xenografts in severe combined immunodeficiency disease (SCID) mice. In a separate experiment, animals bearing previously untreated xenografts were treated for two weeks with G3139 (100 μg/day). Lymphoma growth was abolished in 83% (50 of 60 animals). Similar results were obtained in non-obese diabetic (NOD) SCID animals, which lack NK-, B- or T-cell activity, but which do, however, retain plasmacytoid dendritic cells. The significance of this, and its confounding effects on the interpretation of experiments of this type, is described below.

Based on these preclinical data, Webb *et al.*[10] initiated a Phase I trial with G3139 in nine patients with lymphoma whose cell Bcl-2 protein was 'elevated'. All patients had failed at least two therapeutic interventions. Patients were treated at two doses, 4.6 and 73.6 mg m^{-2} day^{-1}. The maximum tolerated dose (MTD) was not achieved. There was one complete response (CR), three patients with stable disease (SD), and five with progressive disease (PD). A subsequent study administered G3139 via continuous infusion to 21 patients at doses of 4.96–196 mg m^{-2} day^{-1} for 2 weeks. One patient attained a CR, two patients a minor response (MR), nine had SD, and the remaining nine had PD. In seven of 16 evaluable patients, Bcl-2 protein was diminished in cells derived from lymph nodes (two patients) or in samples of peripheral blood or bone marrow (five patients).

In retrospect, given the general paucity of controls in the early pre-clinical work, coupled with our current understanding of the biological behavior of phosphorothioate oligos, it is difficult to accept that the results described above were not substantially the result of non-sequence specific events. This, indeed, is the problem in the interpretation of the vast majority of experiments in which

G3139 has been employed, *i.e.*, the extent to which a biological readout (*e.g.*, apoptosis) is actually related to the assumed mechanisms of action of G3139, which, of course, is silencing of Bcl-2 expression. This question is further complicated by uncertainties that surround the actual function of Bcl-2 protein in cancer cells. For example, it is no longer clear that Bcl-2 protein is both necessary and sufficient for the maintenance of the neoplastic phenotype in follicular lymphoma cells bearing the t14:18 translocation, which produces a fused Bcl-2/immunoglobulin mRNA in about 65–70% of cases.[11] Similarly, despite several clinical correlations between the expression (or 'overexpression') of Bcl-2 protein and a poor prognosis in cancer patients with tumors,[12–15] it is still possible that the elevated expression of Bcl-2 protein in many tumor cells is an epiphenomenon, and not directly related to chemo-resistance. Given the plethora of pro- and anti-apoptotic signals that can bombard the mitochondria of solid tumor cells (certainly over 100 proteins may be contributory to the process), additional questions must be raised about the value of Bcl-2 as a stand-alone target.

There is no question whatsoever that the forced over-expression of Bcl-2 protein produces drug resistance in numerous human tumor cell lines,[16,17] and that this resistance occurs through a variety of mechanisms.[18] However, the reverse is not necessarily so. Indeed, forced downregulation of Bcl-2 protein expression from its basal level need not *necessarily* lead to chemo-sensitization.[19,20] This appears to be true in those melanoma cell lines that have been studied, as is described below. It is also true in PC3 prostate cancer cells (Benimetskaya and Stein, unpublished observations), and may well turn out to be true in other tumor types, were they to be extensively studied.

Also, the reasoning that has led to the idea that forced downregulation of basal levels of Bcl-2 protein expression *necessarily* leads to chemo-sensitization appears to be, in our opinion, somewhat circular. The idea seems to have been derived, at least in part, from preconceptions of what ought to have occurred given the chemo-resistance induced by Bcl-2 forced overexpression, coupled with data obtained from the use of G3139 or other phosphorothioate ONs to silence Bcl-2 expression successfully. However, after G3139 treatment, specific gene-silencing occurs during the same time and at the same oligo concentration as numerous other events. Some events are involved in the process of apoptosis independently of Bcl-2 status. The complexity of these processes is perhaps best exemplified by studies of G3139 in melanoma.

2.2 G3139 in Melanoma: Clinical and Preclinical Results

In earlier studies, treatment of 518A2 melanoma cells *in vitro* with G3139 complexed with Lipofectin led to as much as a 60% downregulation of Bcl-2 protein expression (200 nM, 48 h).[21] 518A2 cells were then xenografted into SCID mice, and treated with G3139 for 14 days at 5 mg kg^{-1} day^{-1} via Alzet minipumps. Dacarbazine, the most active drug in clinical melanoma, was injected at 80 mg kg^{-1} intraperitoneally (i.p.) from days 12 to 16. Apoptosis, as assessed by terminal uridine deoxynucleotidyl transferase nick end labeling (TUNEL) assay, increased from 0.7% in the control, saline or mismatched ON group to 3.3% in the antisense group. Levels of Bcl-2 protein expression were stated as being reduced by 66–72% in the G3139-treated group, but only a single Western blot example was shown. Furthermore, the reverse control oligo altered the flanking sequences of the CpG motifs relative to G3139, while the mismatched oligo eliminated them entirely. The authors of this work were aware of these problems, and in a follow-up[22] the issue of immunostimulation by the two CpG motifs of G3139 was re-examined. SCID mice bearing 518A2 melanoma xenotransplants were treated with G3139 or with G4232, a G3139 variant that contains C5-methylcytosine at each CpG motif. This substitution is purported to abolish immune stimulation by CpG phosphorothioates, but the extent to which it does so is somewhat controversial. Tumor growth after either G3139 or G4232 treatment was decreased by up to 40% *vs.* untreated animals after 14 days continuous intravenous (i.v.) infusion, although from the data it is not clear if statistical significance *vs.* the control was achieved. Bcl-2 protein downregulation of 60% was claimed for both treatments, but given the variability of Western blots and the single example shown, these data must be considered incomplete. Perhaps of greater significance, both G4232 and G3139 caused an increase in TUNEL staining in the tumors of magnitude equivalent to that previously reported.[21] Of interest, too, is the substantial immunostimulatory effect produced by G3139, characterized by splenomegaly and a six-fold increase in serum levels of interleukin-12 (IL-12). [However, IL-12 at concentrations higher than measured in the serum of G3139-treated animals did not produce any direct slowing of the growth of 518A2 cells in tissue culture (J. Lai and C.A. Stein, unpublished results).] Smaller increases (1.5–2-fold normal) in spleen weight and IL-12 levels were observed after G4232 treatment. The authors concluded that the antitumor effects of G3139 and G4232 were due to ' . . . a Bcl-2 antisense effect independent of CpG immune stimulation'.

However, a somewhat different set of conclusions was drawn by Gekeler *et al.*,[23] although these authors did not perform experiments in melanoma cells. Prior to this work, it had been believed that in animals devoid of T cells, human tumor xenograft responses to CpG ONs required cytotoxic chemotherapy.[24,25] However, treatment of nude mice bearing the H69 SCLC xenograft with G3139 produced tumor growth arrest, which was similar in scale to the effects of cisplatin, and to H1826, an optimized immune stimulatory oligo.[26] A G3139 derivative in which all the deoxycytidine residues were C5-methylated did not produce any inhibition of tumor growth in this model, although its ability to downregulate Bcl-2 expression *in vitro* was essentially equivalent to that of G3139 itself. The A2780 ovarian carcinoma and the 549 lung adenocarcinoma xenograft models essentially behaved similarly. It was this lack of activity after cytosine C5-methylation that convinced Gekeler *et al.*[23] that immunostimulation was the origin of the observed antitumor effect.

Some additional data that comment on this problem in melanoma cells were provided by Lai *et al.*[27] The nucleosides of G3139 are derived from the naturally occurring D-deoxyribose stereoisomer at C1 of the sugar ring. The behavior of D-G3139 was compared with that of its unnatural, but isosequential, mirror image, L-G3139, which contained L-deoxyribose (2'-deoxy-β-L-erythro-pentafuranose). Unlike D-G3139, L-G3139 is virtually impervious to nucleases[28,29] although both contain only phosphorothioate linkages. Most significantly, L-G3139 does not elicit ribonuclease H (RNase H) activity, and thus does not cause downregulation of Bcl-2 mRNA or protein expression.

With respect to non-specific properties, *i.e.*, binding to the heparin-binding protein bFGF, uptake in 518A2 or A375 melanoma cells and release of cytochrome c from the mitochondrion (see below), D-G3139 and L-G3139 are virtually indistinguishable. However, the induction of apoptosis in melanoma cells, as measured by procaspase-3 and PARP-1 cleavage, was much more rapid for D-G3139 than for L-G3139 (9.5 hours *vs.* 24 hr after the initiation of the transfection). The reason for this highly reproducible observation is currently uncertain. Most importantly, when SCID mice bearing A375 human tumor xenografts were treated with D-G3139 either at $10 \, mg \, kg^{-1} \, day^{-1}$ for 14 days or at $20 \, mg \, kg^{-1}$ every other day for 14 days, tumor growth was markedly reduced. In contrast, no tumor growth inhibition at all was observed when the same model was treated with L-G3139. Moreover, D-G3139, but not L-G3139, caused increased expression of IL-6 and IL-12 in BALB/c mouse splenocytes in culture, and increased production of IL-12 and interferon-γ (IFN-γ) in SCID mice. These experiments seem to suggest strongly that immune stimulation caused the observed reductions in tumor size, even though neither IL-12 nor IFN-γ may be the responsible tumoricidal agent. However, as a counter argument, the differential silencing of Bcl-2 expression by D-G3139 (excellent) *vs.* L-G3139 (completely inactive) has still not been accounted for. It still remains possible that the downregulation of Bcl-2 expression either directly contributes to antitumor effects of D-G3139, or synergizes with TLR9-mediated effects.

2.3 The Role of Bcl-2 in Melanoma

How are we to understand the role of Bcl-2 in melanoma? The general question of the relevance of Bcl-2 expression to cell death has been discussed by Blagosklonny,[30] who notes that it is not a direct one. He states that it would be expected that cells which express Bcl-2 would be apoptosis resistant, and vice versa, yet Bcl-2 appears to be unnecessary for many cell types. Furthermore, Bcl-2 is often expressed at low levels in highly apoptosis-resistant cells, and its expression in colorectal, breast and lung carcinomas is associated with 'increased apoptotic index, lower risk of distant metastases, and improved prognosis'. In addition, cell lines ' . . . become resistant due to a strong selection during establishment of cells in culture, over-expression of Bcl-2 simply cannot further increase resistance and [the] effects of Bcl-2 are undetectable'. Tumor cells may also downregulate proteins involved in the apoptotic cascade. An example in melanoma is Apaf-1, which is downstream of Bcl-2, and in whose absence the level of Bcl-2 protein expression would appear to be irrelevant.[31]

Bcl-2 protein is present in normal melanocytes, benign nevi, primary melanomas and melanoma metastases,[32] and its role in the pathogenesis and prognosis of clinical melanoma is a matter of some controversy.[33] In older studies, Bcl-2 expression was noted to be decreased in melanoma cells *vs.* normal melanocytes.[34–37] Other studies[32,38,39] have observed minimal differences in the expression of Bcl-2 in this comparison. In advancing melanoma, about one-third of the data suggest an increase in Bcl-2 expression, while one-third actually suggest a decrease.[33] In contrast, one study with an insufficient number of patients has demonstrated that advanced melanoma patients whose lymph node deposits express Bcl-2 have a poorer prognosis than those who do not.[33] The very complexity and lack of clarity that the sum total of these observations represent suggest that the role of Bcl-2 in chemoresistance in melanoma is murky, as suggested above by Blagosklonny,[30] and as further explained below.

To begin to address the question of the relevance of Bcl-2 as a target in melanoma, Benimetskaya *et al.*[20] studied 518A2 cells after treatment with G3139 or with a small interfering RNA (siRNA) called D6 (both as complexes with a lipidic transfection reagent). All treatments successfully, dramatically, reduced expression of Bcl-2 mRNA and protein. However, treatment with G3139 did not lead to any sensitization to the cytotoxic agents docetaxel, gemcitabine, cis-platin or thapsigargin (chosen because its mode of action is independent of cell cycle). Treatment with the D6 siRNA also did not lead to any diminution in cell viability, nor to sensitization to either docetaxel or thapsigargin. Remarkably, when 518A2 cells were treated with G3139, they were dramatically protected against the cytotoxic effects of light-activated dacarbazine.[40] Cytoprotection to dacarbazine toxicity was not observed with a phosphodiester congener of G3139, with the D6 siRNA or an 18-mer homopolymer of thymidine, nor was it seen in several other melanoma cell lines. However, both cis-platin and gemcitabine were also protective to dacarbazine cytotoxicity. Cytoprotection by G3139 to dacarbazine was abolished by O6-benzylguanine, a rather non-specific inhibitor of O6-guanosine alkyltransferase (OGAT) activity. However, this is an enzyme of which 518A2 cells were shown to be devoid, and thus the mechanism of cytoprotection by C3139 (or gemcitabine or cis-platin) remains uncertain.

2.4 G3139 Causes Bcl-2 Independent Apoptosis in Melanoma Cells

Older work in a SCID-hu melanoma model had hinted that the effects of phosphorothioate ONs were non-specific. Subsequently, a much more detailed study of the effects of G3139 treatment of melanoma cells was conducted by Lai *et al.*[41]

The viability of 518A2 melanoma cells, when treated with concentrations of G3139 as high as 40 µM in the absence of a lipidic delivery agent, was not affected. In contrast, treatment with 100 nM G3139 in complex with Lipofectamine 2000 caused extensive apoptosis, as measured by cell-surface annexin V binding and 4',6-diamidino-2-phenylindole (DAPI) staining. This apoptotic process, after two days, was associated with extensive mitochondrial membrane depolarization.

Apoptosis of the 518A2 melanoma cells led to cleavage of PARP-1, cleavage of pro-caspase-3 to caspase-3 (and increased DEVDase activity) and the cleavage of Bid to tBid, although no activation of pro-caspase-8 was observed. These changes were initiated as early as 9.5 hours after initiation of the G3139 transfection, at a time when neither Bcl-2 protein expression nor $\Delta\Psi$m were decreased (Figure 2.1). Similar events were observed in the 346.1, 201.2, 333.1 and A375 melanoma cell lines. Contemporaneously (*i.e.*, 9.5 hours after the initiation of the transfection) with the initiation of these molecular events, cytochrome c was observed in the cytoplasm of treated cells (Figure 2.2). Moreover, the release of cytochrome c from the mitochondria was essentially independent of Bcl-2 protein expression. In a key set of experiments (Figure 2.3), in total cellular lysates of 518A2 cells, Bcl-2 protein expression was decreased by approximately 34% after 24 hours and 91% after 48 hours. However, mitochondrial levels of Bcl-2 protein were only minimally released by G3139 after 24 hours, and only by 40% after 48 hours. By this time substantial apoptosis is observed in treated cells. In contrast, in 518A2 cells treated with the D6 anti-Bcl-2 siRNA, mitochondrial-associated Bcl-2 (actually, a combination of mitochondrial and endoplasmic reticulum-associated) is reduced by 90–99% by three days, and no apoptosis whatsoever was observed. In addition, forced overexpression of Bcl-2 did not block the induction of apoptosis by the identical concentration of G3139, but did seem to delay the initiation of pro-caspase-3 cleavage from 9.5 hours after the initiation of the transfection to about 15–18 hours.

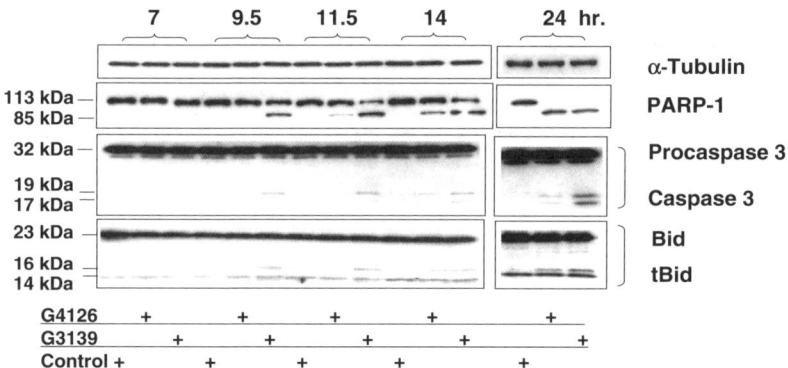

Figure 2.1 Representative Western blot analyses showing the time course of protein processing by G4126 and G3139 in 518A2 cells, as demonstrated by cleavage of PARP-1, Pro-caspase 3 and Bid. Cells were treated with 100 nM of G4126 or G3139 complexed with Lipofectamine 2000 (1.9 µg/mL) for five hours. Total cellular protein was subjected to Western blotting at the indicated time points. The average error in Western blot measurements is approximately 20–25%. Reproduced by permission of the American Association for Cancer Research.

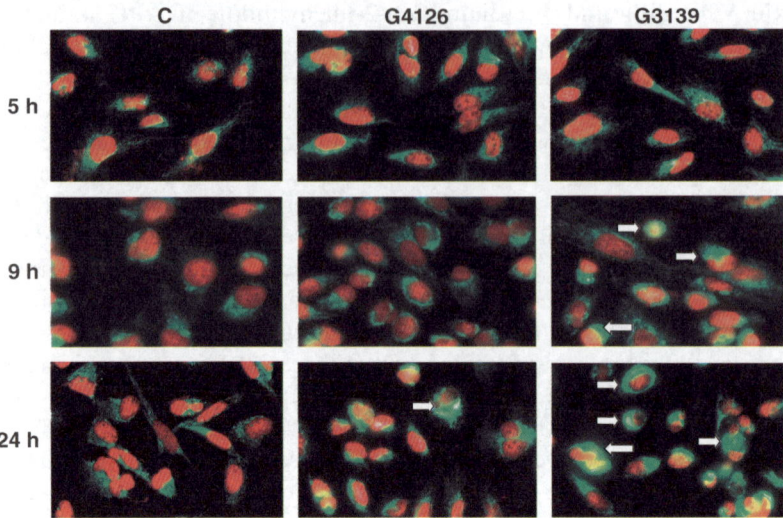

Figure 2.2   Immunohistochemical staining of cytochrome c in 518A2 cells treated with
100 nM of G4126 or G3139 complexed with Lipofectamine 2000 (1.9 µg/
mL). At the indicated times post-transfection, cells were co-stained with
anti-cytochrome c antibody and DAPI and analyzed by fluorescence microscopy. Cytochrome c release could be detected as early as nine hours
post-transfection in G3139-treated cells, as indicated by the white arrows.
By 24 hours, marked diffuse cytoplasmic staining of cytochrome c was
evident in both G4126- and G3139-treated cells. Reproduced by permission of the American Association for Cancer Research.

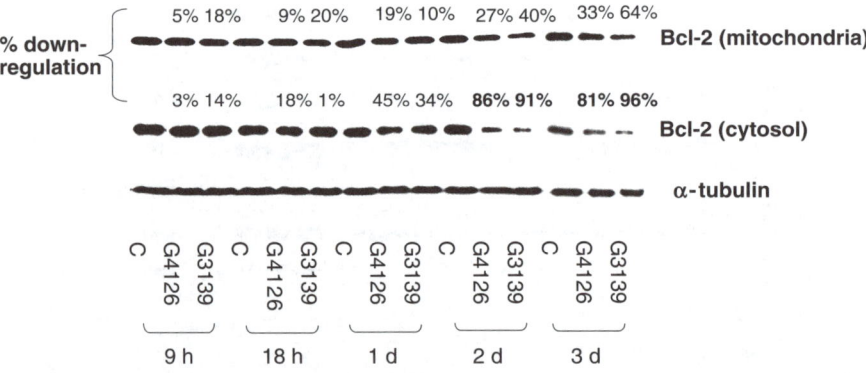

Figure 2.3   Bcl-2 protein downregulation is not required for the induction of apoptosis
in 518A2 melanoma cells. Untreated cells and cells treated with 100 nM of
G4126 or G3139, for the indicated times, were fractionated as described
previously. Bcl-2 protein expression in mitochondria *vs.* cytoplasmic lysate
was determined by Western blotting, as described in Figure 2.1 caption,
and quantitated by laser-scanning densitometry. The average error in this
measurement is approximately 20–25%. Reproduced by permission of the
American Association for Cancer Research.

The early activation of the intrinsic apoptotic pathway by G3139 relative to its activation by cytotoxic chemotherapy (minimum time required = 24 hours) suggests why chemo-sensitization is not, and cannot be, observed in these cells. While extrapolation of these experiments to the *in vivo* setting is also not possible, the data, in the relatively copious amounts in which it exists, strongly suggests that, although G3139 can downregulate Bcl-2 protein expression in several melanoma cell lines, this ability is not relevant to the observed apoptotic phenotype. However, this interpretation does not necessarily rule out any role whatever for Bcl-2 in melanoma. Benimetskaya et al.[42] constructed a plasmid that when transfected into 518A2 melanoma cells, constitutively expressed a short single-stranded ON complementary to the initiation codon region of the Bcl-2 mRNA. This produced long-term suppression of the expression of Bcl-2 mRNA and protein. Bcl-2 silencing was also accomplished via transfection of a stably expressed anti-Bcl-2 shRNA. *In vitro*, the clones with dramatically downregulated Bcl-2 expression behaved virtually almost identically to control clones with respect to growth rate, chemo-sensitivity and release of cytochrome c from isolated mitochondria. *In vivo* (SCID mice), however, xenografted cells that lacked Bcl-2 expression either grew to form very small tumors, or did not grow at all. In previous work, it had been demonstrated[43] in the M14 human melanoma cell line that Bcl-2 overexpression increased hypoxia-stimulated expression of vascular endothelial growth factor (VEGF) protein and mRNA. Expression of the HIF-1α transcription factor, which upregulates VEGF expression, was also increased. Similar observations were also seen in a breast cancer cell line.[44] However, in the transfected 518A2 melanoma cell lines, downregulation of VEGF protein expression was not observed. It has also been suggested that increased Bcl-2 expression in melanoma could increase its aggressiveness by augmenting the activity of MMP-2 or -9,[45] but Benimetskaya et al.[42] did not observe downregulation of MMP-2 or -9 activity after zymography of conditioned media from the transfected cells. Finally, conditioned media from Bcl-2-transduced human mammary epithelial cells (HMECs) can promote neovascularization in a rat corneal assay. This was thought to occur because of Bcl-2-induced increases in the activity of IL-8 and CXCL1, both pro-angiogenic molecules.[46] Unfortunately, the xenografts examined in the work of Benimetskaya et al.[42] contained extensive central necrosis in the setting of poor vascularity, vitiating any possible comparisons.

How could G3139 cause apoptosis in melanoma cells? One possibility was described recently by Lai et al.[47] G3139 has the ability to bind to a protein on the outer mitochondrial membrane known as VDAC (voltage-dependent anion-selective channel). This protein is responsible for the flux of anions, including phosphocreatine, ATP^{4-} and HPO_4^{2-}, across the membrane.[48–50] The binding of G3139 to VDAC causes reversible closure of the channel and a dramatic reduction (six- to seven-fold) in the permeability of the channel to ADP.[51] The resulting failure to exchange metabolites between the cytosol and the mitochondrion results, via an unknown mechanism, in the re-permeabilization of the mitochondrial outer membrane and the release of cytochrome c, which initiates the apoptotic process.

However, it is not entirely clear that VDAC closure is the mechanism whereby apoptosis is induced in melanoma cells by G3139. First, the ability to bind to and close VDAC by G3139 is essentially non-sequence specific, and in cell-free models depends entirely on ON length and phosphorothioate content. Second, while the ability of mirror-image D- and L-G3139 to close VDAC is identical in the same cell-free models, their ability to cause apoptosis in living cells differs. As mentioned previously, apoptosis is initiated within 9.5 hours after D-G3139 transfection, but only after about 24 hours after L-G3139 transfection. Third, the K_i of permeability reduction in isolated mitochondria is approximately 0.18 µM, which may be a higher ON concentration than can be obtained within cells, and release of cytochrome c from isolated mitochondria requires approximately 5 µM. However, these measurements in isolated mito-chondria may well not mimic the situation in intact cells. Apropos, the K_i of VDAC closure by G3139, when the protein is inserted into an artificial lipid membrane, is 9.6 µM, over 50-fold higher than that observed in isolated mito-chondria. Furthermore, we (A. Santal and C.A. Stein, unpublished data) have obtained confocal micrograms in 518A2 melanoma cells that demonstrate mitochondrial co-localization of fluoresceinated G3139 and a Cy3-labeled anticytochrome c mAb, which is mitochondria-specific. Thus, the fact that all phosphorothioate oligos of length greater than or equal to 16-mer induce apoptosis after 24 hours may, indeed, be a reflection of VDAC closure at this time. However, the induction of apoptosis at 9.5 hours after the transfection, which seems to be specific to D-G3139, has at the moment no known cause.

2.5   Clinical Trials in Advanced Melanoma

A small Phase I/II trial of G3139 (brand name oblimersen) in combination with dacarbazine in advanced melanoma was published by Jansen et al.[52] Dacar-bazine is apparently the most active single agent in this disease, although its response rate is approximately 7%. It was approved by the Food and Drug Agency (FDA) in 1975, which marked the last time that the agency has approved any agent for this indication.

Oblimersen was administered via continuous infusion for 14 days. The initial dose was 0.6 mg kg^{-1} day^{-1}, which was rapidly increased in subsequent patients in the absence of toxicity, with a dose maximum of 6.5 mg kg^{-1} day^{-1}. Dacarbazine was administered separately at a dose of 200 mg m^{-2} on days 5–9. Treatment cycles were repeated monthly. Levels of Bcl-2 protein expression were evaluated in biopsy specimens, but the maximum decrease in Bcl-2 expression was highly variable, and the numbers of samples were far too few for any statistically meaningful conclusions to be drawn. Small increases in the number of apoptotic cells, as evaluated by TUNEL assay, were also described. Some evidence of antitumor activity was observed in six patients, some of which had unsuccessfully received prior melanoma treatment. These included one CR, two partial responses (PRs) and two patients whose disease stabilized for a period of at least one year.

The data obtained from this relatively small Phase I/II trial, which was itself based in part on the pre-clinical 518A2 human tumor xenograft melanoma model (as described above), led to the initiation of the largest Phase III trial in advanced melanoma ever, sponsored by Genta, Inc. (Berkeley Heights, NJ). Between July 2000 and February 2003, 771 chemotherapy-naïve patients with advanced malignant melanoma were randomly assigned to receive treatment with dacarbazine alone ($1000 \, \text{mg m}^{-2} \, \text{day}^{-1}$ i.v. for 60 minutes) or oblimersen sodium ($7 \, \text{mg kg}^{-1} \, \text{day}^{-1}$ by continuous infusion for five days) followed by the same dacarbazine dose.[53]

Patients were stratified according to ECOG performance status (0 *vs.* 1–2), presence or absence of liver metastasis and disease site and/or serum lactate dehydrogenase (LDH) level. This latter category included two groups, patients with non-visceral disease (skin, subcutaneous tissue or lymph node disease) *and* normal LDH, and those with visceral disease (excluding liver) *or* elevated LDH [baseline serum level at least $1.1 \times$ the upper limit of normal (ULN)].[53] The baseline characteristics of the groups, including age, sex, time from diagnosis, and prior treatments, were well balanced.

The primary endpoint of the study was a comparison of overall survival between the two treatment groups conducted on an intent-to-treat basis. Secondary endpoints included progression-free survival, overall and durable response, and duration of response. The study design estimated that the addition of Genasense to dacarbazine would increase median survival from 6 months to 8 months, with a 90% power to detect this difference at a two-sided significance level of 0.05. Response was measured according to RECIST (Response Evaluation Criteria in Solid Tumors) standard algorithms assessed by an independent panel blinded to treatment assignment.[54]

With a minimum follow-up of 24 months, the median overall survival in the Genasense–dacarbazine cohort was 9.0 months, compared with 7.8 months observed in the dacarbazine-alone group (Figure 2.4; hazard ratio (HR) 0.87, 95% CI, range 0.75 to 1.01, $p = 0.077$).[53] Overall response rates (complete plus PR) were 13.5% for patients treated with Genasense plus dacarbazine, and 7.5% for patients who received dacarbazine alone ($p = 0.007$). Durable responses (≥ 6 months) were also increased for the Genasense–dacarbazine group (7.3% *vs.* 3.6%; $p = 0.03$). Eleven patients (2.3%) in the combination treatment arm achieved CR in comparison to three patients (0.8%) in the dacarbazine-alone group. Median progression-free survival was also significantly longer among patients who received Genasense than in those treated with dacarbazine alone (2.6 *vs.* 1.6 months, HR 0.75, $p < 0.001$).

Outcome data were analyzed according to the LDH-stratification category – LDH has long been recognized as an important independent biomarker of poor prognosis in malignant melanoma.[55] In this trial, multivariate analyses that accounted for differences in baseline prognostic factors revealed an interaction between treatment and LDH activity level. Study patients with LDH values $< 1.1 \times$ ULN who received the Genasense–dacarbazine combination (Figure 2.5; approximately two-thirds (508) of the 771 subjects) were observed to have significantly better treatment outcomes for all efficacy end points. These included

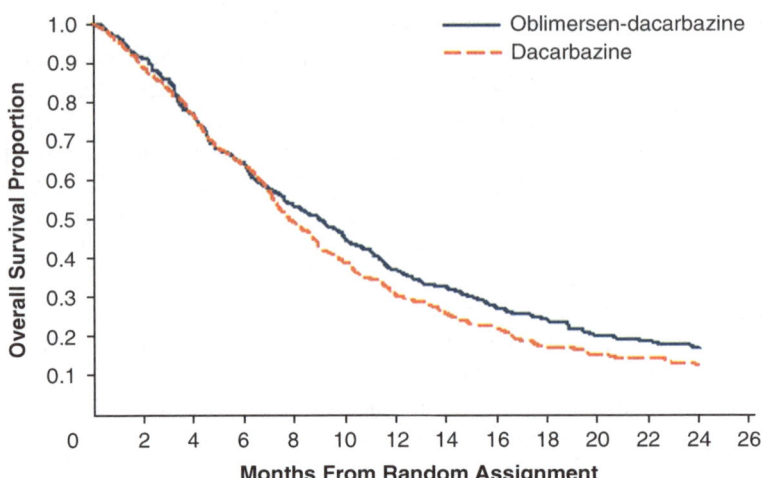

Figure 2.4 Clinical data for oblimersen in melanoma. Depicted are Kaplan–Meier estimates of overall survival. [Median overall survival = 9.0 months for oblimersen–dacarbazine *vs.* 7.8 months for dacarbazine; $P = 0.077$; hazard ratio = 0.87; 95% confidence level (CI), 0.75–1.01.[53]] Reproduced by permission of the American Society of Clinical Oncology.

overall survival (median, 11.4 *vs.* 9.7 months; $p = 0.02$), progression-free survival (median, 3.1 *vs.* 1.6 months, $p < 0.001$), overall response (17.2% *vs.* 9.3%; $p = 0.009$), CR (3.4% *vs.* 0.8%) and durable response (9.6% *vs.* 4.0%; $p = 0.01$). In sharp contrast, significant differences between treatment groups were not observed for patients with elevated ($> 1.1 \times$ ULN) baseline LDH levels (Figure 2.6).

New, unpublished data have revealed the dramatic extent to which LDH stratification, *even within the 'normal' range*, is predictive of prognosis in advanced melanoma. As shown in Figure 2.7, which is a retrospective examination of a recently completed large study in advanced melanoma (EORTC 18951, $N = 330$), there is an almost linear progression to improved prognosis as the value of LDH descends. Moreover, this improved prognosis with diminishing LDH could also be observed in the current trial. As shown in Figure 2.8 for patients ($N = 108$) whose LDH is $< 0.8 \times$ ULN, the median survival (at 24 months, Genasense–dacarbazine *vs.* dacarbazine alone) was 12.3 months *vs.* 9.9 months ($p = 0.0009$, HR 0.64). LDH seems clearly destined to emerge as an important parameter for the identification of patients most likely to benefit from Genasense–dacarbazine treatment in the future.

What could be the origin of this remarkable dependence of treatment success and overall prognosis on LDH? LDH is, of course, found in all cells, but may be seen in higher levels in tumor cells because of their greater reliance on glycolysis secondary to relatively poor vascularization and diminished oxygen delivery. The enzyme is released from dying cells during the process of necrosis, but it is not released after apoptosis. The induction of necrosis of tumor cells may often depend on the balance between their rate of proliferation *vs.* the rate of vascularization of the growing tumor. Therefore, high LDH levels in patients

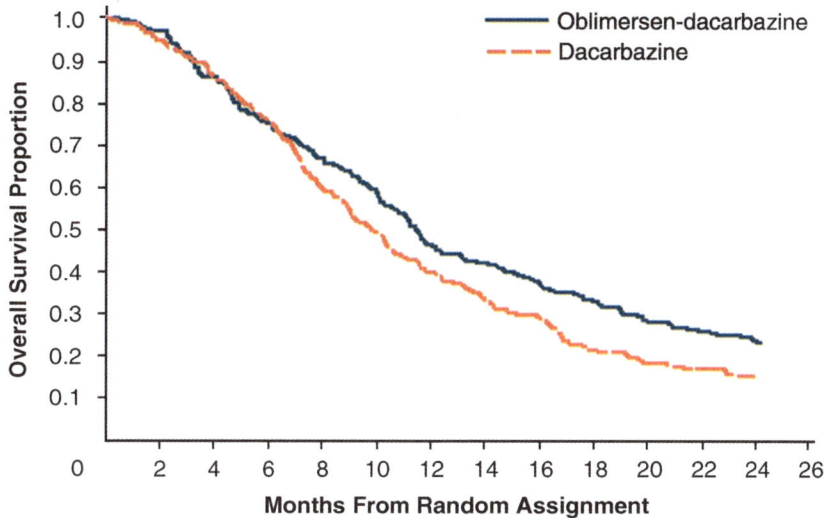

Figure 2.5 Clinical data for oblimersen in melanoma: Depicted are Kaplan–Meier estimates of overall survival in patients with normal ($< 1.1 \times$ ULN) baseline serum LDH ($n = 508$). (Median overall survival, 11.4 months for oblimersen–dacarbazine *vs.* 9.7 months for dacarbazine; $P = 0.02$; hazard ratio $= 0.79$; 95% CI, 0.65–0.96.[53]) Reproduced by permission of the American Society of Clinical Oncology.

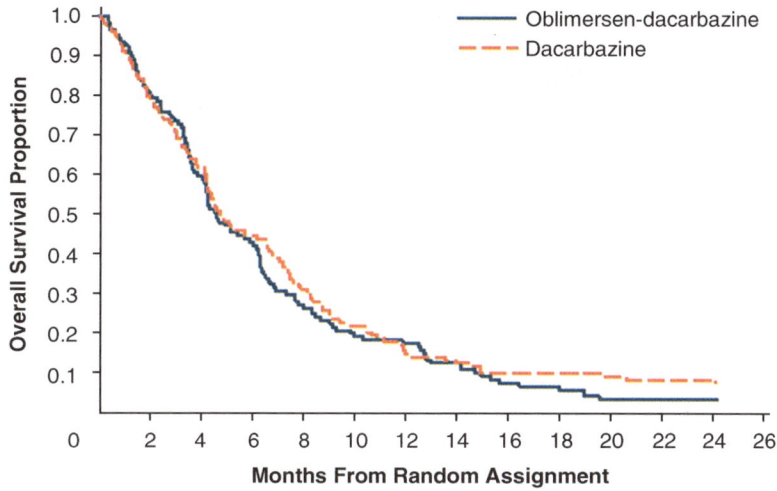

Figure 2.6 Clinical data for oblimersen in melanoma: Depicted are Kaplan–Meier estimates of overall survival of patients with elevated ($>1.1 \times$ ULN) baseline serum LDH ($n = 252$). (Median overall survival, 4.6 months for oblimersen–dacarbazine *vs.* 4.7 months for dacarbazine; $P = 0.41$.[53]) Reproduced by permission of the American Society of Clinical Oncology.

Kaplan-Meier Survival Curve
by LDH Category

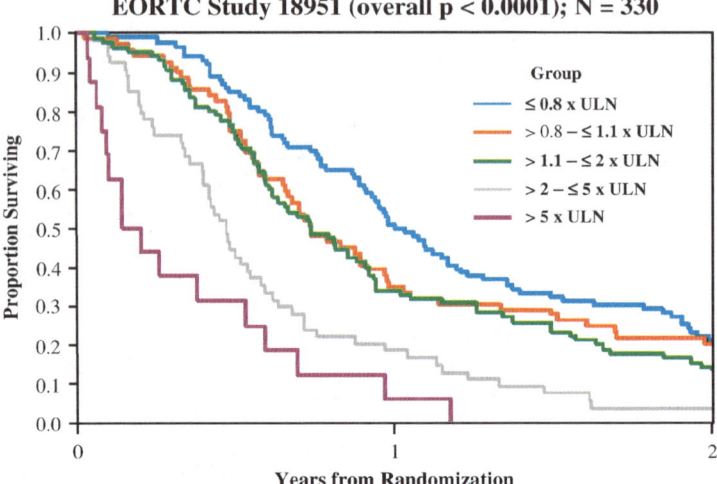

EORTC Study 18951 (overall p < 0.0001); N = 330

Figure 2.7 Depicted are Kaplan–Meier survival curves analyzed by LDH decile in melanoma patients receiving chemotherapy unrelated to oblimersen as part of EORTC 18951 (unpublished data courtesy of L. Itri, Genta Inc., Berkeley Hills, NJ).

Overal Survival
Baseline LDH ≤ 0.8 x ULN

	G + DTIC	DTIC
Median	12.3	9.9
Events/N	108/149 (73%)	106/125 (85%)
Logrank p	0.0009	
HR	0.64	
95% C.I.	0.49–0.83	

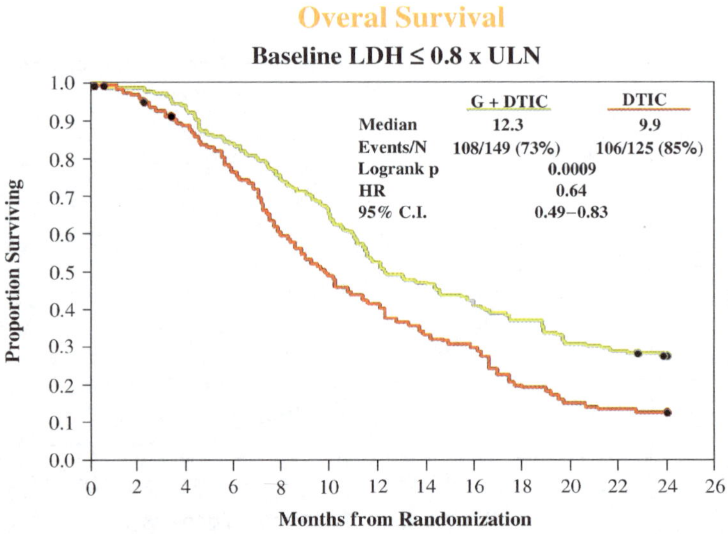

Figure 2.8 Clinical data of oblimersen in melanoma: Kaplan–Meier estimates of overall survival in patients with normal (< 0.8 × ULN) baseline serum LDH (*n* = 274). (Median overall survival, 12.3 months for oblimersen–dacarbazine *vs.* 9.9 months for dacarbazine; *P* = 0.0009; hazard ratio = 0.64; 95% CI, 0.49–0.83) (unpublished data courtesy of L. Itri, Genta Inc., Berkeley Hills, NJ).

may reflect disease that is rapidly growing and poorly vascularized. These types of tumors are frequently highly resistant to chemotherapy because, in addition to poor oxygen delivery, drug delivery may also be quite poor, and hence the lack of response to treatment.

2.6 Safety and Tolerability

The most significant adverse events were neutropenia and thrombocytopenia (Table 2.1). Grade 3 to 4 neutropenia with infection occurred at a rate of 4.3% for the Genasense–dacarbazine group, compared with 2.8% for the dacarbazine-alone cohort.

The incidence of bleeding events was increased in the Genasense–dacarbazine group at 13.7%, an increase from 9.2% observed for the dacarbazine-alone group. However, these events were primarily grade 1 to 2 epistaxis or hematuria. In fact, more grade 3 to 4 bleeding events (mostly gastrointestinal events) occurred in the dacarbazine-alone group (3.1% *vs.* 2.2%). Also of note was an apparent increased rate of catheter-related events (venous thrombosis, infection, occlusion) observed for the combination Genasense–dacarbazine group (19.1%), in comparison to the dacarbazine-alone group (8.6%). Patients without elevated baseline LDH values appeared to fare better with respect to safety findings, and had lower rates of adverse events that result in treatment discontinuation or death.

These safety data must be interpreted with caution, and should be considered in the appropriate context. Although treatment with Genasense was associated with an increased incidence of grade 3 to 4 neutropenia (21%) and thrombocytopenia (16%), these rates are still substantially lower than those associated with other drugs and drug regimens used for the treatment of advanced melanoma.[56–58] Moreover, the rates of serious bleeding events and the incidence of neutropenic infections were both low for patients who received Genasense. However, catheter-related problems remain an issue, which might be expected given that the treatment was provided via continuous i.v. infusion. Undoubtedly, future trials, if they happen, will evaluate more convenient routes of administration, such as daily subcutaneous infections or short i.v. infusions.

To date, the clinical experience with Genasense in combination with dacarbazine for the treatment of advanced malignant melanoma has been encouraging, despite its rejection for marketing by the FDA in 2003. However, this rejection occurred at a time when the benefits of LDH stratification were not clear, as they became several years later. As outlined above, the Genasense–dacarbazine has resulted in significant improvements in multiple clinical outcomes, and appears to be safe. Patients with normal baseline LDH ($<1.1 \times$ ULN), an important biomarker in melanoma, may benefit most from treatment with these agents, with benefits increasing in patients with even lower levels of LDH. Given the critical status of LDH as a stratification factor, the observations to date argue that the optimum treatment place for Genasense belongs in patients with either minimal disease, or in the adjuvant setting.

Table 2.1 Oblimersen in melanoma: adverse events occurring in ≥15% by treatment group.

	Adverse Events Occurring in ≥ 15% of Patients											
	Any Grade				Grade 3				Grade 4			
	Oblimersen-Dacarbazine (n = 371)		Dacarbazine (n = 360)		Oblimersen-Dacarbazine (n = 371)		Dacarbazine (n = 360)		Oblimersen-Dacarbazine (n = 371)		Dacarbazine (n = 360)	
Adverse Event	No.	%	No.	%	No.	%	No.	%	No.	%	No.	%
Gastrointestinal												
Nausea[a]	231	62.3	169	46.9	25	6.7	8	2.2	0	0	1	0.3
Vomiting[b]	139	37.5	75	20.8	16	4.3	6	1.7	0	0	1	0.3
Constipation	103	27.8	93	25.8	8	2.2	3	0.8	0	0	0	0
Diarrhea	100	27.0	63	17.5	6	1.6	2	0.6	0	0	1	0.3
Hematologic/lymphatic												
Thrombocytopenia[a]	107	28.8	40	11.1	49	13.2	19	5.3	10	2.7	4	1.1
Neutropenia[a]	103	27.8	56	15.6	40	10.8	30	8.3	39	10.5	15	4.2
Anemia[b]	86	23.2	61	16.9	24	6.5	14	3.9	3	0.8	3	0.8
Leukopenia[a]	62	16.7	31	8.6	23	6.2	13	3.6	5	1.3	1	0.3
Infection (any)[a]	123	33.2	65	18.1	35	9.4	11	3.1	7	1.9	2	0.6
Other												
Pyrexia[b]	197	53.1	63	17.5	16	4.3	6	1.7	0	0	1	0.3
Fatigue	171	46.1	142	39.4	16	4.3	9	2.5	1	0.3	2	0.6
Anorexia[a]	114	30.7	56	15.6	9	2.4	1	0.3	0	0	0	0
Headache[a]	97	26.1	47	13.1	10	2.7	1	0.3	0	0	0	0
Rigors[b]	76	20.5	16	4.4	3	0.8	0	0	0	0	0	0
Dizziness (excluding vertigo)[b]	56	15.1	22	6.1	1	0.3	0	0	1	0.3	0	0
Faint	56	15.1	32	8.9	13	3.5	6	1.7	1	0.3	0	0

NOTE. In the event that a patient had both grade 3 and grade 4 occurrences of an event, that patient was included in the grade 4 column.
[a]Denotes a statistically significant difference between treatment groups in the number of patients with any grade of the event, as well as in the number of patients with grade 3 or grade 4 occurrences of the event.
[b]Denotes a statistically significant difference between treatment groups in the number of patients with any grade of the event.

2.7 Conclusion

G3139 (Genasense), while initially designed as an anti-Bcl-2-specific drug, clearly has a complex, multimodal mechanism of action that is not completely understood. In this respect, Genasense is similar to virtually every other clinically useful cancer chemotherapeutic agent. Indeed, if Genasense had but a single mechanism of action, it is doubtful that it would be clinically active in a disease as complicated and aggressive as advanced melanoma, a disease that cannot be treated effectively by any of our current therapeutic strategies. These facts alone trump the controversy surrounding the *in vitro* and *in vivo* mechanism of action of G3139, which is essentially a question for those primarily interested in basic questions of molecular pharmacology.

In Memoriam Aeternam

Herbert Stein (September 21, 1919 to January 27, 2007), who would have greatly appreciated this discussion and its nuances.

References

1. R. Klasa, A. Gillum, R. Klem and S. Frankel, *Antisense Nucl. Acid Drug Devel.*, 2002, **12**, 193.
2. J. C. Reed, C. A. Stein, C. Subasinghe, S. Haldar, C. M. Croce, S. Yum and J. Cohen, *Cancer Res.*, 1990, **50**, 6565.
3. S. Kitada, T. Miyashita, S. Tanaka and J. C. Reed, *Antisense Res. Dev.*, 1993, **3**, 157.
4. S. Kitada, S. Takayama, K. D. Riel and J. C. Reed, *Antisense Res. Dev.*, 1994, **4**, 71.
5. W. Stec, G. Zon and V. Egan, *J. Am. Chem. Soc.*, 1984, **106**, 6077.
6. L. Campos, O. Sabido, J. P. Rouault and D. Guyotat, *Blood*, 1994, **84**, 595.
7. F. Durrieu, M. A. Belaud-Rotureau, F. Lacombe, P. Dumain, J. Reiffers, M. R. Boisseau, P. Bernard and F. Belloc, *Cytometry*, 1999, **36**, 140.
8. F. J. Keith, D. A. Bradbury, Y. M. Zhu and N. H. Russel, *Leukemia*, 1995, **9**, 131.
9. F. E. Cotter, P. Johnson, C. Pocock, N. Mahdi, J. K. Cowell and G. Morgan, *Oncogene*, 1994, **9**, 3049.
10. A. Webb, D. Cunningham, F. Cotter, P. A. Clarke, F. di Stefano, P. Ross, M. Corbo and Z. Dziewanowska, *Lancet*, 1997, **349**, 1137.
11. L. Weiss, J. Sklar and M. Cleary, *N. Engl. J. Med.*, 1987, **317**, 1185.
12. J. C. Reed, S. Kitada, S. Takayama and T. Miyashita, *Ann. Oncol.*, 1994, **5**(Suppl. 1), 61.
13. C. A. Schmitt, C. T. Rosenthal and S. W. Lowe, *Nat. Med.*, 2000, **6**, 1029.

14. E. Tian, W.-X. Hu and Y. Gazitt, *Int. J. Oncol.*, 1996, **9**, 165.
15. M. E. Gleave, H. Miayake, J. Goldie, S. Nelson and A. Tolcher, *Urology*, 1999, **54**, 36.
16. A. J. Raffo, H. Perlman, M. W. Chen, M. L. Day, J. S. Streitman and R. Buttyan, *Cancer Res.*, 1995, **55**, 4438.
17. J. C. Reed, *J. Clin. Oncol.*, 1999, **17**, 2941.
18. R. Kim, M. Emi, K. Matsuura and K. Tanabe, *Cancer Gene Ther.*, 2007, **14**, 1.
19. A. Raffo, J. Lai, P. Miller, C. A. Stein and L. Benimetskaya, *Clin. Cancer Res.*, 2004, **10**, 3195.
20. L. Benimetskaya, J. C. Lai, A. Khvorova, S. Wu, E. Hua, P. Miller, L.-M. Zhang and C. A. Stein, *Clin. Cancer Res.*, 2004, **10**, 8371.
21. B. Jansen, H. Schlagbauer-Wadl, B. D. Brown, R. N. Bryan, A. van Elsas, M. Muller, K. Wolff, H. G. Eichler and H. Pehamberger, *Nat. Med.*, 1998, **4**, 232.
22. V. Wacheck, C. Krepler, S. Strommer, E. Heere-Ress, R. Klem, H. Rehamberger, H. G. Eichler and B. Jansen, *Antisense Nucl. Acid Drug Devel.*, 2002, **12**, 359.
23. V. Gekeler, P. Gimmnich, H.-P. Hofmann, C. Grebe, M. Rommele, A. Leja, M. Baudler, L. Benimetskaya, B. Gonser, U. Pieles, T. Maier, T. Wagner, K. Sanders, J. F. Beck, G. Hanauer and C. A. Stein, *Oligonucleotides*, 2006, **16**, 83.
24. G. Pratesi, G. Petrangolini, M. Tortoreto, A. Addis, S. Belluco, A. Rossini, S. Selleri, C. Rumio, S. Menard and A. Balsari, *Cancer Res.*, 2005, **65**, 6388.
25. A. Balsari, M. Tortoreto, D. Besusso, G. Petran-Golini, L. Sfondrini, R. Maggri, S. Menard and G. Pratesi, *Eur. J. Cancer*, 2004, **40**, 1275.
26. G. Hartmann, R. D. Weeratna, Z. K. Ballas, P. Payette, S. Blackwell, I. Suparto, W. L. Rasmussen, M. Waldschmidt, D. Sajuthi, R. H. Purcell, H. L. Davis and A. M. Krieg, *J. Immunol.*, 2000, **164**, 1617.
27. J. C. Lai, B. D. Brown, A. Voskresenskiy, S. Vonhoff, S. Klussman, W. Tan, M. Colombini, R. Weeratna, P. Miller, L. Benimetskaya and C. A. Stein, *Mol. Ther.*, 2007, **15**, 270.
28. D. Eulberg and S. Klussmann, *Chembiochem.*, 2003, **4**, 979.
29. A. Vater and S. Klussman, *Curr. Opin. Drug Discov. Devel.*, 2003, **6**, 253.
30. M. V. Blagosklonny, *BioEssays*, 2001, **23**, 947.
31. M. S. Soengas, P. Capodieci, D. Polsky, J. Mora, M. Esteller, X. Opitz-Araya, R. Mccombie, J. G. Herman, W. L. Gerald, Y. A. Lazebnik, C. Cordon-Cardo and S. W. Lowe, *Nature*, 2001, **409**, 207.
32. U. Leitner, R. Schmid, P. Kaskel, R. U. Petter and G. Krahn, *Arch. Dermatol. Res.*, 2000, **292**, 225.
33. J. A. Bush and G. Li, *Clin. Exp. Metastasis*, 2003, **20**, 531.

34. L. Tang, V. A. Tron, J. C. Reed, K. J. Mah, M. Krajewska, G. Li, X. Zhou, V. C. Ho and M. J. Trotter, *Clin. Cancer Res.*, 1998, **4**, 1865.
35. J. Ramsay, L. From and H. Kahn, *Mod. Pathol.*, 1995, **8**, 150.
36. M. C. Saenz-Santamaria, J. A. Reed, N. S. McNutt and C. R. Shea, *J. Cutan. Pathol.*, 1994, **21**, 393.
37. V. A. Tron, S. Krajewski, H. Klein-Parker, G. Li, V. C. Ho and J. C. Reed, *Am. J. Pathol.*, 1995, **146**, 643.
38. A. Plettenberg, C. Ballaun, J. Pammer, M. Mildner, D. Strunk, W. Weninger and E. Tschachler, *Am. J. Pathol.*, 1995, **146**, 651.
39. L. Cerroni, H. Soyer and H. Kerl, *Am. J. Dermatopathol.*, 1995, **17**, 7.
40. L. Benimetskaya, P. Miller and C. A. Stein, *Oligonucleotides*, 2005, **15**, 206.
41. J. Lai, L. Benimetskaya, A. Khvorova, S. Wu, E. Hua and C. A. Stein, *Mol. Cancer Ther.*, 2005, **4**, 305.
42. L. Benimetskaya, K. Ayyanar, N. Kornblum, D. Castanotto, J. Rossi, S. Wu, J. Lai, B. D. Brown, N. Popova, P. Miller, H. McMicken, Y. Chen and C. A. Stein, *Clin. Cancer Res.*, 2006, **12**, 4940.
43. A. Lervolino, D. Trisciuoglio and D. Ribatti, *FASEB J.*, 2002, **16**, 1453.
44. A. Biroccio, A. Candiloro and M. Mottolese, *FASEB J.*, 2000, **14**, 652.
45. D. Trisciuoglio, M. Desideri, L. Ciuffreda, M. Mottolese, D. Ribatti, A. Vacca, M. Del Rosso, L. Marcocci, G. Zupi and D. Del Bufalo, *J. Cell Physiol.*, 2005, **205**, 414.
46. E. Karl, K. Warner and B. Zeitlin, *Cancer Res.*, 2005, **65**, 5063.
47. J. C. Lai, W. Tan, L. Benimetskaya, P. Miller, M. Colombini and C. A. Stein, *Proc. Natl. Acad. Sci. U.S.A.*, 2006, **103**, 7494.
48. M. Colombini, *Nature*, 1979, **279**, 643.
49. T. Rostovtseva and M. Colombini, *Biophys. J.*, 1997, **72**, 1954.
50. M. Colombini, *Mol. Cell. Biochem.*, 2004, **256/257**, 107.
51. W. Tan, J. C. Lai, P. Miller, C. A. Stein and M. Colombini, *Am. J. Physiol. Cell Physiol.*, 2007, **292**, 1388.
52. B. Jansen, V. Wacheck, E. Heere-Ress, H. Schlagbauer-Wadl, C. Hoeller, T. Lucas, M. Hoermann, U. Hollenstein, K. Wolff and H. Pehamberger, *Lancet*, 2000, **356**, 1728.
53. A. Bedikian, M. Millward, H. Pehamberger, R. Conry, M. Gore, U. Trefzer, A. C. Pavlick, R. DeConti, E. M. Hersh, P. Hersey, J. M. Kirkwood and F. G. Haluska, *J. Clin. Oncol.*, 2006, **24**, 4738.
54. P. Therasse, S. G. Arbuck, E. A. Eisenhauer, J. Wanders, R. S. Kaplan, L. Rubinstein, J. Verweij, M. Van Glabbeke, A. T. Van Oosterom, M. C. Christian and S. G. Gwyther, *J. Natl. Cancer Inst.*, 2000, **92**, 205.
55. J. Manola, M. Atkins, J. Ibrahim and J. Kirkwood, *J. Clin. Oncol.*, 2000, **18**, 3782.
56. M. F. Avril, S. Aamdal, J. J. Grob, A. Hauschild, P. Mohr, J. J. Bonerandi, M. Weichenthal, K. Neuber, T. Bieber, K. Gilde, V. Guillem Porta, J. Fra, J. Bonneterre, P. Saiag, D. Kamanabrou, H. Pehamberger, J. Sufliarsky, J. L. Gonzalez Larriba, A. Scherrer and Y. Menu, *J. Clin. Oncol.*, 2004, **22**, 1118.

57. P. B. Chapman, L. Einhorn, M. L. Meyers, S. Saxman, A. N. Destro, K. S. Panageas, C. B. Begg, S. S. Agarwala, L. M. Schuchter, M. S. Ernstoff, A. N. Houghton and J. M. Kirkwood, *J. Clin. Oncol.*, 1999, **17**, 2745.
58. O. Eton, S. S. Legha, A. Y. Bedikian, J. J. Lee, A. C. Buzaid, C. Hodges, S. E. Ring, N. E. Papadopoulos, C. Plager, M. J. East, F. Zhan and R. S. Benjamin, *J. Clin. Oncol.*, 2002, **20**, 2045.

CHAPTER 3

Antisense Morpholino Oligomers and Their Peptide Conjugates

HONG M. MOULTON[*,1] AND JON D. MOULTON[2]

[1] AVI BioPharma Inc., 4575 SW Research Way, Corvallis, OR 97333, USA;
[2] Gene Tools LLC, One Summerton Way, Philomath, OR 97370, USA

3.1 Introduction

Morpholinos or phosphorodiamidate morpholino oligomers (PMOs) are antisense compounds. Their efficacy, sequence specificity, stability and low toxicity have been demonstrated in many publications (pubs.gene-tools.com). In the first part of this chapter we discuss the chemistry and mechanism of action of Morpholinos, their applications in disease research and the delivery methods used to introduce Morpholinos into cells. The second part of the chapter is devoted to the delivery of Morpholinos by cell-penetrating peptides (CPPs). We discuss the chemistry, sequence specificity, stability and toxicity of Morpholinos when covalently conjugated with cell-penetrating peptides (PPMO), as well as with variants containing non-natural amino acids. We outline applications of PPMO in disease research and their potential for treatment of human diseases. Finally, we briefly summarize some ongoing clinical trials using PMO or PPMO.

3.1.1 What is a Morpholino?

3.1.1.1 Nomenclature and Chemical Structure

Phosphorodiamidate morpholino oligomers are referred to by a number of different names and acronyms. In this chapter, the commonly used terms PMO

RSC Biomolecular Sciences
Therapeutic Oligonucleotides
Edited by Jens Kurreck
© Royal Society of Chemistry 2008

and Morpholino are used interchangeably to describe this class of antisense compounds. In the literature, phosphorodiamidate morpholino oligomers are also referred to as MOs or MORF.

The structure of a Morpholino is superficially similar to that of a nucleic acid oligonucleotide (ON), however, upon closer inspection there are significant differences. Nucleic acids consist of a backbone of organic rings (ribose or deoxyribose) bearing pendant nucleic acid bases and are coupled *via* phosphates. In contrast, Morpholino oligonucleotides (oligos) replace the ribose rings of RNA with morpholine rings which bear the nucleic acid bases (usually A, C, G or T), and the morpholino-base moieties are linked through methylene phosphordiamidates (Figure 3.1). The phosphordiamidate group contains two nitrogen–phosphorous bonds, one with the morpholine nitrogen and one with a dimethylamine. The phosphorodiamidate linkage is non-ionic, in contrast to the anionic phosphate linkage of natural nucleic acids.[1]

The overall structure of the morpholine base with one linkage is called a subunit. Morpholinos are usually 18–31 subunits in length with a typical 25-base Morpholino oligo having a mass ~ 8000 Daltons, dependent on the specific base sequence. The ends of a Morpholino oligo are referred to as the 3′-end and the 5′-end. These numbers do not actually specify atoms of the morpholino subunit; instead, the convention follows the nomenclature of nucleic acids, where 3′ and 5′ refer to carbons of the ribose. Viewed from a perspective at the center of a subunit's ring, the nitrogen of the morpholine ring is toward the 3′-end, and the methylene bound to the morpholine ring is toward the 5′-end.

3.1.1.2 Preparation

Morpholinos are prepared by methods evolved from the method described previously.[1] As shown in Figure 3.2, morpholino subunits are prepared from RNA nucleosides by ring-cleaving oxidation of the ribose 2′- and 3′-hydroxyls, followed by reaction with ammonia and reduction to form a morpholine ring.

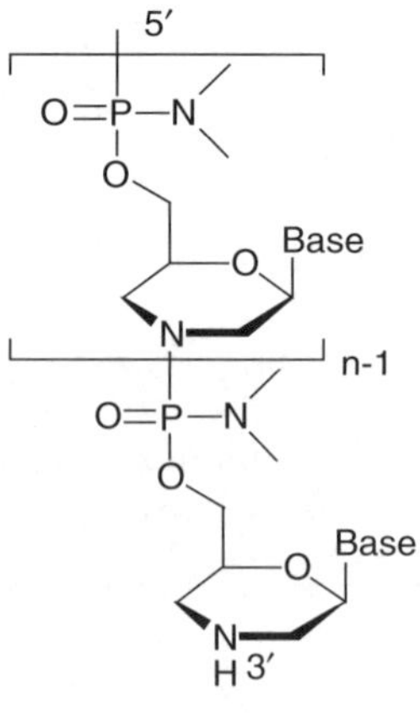

Figure 3.1   Morpholino structure.

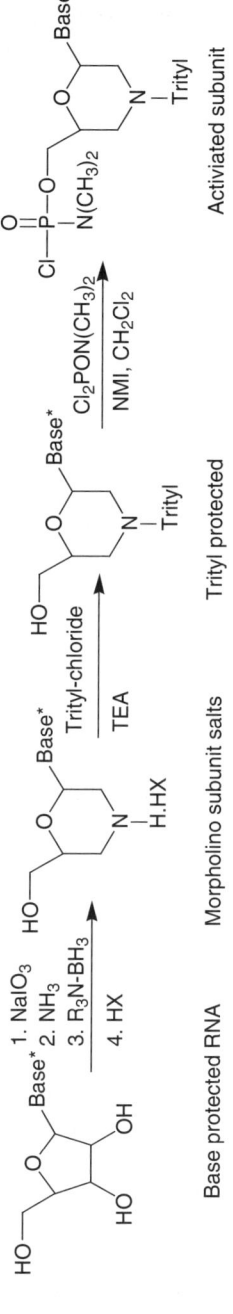

Figure 3.2 Morpholino subunit preparation.

The morpholine nitrogen is protected by tritylation. The hydroxyl on the methylene is reacted with a dichlorophosphoroamidate bearing a dimethyl-amine to form the activated subunit. Morpholino synthesis begins with an activated and base-protected subunit being attached through its 5′-end to a functionalized solid support. The oligo synthesis proceeds opposite to the direction used for standard nucleic acid synthesis, with the 3′-end being the site of further reaction cycles. These consist of detritylation and addition of new activated subunits. At the end of the synthesis, oligos are cleaved from the solid support and the bases are deprotected. Some protocols follow with purification by reverse-phase chromatography. Oligos are analyzed by matrix assisted laser desorption/ionization time-of-flight (MALDI-TOF) mass spectrometry, usu-ally from a sinapic acid matrix. The final product is usually lyophilized.

3.1.1.3   Solubility and Stability

Morpholinos exhibit good water solubility, usually dissolving at over 1 mM, and in some cases dissolving to 10 mM. The more even electrostatic polarity distribution of the uncharged phosphorodiamidate linkage confers better chemical stability on the Morpholinos compared to the anionic phosphodiester linkage of a nucleic acid. Morpholinos are stable in basic conditions and sen-sitive to acidic conditions at pH < 3. In addition to their improved chemical stability, Morpholinos are not degraded by enzyme action[2] and they are bio-chemically stable in human serum and in cells[3] and in tissue lysates.[4]

3.1.2   Mechanism of Action

Binding of a Morpholino to its target RNA sequence prevents other processes from occurring at that target by steric blocking, simply getting in the way of cellular macromolecules (Figure 3.3). Morpholinos bind to sequences of RNA in antiparallel orientation by complementary base pairing. Morpholinos do not require cellular enzymes for their activity.[5]

3.1.2.1   Blocking Translation

A Morpholino can prevent expression of a targeted gene. By binding to a tar-geted messenger RNA (mRNA) somewhere in a region that extends throughout the 5′-untranslated region (UTR) and through the first 25 bases of the coding sequence, a 25-base Morpholino prevents the initiation complex from com-pleting its journey from the 5′-cap to the start codon. If the initiation complex cannot reach the start codon, a complete ribosome will not assemble on that mRNA and the corresponding protein will not be translated (Figure 3.4A). Western blottings or immunohistochemistry are usually used to assess the translation blocking activity of a Morpholino. Translation blocking can some-times be detected by indirect methods such as enzyme activity assays or changes in phenotype during embryonic development. Regardless of the method used to detect the loss of protein activity or decrease in protein concentration, assays

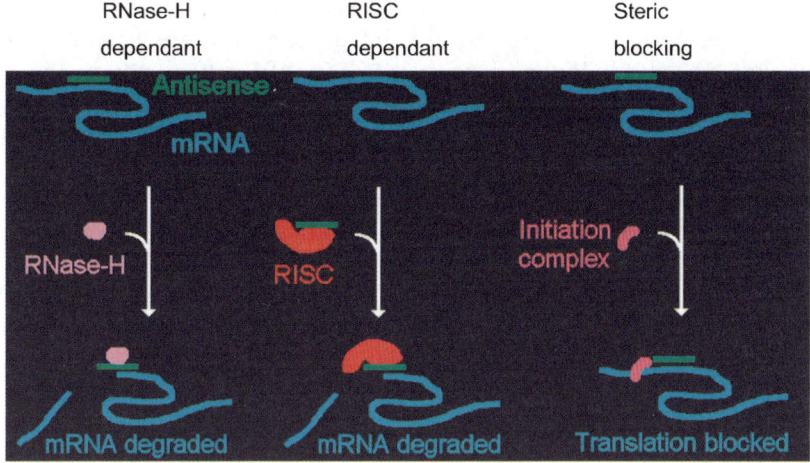

Figure 3.3 Comparison of antisense mechanisms.

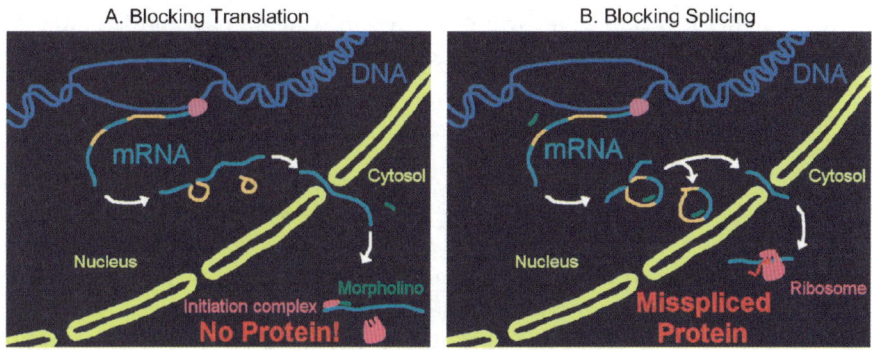

Figure 3.4 Two available mechanisms for Morpholinos.

must be delayed long enough after the start of translation blocking to allow time for enough pre-existing protein to degrade so that a change in protein activity or concentration is detectable. Because a Morpholino does not necessarily accelerate degradation of the RNA to which it is bound, RNA assay techniques such as reverse transcriptase polymerase chain reaction (RT-PCR) or Northern blotting of the target mRNA are unsuitable for detecting activity of translation-blocking Morpholinos – even though protein synthesis is halted, the signal from the corresponding mRNA may persist.

3.1.2.2 Modifying Splicing

By binding to a mostly intronic sequence at a splice donor or splice acceptor site on pre-mRNA, a Morpholino can modify RNA splicing (Figure 3.4B). Targeting the splice junctions of any internal exon usually triggers an exon deletion, while targeting the first or last splice junction in an mRNA usually triggers an intron insertion. Specifically, intron insertions are usually triggered by targeting the splice donor (the exon's 3'-end) of the first exon or the splice acceptor (the exon's 5'-end) of the last exon.[6] When targeting an internal exon, an exon deletion can usually be triggered by targeting either the splice donor or the splice acceptor of the exon to be deleted. However, blocking a splice junction sometimes causes activation of a cryptic splice site, which leads to a partial exon deletion or intron insertion. With the selection of appropriate primers, the activity of a splice-modifying Morpholino can be assayed at the molecular level by RT-PCR. When run on an electrophoretic gel, the RT-PCR product produced after splice modification appears as a different mass from the RT-PCR product of the un-splice-modified mRNA.[7] If splice modification causes a change in protein activity then this altered activity can sometimes be measured by an indirect method, such as an enzyme activity assay or a change in phenotype during embryonic development. As 35–65% of human genes undergo alternative splicing and various known diseases including cancers are caused by aberrant splicing,[8] Morpholinos have potential therapeutic uses either by restoring normal splicing (*e.g.* proposed for β-thalassemia) or by altering splicing to produce a form of a protein which ameliorates the disease [*e.g.* proposed for Duchenne muscular dystrophy (DMD)].

3.1.2.3   Blocking MicroRNA

MicroRNAs (miRNAs) are short RNA sequences transcribed from genomic templates which fold into stem-loops and are processed by two nucleases, Drosha and Dicer, to form short ($\sim$21 base) RNA duplexes with overhanging single-stranded ends. A miRNA binds to a protein complex which includes an Argonaute-family protein. The Argonaute-family protein cleaves one of the RNA strands and incorporates the other (the guide strand) into a mature micro-RNA ribonucleoprotein (miRNP) complex.[9] The mature miRNP complex suppresses the translation and might accelerate the degradation of mRNAs bearing some partially complementary sequences.[10]

Morpholinos can target many stages in the above process. Morpholinos targeting the Drosha cleavage site block nucleolytic release of the stem loop in the primary miRNA (pri-miRNA) transcript, which prevents formation of the free precursor miRNA (pre-miRNA) hairpin. Blocking the Dicer cleavage site prevents formation of the short duplex miRNA.[11] Blocking the guide strand prevents the nucleolytic activity of the mature miRNA–protein complex.[11,12a] A Morpholino may be targeted to bind across several of these sites which have individually been shown to inhibit single stages of miRNA maturation.

Because miRNAs are usually only partially complementary to the sites they block, a given miRNA can recognize a population of sites on mRNA and inhibit translation by interacting with any of those sites. Blocking the target site on an mRNA could potentially prevent recognition by the miRNA–protein complex. This should allow blocking a subpopulation of sites recognized by one miRNA sequence.[12b]

3.1.2.4 Blocking Other RNA Sites – Splice Regulation Sequences, Ribozymes, Slippery Sequences

The ability of Morpholinos to block access of cellular macromolecules to targeted RNA sequences can be exploited to interfere with other cellular processes. Morpholinos have been used to block binding of splice-regulatory proteins such as intronic splice suppressors.[13] The target-recognition motifs of small nuclear ribonucleoprotein particles (snRNPs) can be targeted directly.[14] Morpholinos can protect cleavage sites of ribozymes[15] and those targeted to 'slippery' sequences can trigger translational frame shifts.[16] Blocking viral internal ribosomal entry sites has been shown effective for inhibiting translation of some viral proteins,[17] and blocking cyclization sequences inhibits replication of Dengue virus (DEN).[18]

3.2 Research Applications of Morpholinos

In this section, we discuss research applications of Morpholinos in developmental biology and for studying diseases. We also discuss methods for delivery of Morpholinos to the cytosol and/or nuclei of cells, an important consideration for their research use.

3.2.1 Morpholinos for Developmental Biology

Morpholinos are commonly used tools in developmental biology.[19] When injected into embryos developmental defects unrelated to the targeted gene are caused by siRNA,[20] with both untargeted developmental defects and toxicity typical from embryonic injections of phosphorothioate antisense oligos (AOs).[21–23] In contrast, embryos injected with negative control Morpholino sequences often develop into healthy adults. Embryos are usually injected at one- to few-cell stages. While successive divisions result in progressively smaller cells, the concentration of the Morpholino remains fairly constant until the embryo begins to feed – this is the case for egg-laying organisms like fish and frogs. Dilution of the Morpholino concentration begins more quickly in placental organisms, which in effect begin feeding when nutrient exchange across the placenta begins, yet embryonic Morpholino injections have been useful in studying pre-implantation processes in rodents.[24]

3.2.2 Morpholinos for Studying Diseases: Genetic Disorders, Infection, Cancer, Trauma and Toxins

3.2.2.1 Genetic Disorders

In their review titled 'The zebrafish as a model for human disease', Penberthy *et al.* stated 'The success of morpholino antisense technology in zebrafish potentially opens the door for modeling nearly any inherited developmental defect.'[25] Morpholino knockdowns, often in model embryonic systems, have helped define the functions of genes associated with many human diseases, as shown in Table 3.1.

Potential treatments with Morpholinos for a number of genetic disorders that affect splicing of pre-mRNA have been discussed in reviews by Ryszard Kole, with possible therapeutic applications of splice-modifying oligos including splice-correcting β-globin to relieve β-thalassemia, splice-correcting *tau* to relieve frontotemporal dementia and parkinsonism associated with chromosome 17, blocking splicing to active SMN2 (an inactive form of SMN) and ameliorate spinal muscular atrophy, and modifying dystrophin splicing to ameliorate DMD.[26] More recently there have been additional targets discovered that are amenable to potential treatment by Morpholinos. For example, altering the splicing of a mutated form of lamin-A has reversed a nuclear phenotype typical of cells of elderly humans, also reversing the early onset of the nuclear phenotype when introduced into cells from humans with the genetic disease Hutchinson–Gilford progeria syndrome.[27,28] Cells with splice mutations of ataxia-telangiectasia mutated (ATM) gene have been treated with splice-modifying Morpholinos which resulted in the production of functional ATM proteins.[29] Correction of a mutant splice site restored normal splicing and restored functional protein expression in isolated melanocytes from a human bearing a mutation of the ocular albinism type 1 gene.[30] The use of Morpholinos to correct splicing defects leading to DMD is discussed later in Sections 3.3 and 3.4, as a clinical trial has been started.

3.2.2.2   Infection

While most of the recent reports of viral knockdowns using Morpholinos have used Morpholinos conjugated to CPPs (see later), papers have described viral knockdown or studies of viral components using unconjugated Morpholinos with and without other delivery strategies. Replication of the hepatitis C virus has been knocked down in tissue culture by electroporating the cultures to deliver Morpholinos that targeted internal ribosomal entry sites on viral RNA.[17,31] Morpholinos targeting various sites on Vesivirus RNA caused sequence-specific responses, including increase or decrease of viral titers relative to controls in cultured cells.[32] The ability of the hepatitis C virus helicase to unwind various substrates was investigated. Oligos of peptide nucleic acid (PNA), Morpholino, phosphorothioate or DNA were duplexed with a nucleic acid strand. DNA duplexed with DNA, phosphorothioate or Morpholino unwound with similar rates, while PNA–DNA heteroduplexes unwound 25 times more slowly and PNA–RNA heteroduplexes were not unwound.[33] Dimerization of the RNA genome of Moloney murine sarcoma virus was inhibited by Morpholinos that targeted a palindrome, which has been proposed to mediate genome dimerization – Morpholinos targeting the dimer linkage structure reduced viral titers relative to controls.[34] Endogenous retroviruses were knocked down in the sheep conceptus trophectoderm using *in vivo* delivery of Morpholinos by Endo-Porter in the sheep uterus.[35]

Table 3.1 Morpholino in genetic disease research.

Diseases	Ref.
Alagille syndrome	109
Alzheimer's disease	110
Arthrogryposis–renal dysfunction–cholestasis	111
Inherited atrial septal defect	112
Axenfeld–Rieger syndrome	113
Bardet–Biedl syndrome	114
Barth syndrome	115
Campomelic dysplasia, Ehlers–Danlos syndrome	116
Chorioretinal coloboma	117
Cornelia de Lange syndrome	118
Cranio-lenticulo-sutural dysplasia syndrome	119
Cystic fibrosis	120
Del(5q) syndrome	121
DiGeorge syndrome	122
Down syndrome	123
Fragile X syndrome	124
GLUT1 deficiency syndrome	125
Hematopoietic and cardiovascular disorders	126
Hereditary hemorrhagic telangiectasia 1 and 2	127
Hereditary spastic paraplegia	128
Holt–Oram syndrome	129
Joubert syndrome	130
Lenz microphthalmia, oculofaciocardiodental syndrome	131
Marfan syndrome	132
Multiple sclerosis	133
Congenital muscular dystrophies linked to selenoprotein N mutation	134
Myelodysplastic syndrome	135
Myotonic dystrophy	136
Nephronophthisis	137
Nephrotic syndrome	138
Ocular albinism type 1	30
Okihiro syndrome, Holt–Oram syndrome	139
Opitz G/BBB syndrome	140
Peutz–Jeghers syndrome	141
Polycystic kidney disease	142
Rett syndrome	143
Saethre–Chotzen syndrome	144
Sensorineural hearing loss	145
Simpson–Golabi–Behmel syndrome	146
Smith–Lemli–Opitz syndrome	147
Spinal muscular atrophy	148
Tangier disease	149
Two hereditary leukemia syndromes	150
Velo-cardio-facial syndrome	151
Waardenburg–Shah syndrome	152
Williams–Beuren syndrome	153

Morpholinos targeted to various sites in the RNA of the Ebola virus have been tested at the U.S. Army Medical Research Institute of Infectious Diseases (USAMRID) for efficacy in cell culture, mouse, guinea pig and Rhesus Macaque models. Some targets were effective in rodents in both prophylactic and post-exposure treatments, and after prophylactic treatment 75% of Rhesus Macaques survived Ebola infection.[36] In 2004, the US Dept. of Defense requested AVI BioPharma Inc. to synthesize Morpholinos for the emergency treatment of a scientist exposed to Ebola at USAMRIID. However, no infection developed and so the delivered Morpholinos were not administered.[37]

Another approach using Morpholinos to study infectious disease is to knock down host factors required for the invasiveness or growth of the pathogens. Morpholinos have been used to study the requirement for host factors in pathogenesis by Kaposi's sarcoma-associated herpes virus[38] and *Legionella pneumophila*.[39]

3.2.2.3   Cancer

Morpholinos have inhibited proliferation of some cancers in model systems and have been used to investigate contributions of specific proteins to cancer pathogenesis. Morpholinos inhibited division of human glioblastomas multi-forme cells in both cultures and rats,[40] and human prostate cancer cells in both cultures and mice.[41,42] Suppression of haptotaxis in a human pancreatic cancer cell line resulted from knockdown of ROCK-1 by Morpholinos.[43]

When Lewis lung carcinoma cells were established in mice, followed by the mice being treated with cisplatin or taxol and then by Morpholinos that targeted c-myc, the tumors grew much more slowly than those in control mice.[44] The toxicity of cisplatin to androgen-insensitive prostate cancer cultures and tumor xenografts in mice was enhanced by treatment with a Morpholino that targeted the X-linked inhibitor of apoptosis protein.[45] Treatment by intraperitoneal (i.p.) injection with a Morpholino that targeted MMP-9 in nude mice bearing human prostate xenograft tumors decreased tumor growth and in 50% of animals treated tumor regression was observed.[46] Knockdown of a SNAIL transcription repressor in MIN mice by injection of a Morpholino that targeted SNAIL resulted in a somewhat decreased rate of tumorigenesis.[47] Morpholinos targeting midkine, a heparin-binding growth factor, decreased growth of human colon carcinoma cells and human prostate carcinoma cells in a mouse xenograft model.[48] Knockdown of the ATM protein rendered prostate cancer cell lines more susceptible to radiation-induced apoptosis.[49] Knockdown of B7-H4 on human macrophages decreased the ability of the macrophages to suppress the antigen-specific T cell response to tumors.[50] Angiogenesis in prostate tumors was increased by modification of the ratio of splice forms of VEGF using splice-modifying Morpholinos.[51] VE-cadherin knockdown by Morpholinos in zebrafish embryos with subintestinal tumor xenografts suppressed tumor neovascularization without affecting the development of intersegmental and subintestinal vessels.[52]

Morpholinos have been used to pretarget radioisotopes to tumors as an experimental diagnostic or therapeutic technique. Targeting radioisotopes to tumors by bonding the radioisotopes directly to antibodies suffers from disadvantages conferred by the large size of the antibodies – the large macromolecules diffuse slowly into tumors and clear slowly from the blood, so delivery of a therapeutically useful dose of radiation into the tumor using radiolabelled antibodies requires exposing healthy tissues to potentially damaging doses of radioactivity. Radiation dose to non-tumor tissues can be decreased by pretargeting. A pretargeting scheme developed by Donald Hnatowich's group involves treatment with an antibody bound to a Morpholino, allowing time for the large molecules to bind in the tumor and clear from the blood. This is then followed by treatment with a complementary Morpholino bound to a radioisotope. The relative small Morpholino–isotope conjugate enters tumors relatively rapidly and binds to its complementary Morpholino on the tumor-associated antibody. The Morpholino–isotope conjugate clears from the blood more quickly than an antibody–isotope conjugate, decreasing off-target radiation exposure.[53] The pretargeting signal can be amplified by administering a polymer conjugated with multiple copies of a Morpholino complementary to the Morpholino on the antibody – the polymer then associates with the antibody by Morpholino–Morpholino pairing. This is followed by a radioisotope-linked Morpholino with the same sequence as the original Morpholino on the antibody – many copies of this isotope-linked Morpholino can associate with the remaining unpaired oligos conjugated to the polymer. By using this amplification method, many radioactive atoms can be pretargeted by each antibody molecule.[54] Treatments of tumor-bearing mice with pretargeted Morpholinos have resulted in significantly lower tumor weights relative to controls.[55]

3.2.2.4 *Trauma and Toxins*

Morpholinos may be useful in the treatment of traumatic injury. Several studies have shown that gene knockdowns by Morpholinos can aid healing during tissue reperfusion, as shown by experiments involving ischemic injury in mice[56] and recovery of cultured rat neonatal cardiac muscle after hypoxia.[57]

Toxicological studies have generally used Morpholinos to study the effect of knocking down proteins involved in the transport or metabolism of toxins. Studies of the effects of environmental toxins in fish have used Morpholino knockdowns of genes, such as the various aryl hydrocarbon receptors,[58,59] the aryl hydrocarbon receptor nuclear translocators,[58,59] and cytochrome P450,[60] to determine the relevance of these genes to toxin uptake and metabolism. This technique has also been applied to study the participation of specific genes in the toxicity of environmental mixtures.[61] Biomolecular factors influencing the toxicity of MPTP, a chemical that can induce Parkinson's disease, have been explored with Morpholinos in the zebrafish model.[62] Formation of DNA adducts associated with metabolites of the chemotherapeutic drug tamoxifen can be reduced in rats by Morpholino-mediated knockdown of CYP3A2.[63]

3.2.3 Methods for Delivery of Morpholinos

In most cases, Morpholinos require the use of a delivery technique to enter cells in culture in useful concentrations. Once Morpholinos have been delivered to the cytosol they freely diffuse into the nucleus, shown by the efficacy of splice-modifying oligos when microinjected into the cytosol.[7] Therefore delivery of Morpholinos to the cytosol is sufficient for blocking both cytoplasmic processes (translation blocking, blocking mature miRNA) and nuclear processes (splice modification, blocking maturation of pri-miRNA). However, delivery into endosomes may not be sufficient to achieve Morpholino antisense activity, as the oligos must still cross the endosomal membrane to reach the cytosol. It is important when assessing delivery to distinguish between delivery to endosomes and delivery to the cytosolic and/or nuclear compartment.[64]

Single-stranded unmodified Morpholinos diffuse into some permeable cell types and tissue explants, resulting in experimentally useful cytosolic concentrations without employing any special delivery procedures beyond bathing cells in a solution containing dissolved Morpholinos. Such simple delivery has been reported in dystrophic muscle,[65,66] in Ebola virus-infected mice,[36] in adult rat liver,[67] in the retina of later-stage developing zebrafish,[68] in the epithelial cells of the explanted pancreas of the stage E11 through E13 mouse embryo[69] and in the explanted liver of the E10 mouse embryo.[70] Successful cytosolic delivery in the brain by intracerebroventricular infusion of single-stranded unmodified Morpholinos has been reported.[71] However, most cell types do not readily take up bare Morpholinos from solution and so some assistance is required to achieve experimentally useful delivery.

3.2.3.1 Microinjection and Electroporation

Delivery to embryos is usually accomplished by microinjection. Morpholinos can be injected into zygotes or early embryos and will be divided between daughter cells with each division. Some embryos, such as zebrafish, allow large molecules such as Morpholinos to diffuse between cells through the first few cell divisions[72,73] while others, such as *Xenopus* frogs, immediately establish permeability barriers between cells so that only daughter cells of the injected cell contain significant concentrations of Morpholino.[74]

Although Morpholinos are not charged, electroporation methods have been used successfully to deliver Morpholinos. Electroporation renders cells transiently permeable, which allows Morpholinos to diffuse into the cytosol. Successful delivery of Morpholinos by electroporation has been reported in cells,[31] chick embryos,[75] zebrafish,[76] *Xenopus*[77] and rodents.[78,79]

3.2.3.2 Reagent-based Methods: Endo-Porter and Ethoxylated Polyethylenimine

Reagent-based delivery methods have been developed to deliver Morpholinos. Endo-Porter is a peptide-based reagent developed to deliver Morpholinos into cells by co-endocytosis followed by release of the endocytosed materials by permeabilization of the acidic endosome. Endo-Porter and a Morpholino are delivered as independent molecules, without conjugation or complexation. This endocytotic entry is more gentle to cells than are lipofection reagents that render the plasma membrane leaky.[80] Uncharged Morpholinos do not electrostatically complex with cationic lipid transfection reagents, so these reagents cannot efficiently deliver them. However, the cationic transfection reagent EPEI has been successfully used to 'indirectly' deliver Morpholinos as Morpholino–DNA heteroduplexes, in which the DNA strand acts as an adapter, associating with the Morpholino by Watson–Crick pairing and to the EPEI by electrostatic complexation.[81]

3.2.3.3 Particle-based Methods

Perfluorocarbon gas microbubble carrier (PGMC) has been used for systemic delivery of the anti-c-myc Morpholino, AVI-4126. A study was performed in pigs to test the ability of PGMC AVI-4126 to prevent restenosis from developing after stent implantation. Perfluorobutane containing 2 mg AVI-4126 was administered to each pig by intravenous (i.v.) injection. AVI-4126 was detected in treated arteries by high performance liquid chromatography (HPLC) and modest inhibition of c-myc was detected by Western blot. Neointimal area was reduced significantly in the treated group when compared with controls at 28 days.[82]

A nanoparticulate conjugate of Morpholinos with poly(beta-l-malic acid) purified from *Physarum polycephalum*, called Polycefin, has been developed to deliver Morpholinos into tumors. After human breast cancer cells were delivered subcutaneously in nude mice and tumors established, Polycefin delivered by i.v. injection selectively accumulated in the tumor cells. After tumors established by intracranial injection of a human glioma cell line grew in nude mice, again Polycefin delivered by i.v. injection selectively accumulated in the tumor cells.[83]

With the possible exception of PGMC and Polycefin, other delivery methods are difficult to implement for enhancing *in vivo* delivery of Morpholino. Since 2003, efforts to develop a CPP-based delivery method suitable for therapeutic applications have generated over 20 publications, especially in the DMD and antiviral fields. We discuss this technology in detail in Section 3.3.

3.3 Cell-penetrating Peptide–Morpholino Conjugates

Cationic CPPs are promising carriers to enhance the cytoplasmic and nuclear delivery of Morpholinos, referred to hereon as PMOs, for several reasons. First, negatively charged AOs might electrostatically interact with conjugated cationic CPPs with loss of antisense or delivery activity – because of the uncharged nature of PMOs undesired inter- and intra-molecular charge interactions between a CPP and PMO are not a concern. Second, CPP–PMO (PPMO) conjugates are simple to use as they can be directly added to cultures or injected into animals without complexation with transfection agents.

Lastly, the short CPPs and their PMO conjugates can easily be manufactured and characterized in a quality-controlled manner, which is an important consideration for their use in therapeutic settings. In this section, we discuss PPMO in terms of their chemistry, internalization mechanism, effect of CPP sequences on PPMO efficacy, specificity, stability, applications and toxicity.

3.3.1   Chemistry

Covalent conjugation of a CPP to PMO is required to enhance the delivery of the PMO. Table 3.2 lists the names and sequences of CPPs that have been conjugated to PMOs and tested in various biological systems. Uncharged PMOs do not form a complex with CPPs, so simply mixing of PMOs and CPPs failed to produce antisense effects.[84,85]

Three types of linkers – amide, thioether and disulfide (Figure 3.5) – have been tested to determine effects of linkage on the activity[86,87] and stability[3] of PPMO conjugates. In cell culture, the nuclear antisense activities of the PPMO were similar for the three linkage types (Moulton *et al.*[86]and unpublished results). Synthetically the amide linkage has two advantages over disulfide and thioether linkages. First, a PMO can easily be conjugated to a CPP via an amine at either end of the oligo. Second, formation of an amide bond requires only a one-step synthetic and purification scheme as compared to a two-step strategy for other linkages. Therefore, the amide linkage leads to an overall greater yield of PPMO. In addition, amide bonds are more effective *in vivo*, as unhindered disulfide linked PPMO was found to be unstable in human serum.[3]

Although a CPP can be conjugated at either end of a PMO, the position of the CPP affects the activity of PPMO. A PPMO having a CPP conjugated at the PMO's 5′-end was more active than with the CPP conjugated at the PMO's 3′-end.[86] Adding a bulky fluorochrome, such as carboxyfluorescein, at the 3′-end of PPMO decreased the activity of the PPMO by about 30% (unpublished data).

3.3.2   Internalization Mechanism

Study of the mechanism of uptake of PPMOs revealed that cell-surface proteoglycans were involved in binding the conjugates, which were subsequently

Table 3.2   Names and sequences of CPPs.

Names	Sequences
R_9F_2C	RRRRRRRRRFFC
$(RXR)_4XB$	RXRRXRRXRRXRXB
$R_5F_2R_4C$	RRRRRFFRRRRC
$(RFF)_3$	RFFRFFRFFXB
Tat	CYGRKKRRQRRR
RTR	RTRTRFLRRTXB
$(RFF)_3R$	RFFRFFRFFRXB
$(KFF)_3K$	KFFKFFKFFKXB

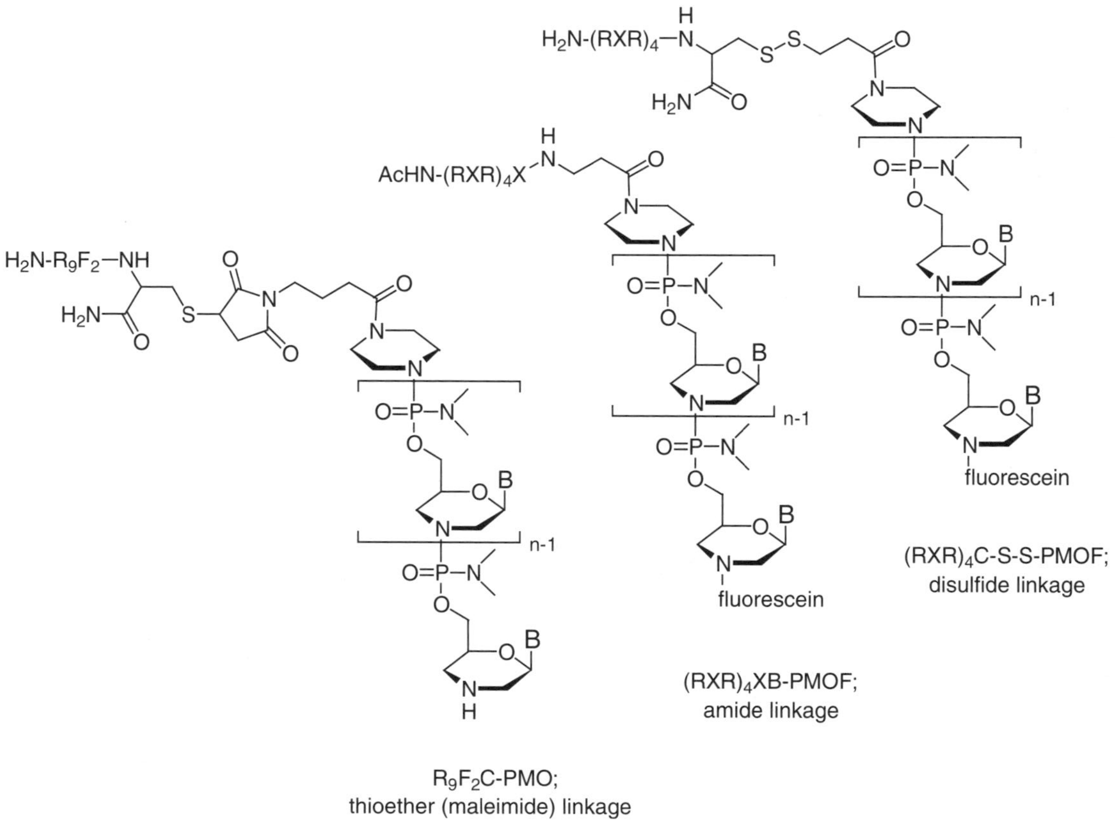

Figure 3.5 Structures of PPMO: R_9F_2C–PMO with thioether (left), $(RXR)_4XB$–PMOF with amide (middle) and $(RXR)_4C$ –PMOF (right) with disulfide linkers.

taken up by endocytosis. When R_9F_2C, Tat and $(RXR)_4XB$ PPMO were studied, the internalization mechanism was found to be energy and temperature dependant, as expected for endocytotic uptake. The majority of conjugate co-localized in cells with a marker for endocytotic and/or lysosomal vesicles. A small amount of the conjugates escaped the vesicles to enter the cytosol and nucleus by an unknown mechanism. The degree of endosomal escaping is related to structural features of the CPPs, with details of this structure–activity relationship yet to be defined. A significant portion of the conjugates was found to be trapped within vesicles and could be released by chloroquine treatment.[87]

3.3.3   Effects of CPP Sequences on PPMO Efficacy

Length, sequence and amino acid types of a CPP affect efficacy of the resulting PPMO conjugates. A splice-correction reporter assay[88] was used to determine the effects of these properties on the effectiveness of CPP to deliver PMO to cell nuclei. This assay uses the ability of steric-blocking AOs to block a splice site created by a mutation to restore normal splicing. The reporter luciferase's coding sequence was interrupted by the human β-globin thalassemic intron 2, which carried a mutated splice site at nucleotide 705. HeLa cells were stably transfected with the plasmid and therefore named pLuc705 cells. In the pLuc705 system, steric-blocking AOs must be present in the cell nucleus for splicing correction and reporter expression to occur. Advantages of this system include the positive readout and high signal-to-noise ratio. With this system the relative efficiencies of various CPPs to deliver a splice-correction AO to cell nuclei can be easily compared.

Using this assay, we first investigated the ability of several well-known CPPs, such as penetratin, Tat, nuclear localization signal and oligoarginine peptides, to deliver PMO into the cell nucleus. We found that oligoarginine[86] and Tat[84] peptides were more efficient than others. However, their instability in serum[3] and large endosomal entrapment[87] limit their use in therapeutic settings. Insertion of non-α amino acids, 6-aminohexanoic acid or β-alanine enhances the serum stability of resulting CPPs[3] and the release of PPMO conjugates from endosomal entrapment.[87]

Further investigation of the effect of these non-α amino acids on the activities of the resulting PPMO conjugates revealed that several structural features of CPPs affect apparent cellular uptake and antisense activity of the PPMO, including the number of arginine residues and the charge density of a CPP. The apparent cellular uptake of PPMO, usually measured as fluorescence of cells treated with a fluorochrome-tagged PPMO, increases with increased number of arginines and charge density.[86,89] The antisense activity of a PPMO increases with increased number of arginines, but not necessarily with higher charge density. A polyarginine CPP conjugate was compared in the splice correction assay to a panel of CPP conjugates that contained the same number of arginines, but with the arginine residues separated by 6-aminohexanoic acid or β-alanine residues. It was found that the PPMO containing these non-α amino acids had higher antisense activity than the polyarginine PPMO,[89] which suggests that PPMO with lower charge density had higher activities.

Murine hepatitis virus (MHV) cell culture and mouse models were used to compare the PMO delivery efficacy of R_9F_2C, $(RXR)_4XB$ and their derivatives. $(RXR)_4XB$ PPMO caused statistically significant reduction of MHV viral titer in the target organs of the MHV-infected mice, but R_9F_2C PPMO did not, although both PPMOs were equally effective in a cell-culture model. Shortening the $(RXR)_4XB$ CPP or modifying the CPP with mannose significantly reduced the activity of the resulting PPMOs.[90]

3.3.4 Specificity

Off-target effects of PPMOs depend on the intracellular stability of the CPPs and the concentrations of PPMO available in the cytosol and nuclei of cells. Effects of CPPs [R_9F_2C, $R_4F_2R_4C$, $(RXR)_4XB$] on the specificity of PPMO have been investigated in various models from cell-free translation assays to cell cultures and *in vivo* models. Cell-free translation studies showed that these PPMOs had minimum off-target effects at $<1\,\mu M$ but off-target effects were observed at higher concentrations.[91–94] The non-specificity was caused by the cationic charged nature of PPMO, and the degree of non-specificity of PPMO appears to depend on the PMO sequences.[91] However, the non-specificity was not observed either in cell culture[18,90–97] up to $20\,\mu M$ treatment concentrations or in mouse[90,93,96,98] models.

Two major reasons give rise to the discrepancy between cell-free and cell and/or animal models. First, the intracellular instability of CPP reduces off-target effects. In cells most of the CPP was degraded from the PMOs and intact conjugates were not detected.[3] Non-specificity in cells was observed for the non-degradable D-isomer of R_9F_2C PPMO.[91] Second, low concentration of a PPMO in cytosol and nuclei reduces off-target effects. Although the treatment concentrations were $>1\mu M$ in most applications, the concentration of PPMO available in the cytosol and nucleus is only a small fraction of the treatment concentration because only small amounts of PPMO entered cells (unpublished data) and the majority of the PPMOs were trapped within endosomal and/or lysosomal vesicles.[87]

3.3.5 Stability

Conjugates with 12 peptides having different sequences, amino acid compositions and linkers were studied in cells and in human serum to investigate the effects of the sequence, the linker and the stereochemistry of the amino acids. In cells and in human serum, the PMO portions of PPMOs were completely stable. The stability of the peptide moiety in serum ranked in the order D-CPPs $= (RX)_8B = (RB)_8B > (RXR)_4XB > R_9F_2C = $ Tat. Stabilities of peptide moieties within cells were found to be D-CPPs $> (RB)_8B > (RX)_8B = (RXR)_4XB = R_9F_2C = $ Tat. Stabilities of linkers in serum or in cells were found to be amide $=$ maleimide $>$ disulfide.[3]

In addition, the stability of a (RXR)$_4$XB PPMO was determined in rat serum and organ lysates. Again, the PMO portion was completely stable. The CPP portion of the conjugate exhibited time- and tissue-dependent degradation, with biological stability ranked in the order of liver > heart = kidney > plasma. Up to 6 hours, the CPP portion of PPMO was partially degraded in the plasma but had no apparent degradation in tissues.[4]

3.3.6 Applications

We discuss here recent publications of PPMO that involve genetic diseases and antiviral research. More detailed information about cell lines, disease types, PPMO treatments and references in cell culture and animal models are given in Tables 3.3 and 3.4.

3.3.6.1 Genetic Disorders

DMD is an X-linked disease usually caused by mis-splicing of the human dystrophin transcript, leading to muscle degeneration and decreased span and quality of life in affected humans. Using steric-blocking oligos to further modify splicing, resulting in additional exon exclusion and restoration of the correct reading frame, has been shown to restore much dystrophin function, offering hope of changing Duchenne's pathology to resemble the milder Becker muscular dystrophy.

A dog model of muscular dystrophy in golden retrievers (GRMD) was the source of primary muscle cells used to test the efficacy of a splice-modifying (RXR)$_4$XB PPMO for restoring dystrophin function. These dogs have a mutation in intron 6 that causes altered splicing and exclusion of exon 7, which results in disruption of the reading frame. Excision of both exons 6 and 8 should restore the reading frame error caused by the GRMD mutation and produce a shorter, mostly functional dystrophin protein. PPMOs were found to be the most effective compounds to cause the desired exon-skipping among the three types of chemistries tested. PPMO treatment of GRMD muscle cells caused the RT-PCR product of the mis-spliced GRMD transcript to disappear from gels. This was a much more complete splice modification than afforded by 2'-*O*-methyl oligos delivered by lipoplexes. Ten days after a single treatment, PPMO-treated cells continued to show abundant induced in-frame transcript.[99]

A mouse model of muscular dystrophy (*mdx* mice) carries a nonsense mutation in exon 23. Excision of exon 23 in *mdx* mouse cells by a splice-modifying oligo can restore mostly functional dystrophin expression.[66] PPMOs caused the desired exon-skipping in both *mdx* mouse muscle explants[100] and *mdx* mice.[101] When *mdx* mouse-muscle explants were treated with PPMO that targeted a donor site of exon 23, almost all of the RT-PCR product was splice modified, mostly as an exon 23 deletion. To restore dystrophin expression in *mdx* mouse muscles PPMO was more effective than PMO.[100]

Dystrophin level in diaphragm muscles of *mdx* mice, detected two weeks after a single i.p. injection of (RXR)$_4$XB PPMO, was similar to the levels in

Table 3.3 Antisense effects of PPMO observed in cell Culture and *ex vivo* models.

Cell lines; disease or system	Typical treatment	Ref.
R$_9$F$_2$C PPMO		
DBT; MHV-2, 3, 4, A59 (mouse hepatitis virus), 17Cl-1; MHV-1 Vero-E6; SARS-CoV-Tor2	PPMO ranged 2 μM to 20 μM, treatment time varied through pre-infection and post-infection. Assayed by plaque size, formation of syncytium, cell morphology and viral titer.	90, 102, 103
HeLa pLuc705; upregulation splice correction assay	PPMO ranged 0.1 to 10 μM for 4 or 24 hours. Assayed for luciferase production.	86, 87
Grown on C6/36, titered on BHK-21; Dengue DEN-1, -2, -3, -4	PPMO treatment at 5 μM 2–3 hours before infection. 2 hours after infection, the culture medium was replaced by the medium containing PPMO. Viral titer was assayed 24 hours post-infection.	92
Vero or Vero E6; Ebola Zaire or Marburg Musoke	PPMO treatment at 0.1 to 5 μM for 3 hours before infection. After infection, PPMO was re-applied for 24 to 72 hours post-infection when the tissue culture infectious dose was determined.	93
(RXR)$_4$XB PPMO		
DBT; MHV A59	PPMO treatment at 20 μM for 3 hours before infection. Assayed for viral yield 24 hours post-infection.	90
HeLa, HL-1 (murine cardiac muscle); Coxsackievirus B3	PPMO pretreatment for 4 hours at 10 or 20 μM before infecting cells for 1 hour; or 4 hours PPMO treatment 1 hour after infecting cells. Assayed for viral titer 8 or 40 hours later.	96
MDCK; H1N1, H3N2, H3N8, H7N7, H5N1 (influenza A)	Pretreatment with PPMO up to 20 μM, 4 or 6 hours before infection. 15 μM PPMO treatment at 1, 2 or 3 hours post-infection.	95
BHK-21 and Vero; West Nile virus 3356	PPMO treatment at up to 10 μM immediately before infection. Assayed for viral titer 48 hours later.	97

(*Continued*)

Table 3.3 *(Continued).*

Cell lines; disease or system	Typical treatment	Ref.
Primary myoblast cultures from Golden Retriever Muscular Dystrophy dogs	Single treatment of PPMO at up to 10 µM and assayed by RT-PCR at day 1, 4, 7 and 10. 20 µM treatments were assayed by Western blots 7 days later.	99
Explants of tibialis anterior of muscular dystrophy mice	Single treatment of PPMO at 10 µM targeting exon 23, followed by RT-PCR assay at day 7.	100
Muscle explant from normal human	Single treatment of PPMO at 10 µM targeting exons 6 and 8, followed by RT-PCR assay at day 7.	100
(RXR)₄C, Mannose (RXR)₄XB, (RXR)₃C, (RXX)₂C, (RX)₄C and Mannose (RX)₄C PMO		
DBT; MHV-A59	PPMO treatment at 20 µM for 3 hours before infection. Assayed for viral yield 24 hours post-infection.	90
R₅F₂R₄C PMO		
Vero; H1N1	Pretreatment with PPMO up to 20 µM, 6 hours before infection.	95
Vero; Dengue 1, 2, 3,4	Up to 20 µM of PPMO treatment 12 hours before infection and re-treatment after the infection. Viral titer was assayed up to 10 days later.	18
BC-1, BL-1, BJAB: Kaposi's sarcoma associated herpesvirus	Range of treatments including pre-infection, post-infection and combined protocols. PPMO concentrations ranged up to 20 µM.	94
Tat PPMO		
HeLa pLuc705; upregulation splice correction assay	PPMO ranged 0.1 to 10 µM for 4 or 24 hours. Assayed for luciferase production.	84,87
(RFF)₃XB, RTRTRFLRRTXB, (RXX)3B, (KFF)₃K PMO		
E. coli W3110, E. coli E2348/69, Salmonella enterica serovar Typhimurium TA1535; bacterial pure culture and inhibition of infection of Caco-2 cell models	Pure culture model: PPMO treatment up to 80 µM. Viable bacterial cells were measured over the 24 hour period. Caco-2 model: 10 µM PPMO treatment was immediately after the E2348/69 infection.	85

Table 3.4 Antisense effects of PPMO observed in mouse models.

Disease	Treatment	Ref.
R₉F₂C PPMO		
Mouse hepatitis virus: MHV-A59	Mice received three doses of PPMO at 8 nmol/dose: 5 hours before, 24 and 48 hours after infection.	90
Mouse-adapted EBOV	Mice IP-injected with 500 μg/dose PPMO either 24 and 4 hours prior to infection or 24 hours post-infection with either ~ 1000 pfu or ~ 30 000 pfu EBOV.	93
(RXR)₄XB PPMO		
Mouse hepatitis virus: MHV-2, MHV-3, MHV-4, MHV-A59, MHV Alb139	Mice received three doses of PPMO at 8 nmol/dose: 5 hours before, 24 and 48 hours after infection with 1000 or 100, 10 pfu.	90
Coxsackievirus B3	Mice were injected with 200 μg PPMO 3 hours before and again 48 hours after infection with the virus at 10 000 pfu.	96
Duchenne muscular dystrophy	*Mdx* mice received 4 weekly doses of PPMO at 5 mg/kg. The expression of dystrophin was assayed in different muscle tissues.	101
West Nile virus	Mice received 5 or 10 mg kg⁻¹ PPMO IP injections either nine daily doses before the viral infection or 11 daily doses 5–15 days after viral infection.	98
(RFF)₃RXB PPMO		
Escherichia coli peritonitis	BALB/c mice infected IP with *E. coli* W3110 or a derivative acyl carrier protein mutant, LT1. PPMO injected IP at various times post-infection.	107

normal mice. Four doses of the PPMO at $5 \, \text{mg kg}^{-1}$ per dose, administered to neonatal mice once a week for four weeks, resulted in normal levels of dystrophin being present in the diaphragms of 6-week-old mice. The PPMO effect was more modest in the tibialis anterior, gluteus maximus and triceps brachialis, with a weak effect in the colon and stomach and no dystrophin present in the heart. Western blots detected normal levels of dystrophin in the diaphragm at six weeks, lower levels in the tibialis and gluteus and no dystrophin was detected in cardiac muscle. At 22 weeks post the fourth injection, dystrophin was still detectable in the diaphragm, but was discontinuous with some disruption of muscle architecture. Animals of one day, four weeks or one year of age were treated to test the effect of age. While splice-modified dystrophin expression increased for all ages treated, muscle architecture was best in muscles from mice treated at younger age and was far from normal when older mice were treated. Normalized pathology and central nucleation were observed for the sections taken from the 6-week-old mice treated with the PPMO. The study also revealed that treatment of DMD mice at a young age is clearly more beneficial than at an older age.[101]

Normal human muscle explant without muscular dystrophy was used to test the effectiveness of several antisense structures; an $(RXR)_4XB$ PPMO was found to be the most effective of the chemistries tested.[100] The PPMO targeting exon 6 triggered splice modification and produced predominant exon 6 deletions and the PPMO targeting exon 8 triggered splice modification and produced a mixture of products, including exon 9 deletions and predominantly exon 8 and 9 deletions.

3.3.6.2 Viral and Bacterial Infections

Coxsackievirus. Coxsackievirus is a primary cause of viral myocarditis and accounts for 20% of sudden heart failure in children and adolescents in North America.[96] $(RXR)_4XB$ PPMO targeted to the internal ribosome entry site (IRES) core sequence of Coxsackievirus B3 inhibited replication in a dose-dependant and sequence-dependant manner in the virus-infected cells and mice. When pretreated with PPMO, the viral titers were reduced by 4 and 3 logs in virus-infected HeLa and HL-1 cells, respectively. A viral protein was reduced to a level undetectable in both cell lines when the cells were pretreated with PPMO and infected with the virus. To test post-infection efficacy, cells were treated with PPMO 1 hour after viral infection; a 2 and 4.5 log reduction in viral titer was measured in HeLa and HL-1 cells, respectively. When viral infection was delayed after conjugate treatment, viral inhibition was still 50% effective after a five-day delay between treatment and infection. Seven days post-infection, the viral titers in hearts of PPMO-treated mice were 2 logs below titers in control groups. Myocarditis and tissue damage in pancreas, liver and spleen was milder in the PPMO-treated mice.[96]

Coronavirus. The recently emerged severe acute respiratory syndrome (SARS) coronavirus has shown the threat of worldwide pandemic and demonstrated the need for antiviral therapies. R_9F_2C PPMOs targeting a conserved region of SARS-CoV isolates, the transcription-regulatory sequence (TRS) in the 5'-UTR, was the most effective PPMO found. Treatment with the PPMO reduced cytopathic effects, knocked down viral titers to background levels and halted the spread of the virus, with no new plaques formed. The conjugate was effective if administered at any time before peak viral synthesis. A resistance study to assess the development of viral resistance to submaximal doses of the PPMO that targeted the TRS transcript found that some increases in viral titer had developed, which indicates partial viral resistance to the TRS2 PPMO. However, the partially resistant virus grew more slowly and formed smaller plaques than the wild-type virus.[102]

Several strains of MHV, a coronavirus, were used to investigate the target sites and antiviral effects of PPMOs in cell culture and in mouse models. An R_9F_2C PPMO targeting a conserved sequence of the MHV polyprotein gene inhibited MHV-A59 viral replication, reducing viral titer 10- to 100-fold in a dose-dependant manner in delayed brain tumor (DBT) cells.[103]

Ten R_9F_2C PPMOs targeted to various sites of the MHV viral genome were screened for their antiviral effects and the one targeted to the 5′-terminus of the genomic RNA (5TERM) was found to be effective against six strains of MHV in cell culture. The $(RXR)_4XB$ PPMO reduced viral titers and protected tissue from damage in target organs of MHV-infected mice.[90]

Influenza A virus. Influenza A virus causes many deaths worldwide every year. The recent threat of influenza virus H5N1 to the human population has caused considerable concern. A panel of PPMOs targeted to the AUG translation start-site regions of the mRNAs that involve viral RNA synthesis (PA, PB1, PB2 or NP) or one of the four terminal regions of the NP viral RNA or complementary RNA were tested against several strains of the virus. Treatment of cells with several R_9F_2C PPMOs prior to the viral infection caused a 4.5 log reduction of H1N1 viral titer. Two of the effective PMOs were selected to conjugate to the $(RXR)_4XB$ peptide and the corresponding PPMOs were tested against H7N7, H3N8 and H5N1 in Madin–Darby Canine Kidney (MDCK) cells. Both PPMOs caused >85% inhibition of H5N1 and one of the conjugates caused >85% inhibition of H3N8 and H7N7. The conjugate that inhibited H3N8 was also tested for its effectiveness in a post-infection model. The H3N8 viral titer was inhibited by 70, 40 or 20% after the conjugate was added to cells 1 hour, 2 hours or 3 hours post-infection, respectively. The decrease in the effectiveness of the PPMO in the post-infection model could partially be caused by the degradation of the CPP in the trypsin-containing medium that was required for the production of the flu particles in the cells.[95]

Dengue virus. DEN causes sometimes fatal hemorrhagic fever and 40% of the world's population is exposed to potential infection. $R_5F_2R_4C$ PPMOs effectively inhibited the replication of DEN. Five PPMOs were assayed for their ability to inhibit DEN in Vero cells. The PPMOs were targeted to the highly conserved and/or critical elements of the DEN-2 virus genome in the regions of the 5′-SL (stem loop), AUG, 5′-CS (cyclization sequence), 3′-CS and 3′-SL. Two PPMOs targeted to the 5′-SL and 3′-CS regions were most effective, giving a 5.7 and >4 log reduction in viral titers of DEN-2, respectively. However, the virus rebounded after 6 days following the initial PPMO-treatment. Various PPMO treatment time protocols were explored; treatment both before and after infection was the most effective protocol and pre-infection treatment was more effective than post-infection treatment.[18] In another study, Holden *et al.*, assayed the effect on DEN replication and gene expression of three R_9F_2C PPMOs targeted to 5′-SL, 3′-CS and 3′-SLT (top of stem loop) of the viral RNA. Viral RNA synthesis was suppressed by any of these PPMOs, with 3′-SLT inhibiting viral RNA synthesis most effectively with a 450-fold reduction. Viral translation was inhibited most effectively by 5′-SL.[92]

West Nile virus. West Nile virus has caused considerable morbidity and mortality in the USA. Nine $(RXR)_4XB$ PPMOs were tested against a West Nile virus persistently-replicating replicon with a dual luciferase reporter system in cell culture. Two sequences, 5′-End (5′-terminal) and 3′-CSI (cyclizing interaction), were identified that most effectively inhibited luciferase expression and these were subsequently used to challenge West Nile virus replication. Vero cells were infected with the West Nile virus immediately after the addition of each PPMO. Viral titer was reduced by 6 or 5 logs by the 5′-End or 3′-CSI PPMO treatment, respectively, while the scramble control PPMO did not measurably inhibit viral replication. Based on results from a West Nile virus transient replicon luciferase reporter system, the 5′-End PPMO primarily inhibits viral translation while the 3′-CSI PPMO primarily inhibits RNA replication.[97] The 5′-End PPMO was the most effective compound to protect mice from West Nile virus disease *in vivo*, compared to the 3′-CSI PPMO and the corresponding PMOs.[98]

Ebola virus. Ebola virus causes severe hemorrhage fever with fatality rates up to 90% in humans. PPMO treatment rescued Ebola virus-infected mice. Enterlein *et al*. tested several R_9F_2C PPMOs targeted against the negative strand viral genome and the positive strand replication intermediate.[93] Two of the PPMOs were the most effective inhibitors of viral replication. When pre-incubated with an effective PPMO for 3 hours and then infected with Ebola virus, the PPMO protected Vero cells from cytopathic effects for six days and the cells had lower viral titers and decreased viral RNA compared to control PPMO. Cells pretreated with the PPMO had not supported significant viral replication by 48 hours post-infection, while cells treated with the PPMO at 0, 4 or 8 hours post-infection had low but detectable levels of viral RNA, with much higher levels after a 24-hour post-infection treatment. Mice pretreated with 500 µg of the PPMO all survived through 28 days. Mice treated with a single dose of 500 µg PPMO 24 hours post-infection all survived, while mice similarly treated with control PPMO all died.[93]

Kaposi's sarcoma associated herpes virus (KSHV). KSHV is a large DNA virus. It associates with Kaposi's sarcoma, which is the most common skin malignancy among acquired immunodefiency syndrome (AIDS) patients. Three $R_5F_2R_4C$ PPMO sequences were found to inhibit two key proteins, RTA and LANA, and control replication of KSHV. RTA is an immediate early gene that is expressed at the start of the lytic replication cycle and LANA is required to maintain latency of KSHV. PPMOs targeting three sequences of KSHV were found to knock down KSHV protein expression. If the cells were not induced to enter the lytic cycle with tetradecanoyl phorbol acetate (TPA), less than 1% of cells were RTA-positive, indicating that less than 1% of the cells had spontaneously entered the lytic cycle. After induction of lytic replication by TPA treatment, 20% of cells treated with control PPMO or not treated with PMO were RTA-positive. If cells were treated with RP1 PPMO, the fraction of cells becoming RTA-positive was reduced to 5%. Western blots show treatment with RP1 PPMO reduces RTA expression up to 91%. Flow cytometric analysis found CBLB-1 cells were 25% RTA-positive after TPA induction, while treatment with RP1 PPMO reduced the RTA-positive population to 7%.[94]

Other viruses. R_9F_2C PPMOs have been shown to be effective against the equine arteritis virus,[104] fish infectious hematopoietic necrosis virus[105] and porcine reproductive and respiratory syndrome virus.[106]

Bacteria. The growing emergence of antibiotic-resistant bacteria has highlighted the need for additional antibacterial agents. (RFF)$_3$RXB PPMO reduced bacteremia and promoted survival of mice infected with *Escherichia coli* W3110 in a sequence-specific manner. 100% of the mice survived compared with 20% of mice treated with water or scrambled PPMO when the infected mice were treated with the PPMO twice post-infection. The PPMO targeted a transcript encoding *acpP*, a component of *E. coli* Acyl Carrier Protein (ACP). (RFF)$_3$RXB peptides synthesized from D- and L-isomers of amino acids were compared; the D-isomer was less effective. A mutant strain of *E. coli*, LT1, was engineered with four wobble-base point mutations in the *acpP* transcript within the target site of the PPMO; growth of this strain was not inhibited by the PPMO in the peritonitis model.[107]

PPMOs have also been shown to inhibit growth of *Salmonella enterica* and *E. coli* in pure bacteria culture and to inhibit *E. coli* infection in Caco-2 cells. Several bacterium-permeating peptides were conjugated to the PMO targeting the *E. coli acpP* transcript and tested for the ability to inhibit growth of *E. coli* and *S. enterica* in pure bacterial cultures. The most effective conjugates tested against *E. coli* used either the (RFF)$_3$XB or RTR peptides, while both these and the (KFF)$_3$ peptide were effective at inhibiting growth of pure *S. enterica* cultures. (RFF)$_3$XB PPMO treatment of *E. coli*-infected Caco-2 cells reduced the bacterial count to undetectable levels and rescued 95% of Caco-2 cells from bacterium-induced cell death.[85]

3.3.7 Toxicity

The toxicity of PPMO depends on the concentration, treatment time, serum concentration, CPP sequences, cell types and impurities in the PPMO preparation. PPMO toxicity reported in the literature is summarized in Table 3.5. We will discuss some highlights in this section.

Concentration-dependent toxicity has been observed for PPMOs both in cell culture and in animals. Animals (mouse and rat) tolerated (RXR)$_4$XB PPMO well with repeated doses at <10 mg kg^{-1}.[4,90,96,98,101] At higher doses, animals appear lethargic and experience weight loss accompanied by elevated serum blood urea nitrogen (BUN) and creatinine. The LD$_{50}$ of the PPMO was around 220–250 mg kg^{-1}.[4] Longer treatment time with PPMO decreases cell viability. HeLa cells treated with 40 µM of (RXR)$_4$XB PPMO for 5 hours had near 100% cell viability, while 24 hours of treatment reduced viability to 75%.[89] PPMO treatment in serum-containing medium was less toxic than the treatment in serum-free medium.[89]

Toxicity of PPMO increases with increasing number of arginine residues and is also affected by the nature of other amino acid residues in a CPP sequence.[86,89] R$_9$F$_2$C PPMO was found to be more toxic than the (RXR)$_4$XB PPMO, causing leakage of cell membranes[87] while removing the two F residues reduced the toxicity (unpublished results). The influence of amino acid residues on the toxicity of PPMO was also found in another study where replacement of 6-aminohexanoic acid with β-alanine reduced toxicity significantly.

Table 3.5 Toxicity of PPMOs.

Animals or cells	Results	Ref.
(RXR)₄XB PPMO		
HeLa cells, HL-1 (murine cardiomyocytes)	No cytotoxicity was observed when cells were treated with 10 μM PPMO for 4 hours and assayed for toxicity 8 and 40 hours post-treatment for HeLa and HL-1 cells , respectively.	96
HeLa	At >20 μM with 24 hours treatment, cell viability was reduced.	89
MDCK cells	In absence of virus, cells were treated with 10–400 μM for 6 hours and cell viability was determined 24 hours after the treatment. 15–30% loss in cell viability was observed for the 400 μM treatment. <10% loss in cell viability was observed for cells treated with 20 μM for 24 hours.	95
Vero cells	Viabilities of about 85% for treatments of up to 10 μM PPMO and about 70% viability with changes in cell morphology after treatment with 20 μM PPMO.	97
Mice infected with Coxsackievirus B3	At 10 or 15 mg kg⁻¹ doses, two doses administered 48 hours apart. Monitored 7 days. Mice maintained average body weight, appeared and behaved normal. No histopatological abnormality in heart, liver, kidney, spleen and pancreas.	96
Mice infected with MHV-3 or MHV Alb139 (mouse hepatitis virus)	Infected mice treated daily with conjugates of negative control oligos lost weight more rapidly than saline-injected controls. No toxicity was observed for uninfected mice receiving four daily doses at 8 nanomoles (about 3.7 mg kg⁻¹).	90
MDX mice	PPMO-treated or sham-treated young mice gained weight at rates with no statistically significant difference. Similarly there was no statistically significant difference between PPMO-treated or sham-treated young mice in liver transaminases, alkaline phosphatase, creatinine or general health.	101
C3H/HeN mice	Dose-dependent toxicity: weight loss at 15 mg kg⁻¹ after five daily IP injection. No significant difference in organ weights and gross abnormalities except possible fibrosis in liver.	98
R₉F₂C PPMO		
HeLa	1 μM treatment in serum-free medium caused leaking membrane.	87

Table 3.5 (*Continued*).

Animals or cells	Results	Ref.
DBT (murine astrocytoma)	20 µM reduced viability 10% or 40% in DMEM or VP-SFM medium, respectively. No reduction in viability at 10 µM.	103
MDCK cells	20% loss in cell viability at 20 µM.	95
BHK-21 cells	Viability decreased to 40% at 25 µM PPMO with 24 hours treatment. No effect on viability with 10 µM treatment.	92
Vero	One PPMO was not toxic up to 20 µM. Another PPMO was toxic with cell viability dropping to 40% of controls.	93
Vero-E6	After 6 hours exposure to PPMO in serum-free medium, cells were grown with 8% serum. Assayed 24 hours later, 80 µM PPMO resulted in decreased growth and altered morphology. At 40 µM PPMO and below, no effects on growth or morphology were observed.	104
$R_5F_2R_4C$ PPMO		
BCBL-1 (EBV-negative)	By CellTiter-Blue assay, concentrations up to 16 µM PPMO caused no detectable toxicity.	94
CRL11171	By CellTiter-Blue assay, concentrations up to 16 µM PPMO caused no detectable toxicity.	106
$(RFF)_3R$ PPMO		
BALB/c mice with *E. coli* peritonitis	Mice receiving 30 µg PPMO twice had 100% survival at 48 hours while 300 µg PPMO twice had 75% survival and 1 mg PPMO twice had 0% survival.	107

Replacement of L-arginine with D-arginine in $(RX)_8$ PPMO reduced toxicity but the L→D replacement did not change the toxicity of $(RXR)_4XB$ PPMO.[89] The stereochemistry of arginine, the nature of other amino acid residues, the length of a CPP and the sequence of a CPP all play roles in the toxicity of PPMO.

Based on experiments with MHV, PPMO appears to be more toxic in virus-infected mice than in healthy mice.[90] This finding poses an additional hurdle for the development of PPMO to treat infectious diseases.

If free peptide in a PPMO preparation is not removed, it will cause additional cellular toxicity. MALDI-TOF mass spectrometry can be employed to determine the existence of free peptide, but cannot determine its concentration. HPLC can be used to quantitatively determine free peptide. Because PPMO has a much higher molar absorbance at 210, 260 and 280 nm compared to the free peptide, one needs to take the difference in molar absorbance into account when interpreting HPLC data where UV detection is used. A separate titration standard curve needs to be constructed for the free peptide. HPLC with a cation-exchange column can be used to remove the free peptides.[89]

3.4   Therapeutic Applications of Morpholinos and Peptide–Morpholino Conjugates

AVI BioPharma (www.avibio.com) has conducted number of clinical trials using the PMO technology. We discuss several ongoing clinical trials below.

3.4.1   Cardiovascular Trials

3.4.1.1   AVI-4126

Phase I, Ib and II clinical trials have been performed with AVI-4126, a PMO targeted to block translation of c-myc mRNA and intended to prevent cardiac restenosis following balloon angioplasty. Doses of up to 90 mg were delivered in Phase I trials without resulting in dose-limiting events. Doses of 3 mg or 10 mg were used in the Phase II trial, in which the PMO was delivered locally at the site of balloon angioplasty using infiltrator balloon catheters. The experimental dose of PMO dissolved in 0.4 mL solution volume was delivered for 20 seconds and then the stent was implanted. There were 8, 7 and 10 efficacy-evaluable patients in the placebo, 3 mg and 10 mg groups respectively. The placebo, 3 mg and 10 mg groups had similar pre-procedure mean luminal diameters and they also had similar post-procedure mean luminal diameters. However, at the six-month follow-up the mean luminal diameters of the placebo and 3 mg groups had decreased to about half of the mean post-procedure diameters, while the 10 mg group had, on average, lost less than a third of their mean post-procedure luminal diameters. Stenosis was significantly lower in the 10 mg group at six months compared to the other groups. The counts of patients evaluated with binary restenosis at the 6-month follow-up were 3, 2 and 0 for the placebo, 3 mg and 10 mg groups, respectively.[108]

3.4.1.2   AVI-5126

This is the first clinical trial using PPMO technology. An ongoing safety and efficacy study of PPMO to prevent eventual blockage of a transplanted vein after cardiovascular bypass surgery uses AVI-5126, a Morpholino oligo conjugated to the $(RXR)_4XB$ CPP. After the vein is excised it is immersed in a solution containing 10 μM AVI-5126 and then implanted as a bypass graft. The study is titled 'Clinical Study to Assess the Safety and Efficacy of *ex-Vivo* Vein Graft Exposure to AVI-5126 in Coronary Artery By-Pass Grafting to Reduce Clinical Graft Failure'.

3.4.2 Genetic Disease Trial

3.4.2.1 AVI-4658

A Phase I/II study of PMO to treat DMD uses AVI-4658, an unmodified Morpholino oligo targeted to modify splicing of exon 51 of the human dystrophin gene. In this initial study, PMO will be delivered by injection into the extensor digitorum brevis muscle. The study is titled 'Restoring Dystrophin Expression in DMD: A Phase I/II Clinical Trial Using AVI-4658'.

3.5 Conclusion

PMOs are effective, specific, stable and nontoxic when delivered into the cytosol and nuclei of cells. They are versatile reagents for blocking processes that involve RNA, including mRNA translation, pre-mRNA splicing and miRNA maturation and activity. As therapeutics, the excellent safety record of PMOs in clinical trials facilitates approval for exploring new applications in humans. The advent of PPMO offers potential for enhanced delivery of human therapeutics. Improved CPP sequences and compositions for PPMOs have produced decreasing toxicity and improved delivery efficacy in both cell culture and *in vivo* models. Research is currently focused on developing CPPs that produce PPMOs with more effective endosomal escape and further decreases in toxicity. The PMOs, and now the PPMOs, are drug platform technologies which can be adapted for many different applications by selecting the appropriate base sequence of the PMO (or the PMO component of a PPMO). We hope that clinical trials of next-generation PMO and PPMO will produce improved-efficacy antisense treatments that, by varying the PMO sequence, can address a broad range of human diseases.

Acknowledgements

We thank Dr Susie Hatlevig for her careful editing of this chapter, Drs Patrick Ivensen and Jim Summerton for their support of this writing project and our daughters Zoe and Simone for giving us just enough time to finish this chapter.

References

1. J. Summerton and D. Weller, *Antisense Nucleic Acid Drug Dev.*, 1997, **7**, 187.
2. R. M. Hudziak, E. Barofsky, D. F. Barofsky, D. L. Weller, S. B. Huang and D. D. Weller, *Antisense Nucleic Acid Drug Dev.*, 1996, **6**, 267.
3. D. S. Youngblood, S. A. Hatlevig, J. N. Hassinger, P. L. Iversen and H. M. Moulton, *Bioconjugate Chem.*, 2007, **18**, 50.

4. A. Amantana, H. M. Moulton, M. L. Cate, M. T. Reddy, T. Whitehead, J. N. Hassinger, D. S. Youngblood and P. L. Iversen, *Bioconjugate Chem.*, 2007, **18**, 1325.

5. J. Summerton, *Biochim. Biophys. Acta*, 1999, **1489**, 141.

6. P. A. Morcos, *Biochem. Biophys. Res. Commun.*, 2007, **358**, 521.

7. B. W. Draper, P. A. Morcos and C. B. Kimmel, *Genesis*, 2001, **30**, 154.

8. B. R. Graveley, *Trends Genet.*, 2001, **17**, 100.

9. C. Matranga, Y. Tomari, C. Shin, D. P. Bartel and P. D. Zamore, *Cell*, 2005, **123**, 607.

10. R. J. Jackson and N. Standart, *Sci. STKE*, 2007, **367**, re1.

11. W. P. Kloosterman, A. K. Lagendijk, R. F. Ketting, J. D. Moulton and R. H. A. Plasterk, *PLoS Biol.*, 2007, **5**, e103.

12. (a) A. S. Flynt, N. Li, E. J. Thatcher, L. Solnica-Krezel and J. G. Patton, *Nat. Genet.*, 2007, **39**, 259.
    (b) W. Y. Choi, A. J. Giraldez and F. Schier, *Science*, 2007, **318**, 217.

13. I. G. Bruno, W. Jin and G. J. Cote, *Hum. Mol. Genet.*, 2004, **13**, 2409.

14. N. Matter and H. Konig, *Nucleic Acids Res.*, 2005, **33**, e41.

15. L. Yen, J. Svendsen, J. S. Lee, J. T. Gray, M. Magnier, T. Baba, R. J. D'Amato and R. C. Mulligan, *Nature*, 2004, **431**, 471.

16. M. T. Howard, R. F. Gesteland and J. F. Atkins, *RNA*, 2004, **10**, 1653.

17. R. Jubin, N. E. Vantuno, J. S. Kieft, M. G. Murray, J. A. Doudna, J. Y. Lau and B. M. Baroudy, *J. Virol.*, 2000, **74**, 10430.

18. R. M. Kinney, C. Y. Huang, B. C. Rose, A. D. Kroeker, T. W. Dreher, P. L. Iversen and D. A. Stein, *J. Virol.*, 2005, **79**, 5116.

19. S. C. Ekker and J. D. Larson, *Genesis*, 2001, **30**, 89.

20. J. Gruber, H. Manninga, T. Tuschl, M. Osborn and K. Weber, *RNA Biol.*, 2005, **2**, 101.

21. T. M. Woolf, C. G. Jennings, M. Rebagliati and D. A. Melton, *Nucleic Acids Res.*, 1990, **18**, 1763.

22. L. Yamada, E. Shoguchi, S. Wada, K. Kobayashi, Y. Mochizuki, Y. Satou and N. Satoh, *Development*, 2003, **130**, 6485.

23. J. E. Summerton, *Curr. Top. Med. Chem.*, 2007, **7**, 651.

24. L. S. Siddall, L. C. Barcroft and A. J. Watson, *Mol. Reprod. Dev.*, 2002, **63**, 413.

25. W. T. Penberthy, E. Shafizadeh and S. Lin, *Front Biosci.*, 2002, **7**, d1439.

26. R. Kole, M. Vacek and T. Williams, *Oligonucleotides*, 2004, **14**, 65.

27. P. Scaffidi and T. Misteli, *Nat. Med.*, 2005, **11**, 440.

28. P. Scaffidi and T. Misteli, *Science*, 2006, **312**, 1059.

29. L. Du, J. M. Pollard and R. A. Gatti, *Proc. Natl. Acad. Sci. U. S. A.*, 2007, **104**, 6007.

30. F. Vetrini, R. Tammaro, S. Bondanza, E. M. Surace, A. Auricchio, M. De Luca, A. Ballabio and V. Marigo, *Hum. Mutat.*, 2006, **27**, 420.

31. R. Jubin, *Methods Mol. Med.*, 2005, **106**, 309.

32. D. A. Stein, D. E. Skilling, P. L. Iversen and A. W. Smith, *Antisense Nucleic Acid Drug Dev.*, 2001, **11**, 317.

33. A. J. Tackett, L. Wei, C. E. Cameron and K. D. Raney, *Nucleic Acids Res.*, 2001, **29**, 565.
34. H. Ly, D. P. Nierlich, J. C. Olsen and A. H. Kaplan, *J. Virol.*, 2000, **74**, 9937.
35. K. A. Dunlap, M. Palmarini, M. Varela, R. C. Burghardt, K. Hayashi, J. L. Farmer and T. E. Spencer, *Proc. Natl. Acad. Sci. U. S. A.*, 2006, **103**, 14390.
36. K. L. Warfield, D. L. Swenson, G. G. Olinger, D. K. Nichols, W. D. Pratt, R. Blouch, D. A. Stein, M. J. Aman, P. L. Iversen and S. Bavari, *PLoS Pathog.*, 2006, **2**, e1.
37. C. Vander-Linden, *Chem-Bio Defense Quarterly*, 2006, **3**, 8.
38. C. Raggo, R. Ruhl, S. McAllister, H. Koon, B. J. Dezube, K. Fruh and A. V. Moses, *Cancer Res.*, 2005, **65**, 5084.
39. E. K. Wright, S. A. Goodart, J. D. Growney, V. Hadinoto, M. G. Endrizzi, E. M. Long, K. Sadigh, A. L. Abney, I. Bernstein-Hanley and W. F. Dietrich, *Curr. Biol.*, 2003, **13**, 27.
40. M. Fujita, N. M. Khazenzon, A. V. Ljubimov, B. S. Lee, I. Virtanen, E. Holler, K. L. Black and J. Y. Ljubimova, *Angiogenesis*, 2006, **9**, 183.
41. G. R. Devi, J. R. Oldenkamp, C. A. London and P. L. Iversen, *Prostate*, 2002, **53**, 200.
42. P. L. Iversen, V. Arora, A. J. Acker, D. H. Mason and G. R. Devi, *Clin. Cancer Res.*, 2003, **9**, 2510.
43. K. Kaneko, K. Satoh, A. Masamune, A. Satoh and T. Shimosegawa, *Pancreas*, 2002, **24**, 251.
44. D. C. Knapp, J. E. Mata, M. T. Reddy, G. R. Devi and P. L. Iversen, *Anticancer Drugs*, 2003, **14**, 39.
45. A. Amantana, C. A. London, P. L. Iversen and G. R. Devi, *Mol. Cancer. Ther.*, 2004, **3**, 699.
46. C. A. London, H. S. Sekhon, V. Arora, D. A. Stein, P. L. Iversen and G. R. Devi, *Cancer Gene Ther.*, 2003, **10**, 823.
47. H. K. Roy, P. Iversen, J. Hart, Y. Liu, J. L. Koetsier, Y. Kim, D. P. Kunte, M. Madugula, V. Backman and R. K. Wali, *Mol. Cancer Ther.*, 2004, **3**, 1159.
48. Y. Takei, K. Kadomatsu, K. Yuasa, W. Sato and T. Muramatsu, *Int. J. Cancer*, 2005, **114**, 490.
49. J. P. Truman, N. Gueven, M. Lavin, S. Leibel, R. Kolesnick, Z. Fuks and A. Haimovitz-Friedman, *J. Biol. Chem.*, 2005, **280**, 23262.
50. I. Kryczek, L. Zou, P. Rodriguez, G. Zhu, S. Wei, P. Mottram, M. Brumlik, P. Cheng, T. Curiel, L. Myers, A. Lackner, X. Alvarez, A. Ochoa, L. Chen and W. Zou, *J. Exp. Med.*, 2006, **203**, 871.
51. R. Catena, V. Muniz-Medina, B. Moralejo, B. Javierre, C. J. Best, M. R. Emmert-Buck, J. E. Green, C. C. Baker and A. Calvo, *Int. J. Cancer*, 2007, **120**, 2096.
52. S. Nicoli, D. Ribatti, F. Cotelli and M. Presta, *Cancer Res.*, 2007, **67**, 2927.

53. G. Liu, C. Liu, S. Zhang, J. He, N. Liu, S. Gupta, M. Rusckowski and D. J. Hnatowich, *Nucl. Med. Commun.*, 2003, **24**, 697.
54. J. He, G. Liu, S. Zhang, M. Rusckowski and D. J. Hnatowich, *Cancer Biother. Radiopharm.*, 2003, **18**, 941.
55. G. Liu, S. Dou, G. Mardirossian, J. He, S. Zhang, X. Liu, M. Rusckowski and D. J. Hnatowich, *Clin. Cancer Res.*, 2006, **12**, 4958.
56. J. S. Isenberg, F. Hyodo, K. Matsumoto, M. J. Romeo, M. Abu-Asab, M. Tsokos, P. Kuppusamy, D. A. Wink, M. C. Krishna and D. D. Roberts, *Blood*, 2007, **109**, 1945.
57. M. Masaki, M. Izumi, Y. Oshima, Y. Nakaoka, T. Kuroda, R. Kimura, S. Sugiyama, K. Terai, M. Kitakaze, K. Yamauchi-Takihara, I. Kawase and H. Hirota, *Circulation*, 2005, **111**, 2752.
58. A. L. Prasch, R. L. Tanguay, V. Mehta, W. Heideman and R. E. Peterson, *Mol. Pharmacol.*, 2006, **69**, 776.
59. S. M. Billiard, A. R. Timme-Laragy, D. M. Wassenberg, C. Cockman and R. T. Di Giulio, *Toxicol. Sci.*, 2006, **92**, 526.
60. H. Teraoka, W. Dong, Y. Tsujimoto, H. Iwasa, D. Endoh, N. Ueno, J. J. Stegeman, R. E. Peterson and T. Hiraga, *Biochem. Biophys. Res. Commun.*, 2003, **304**, 223.
61. J. P. Incardona, M. G. Carls, H. Teraoka, C. A. Sloan, T. K. Collier and N. L. Scholz, *Environ. Health Perspect.*, 2005, **113**, 1755.
62. E. T. McKinley, T. C. Baranowski, D. O. Blavo, C. Cato, T. N. Doan and A. L. Rubinstein, *Brain Res. Mol. Brain Res.*, 2005, **141**, 128.
63. B. Mahadevan, V. Arora, L. J. Schild, C. Keshava, M. L. Cate, P. L. Iversen, M. C. Poirier, A. Weston, C. Pereira and W. M. Baird, *Mol. Carcinog.*, 2006, **45**, 118.
64. H. M. Moulton and J. D. Moulton, *Curr. Opin. Mol. Ther.*, 2003, **5**, 123.
65. S. Fletcher, K. Honeyman, A. M. Fall, P. L. Harding, R. D. Johnsen and S. D. Wilton, *J. Gene Med.*, 2006, **8**, 207.
66. J. Alter, F. Lou, A. Rabinowitz, H. Yin, J. Rosenfeld, S. D. Wilton, T. A. Partridge and Q. L. Lu, *Nat. Med.*, 2006, **12**, 175.
67. V. Arora, M. L. Cate, C. Ghosh and P. L. Iversen, *Drug Metab. Dispos.*, 2002, **30**, 757.
68. D. L. Stenkamp and R. A. Frey, *Dev. Biol.*, 2003, **258**, 349.
69. K. Prasadan, E. Daume, B. Preuett, T. Spilde, A. Bhatia, H. Kobayashi, M. Hembree, P. Manna and G. K. Gittes, *Diabetes*, 2002, **51**, 3229.
70. S. P. Monga, H. K. Monga, X. Tan, K. Mule, P. Pediaditakis and G. K. Michalopoulos, *Gastroenterology*, 2003, **124**, 202.
71. I. S. Oh, H. Shimizu, T. Satoh, S. Okada, S. Adachi, K. Inoue, H. Eguchi, M. Yamamoto, T. Imaki, K. Hashimoto, T. Tsuchiya, T. Monden, K. Horiguchi, M. Yamada and M. Mori, *Nature*, 2006, **443**, 709.
72. C. B. Kimmel and R. D. Law, *Dev. Biol.*, 1985, **108**, 94.
73. C. B. Kimmel and R. D. Law, *Dev. Biol.*, 1985, **108**, 78.
74. S. L. Nutt, O. J. Bronchain, K. O. Hartley and E. Amaya, *Genesis*, 2001, **30**, 110.

75. R. Kos, R. P. Tucker, R. Hall, T. D. Duong and C. A. Erickson, *Dev. Dyn.*, 2003, **226**, 470.
76. G. A. Cerda, J. E. Thomas, M. L. Allende, R. O. Karlstrom and V. Palma, *Methods*, 2006, **39**, 207.
77. J. E. Bestman, R. C. Ewald, S. L. Chiu and H. T. Cline, *Nat. Protoc.*, 2006, **1**, 1267.
78. G. Mellitzer, M. Hallonet, L. Chen and S. L. Ang, *Mech. Dev.*, 2002, **118**, 57.
79. M. Takahashi, K. Sato, T. Nomura and N. Osumi, *Differentiation*, 2002, **70**, 155.
80. J. E. Summerton, *Ann. N. Y. Acad. Sci.*, 2005, **1058**, 62.
81. P. A. Morcos, *Genesis*, 2001, **30**, 94.
82. N. N. Kipshidze, T. R. Porter, G. Dangas, H. Yazdi, F. Tio, F. Xie, D. Hellinga, J. Fournadjiev, R. Wolfram, R. Seabron, R. Waksman, A. Abizaid, G. Roubin, S. Iyer, M. B. Leon, J. W. Moses and P. Iversen, *Cardiovasc. Radiat. Med.*, 2003, **4**, 152.
83. J. Y. Ljubimova, M. Fujita, N. M. Khazenzon, B. S. Lee, S. Wachsmann-Hogiu, D. L. Farkas, K. L. Black and E. Holler, *Chem. Biol. Interact.*, 2007, in, press.
84. H. M. Moulton, M. C. Hase, K. M. Smith and P. L. Iversen, *Antisense Nucleic Acid Drug Dev.*, 2003, **13**, 31.
85. L. D. Tilley, O. S. Hine, J. A. Kellogg, J. N. Hassinger, D. D. Weller, P. L. Iversen and B. L. Geller, *Antimicrob. Agents Chemother.*, 2006, **50**, 2789.
86. H. M. Moulton, M. H. Nelson, S. A. Hatlevig, M. T. Reddy and P. L. Iversen, *Bioconjugate Chem.*, 2004, **15**, 290.
87. S. Abes, H. M. Moulton, P. Clair, P. Prevot, D. S. Youngblood, R. P. Wu, P. L. Iversen and B. Lebleu, *J. Controlled Release*, 2006, **116**, 304.
88. S. H. Kang, M. J. Cho and R. Kole, *Biochemistry*, 1998, **37**, 6235.
89. R. P. Wu, Y. S. Youngblood, J. N. Hassinger, C. E. Lovejoy, M. H. Nelson, P. L. Iversen and H. M. Moulton, *Nucleic Acids Res.*, 2007, **35**, 5182.
90. R. Burrer, B. W. Neuman, J. P. Ting, D. A. Stein, H. M. Moulton, P. L. Iversen, P. Kuhn and M. J. Buchmeier, *J. Virol.*, 2007, **81**, 5637.
91. M. H. Nelson, D. A. Stein, A. D. Kroeker, S. A. Hatlevig, P. L. Iversen and H. M. Moulton, *Bioconjugate Chem.*, 2005, **16**, 959.
92. K. L. Holden, D. A. Stein, T. C. Pierson, A. A. Ahmed, K. Clyde, P. L. Iversen and E. Harris, *Virology*, 2006, **344**, 439.
93. S. Enterlein, K. L. Warfield, D. L. Swenson, D. A. Stein, J. L. Smith, C. S. Gamble, A. D. Kroeker, P. L. Iversen, S. Bavari and E. Muhlberger, *Antimicrob. Agents Chemother.*, 2006, **50**, 984.
94. Y. J. Zhang, K. Y. Wang, D. A. Stein, D. Patel, R. Watkins, H. M. Moulton, P. L. Iversen and D. O. Matson, *Antiviral Res.*, 2007, **73**, 12.
95. Q. Ge, M. Pastey, D. Kobasa, P. Puthavathana, C. Lupfer, R. K. Bestwick, P. L. Iversen, J. Chen and D. A. Stein, *Antimicrob. Agents Chemother.*, 2006, **50**, 3724.

96. J. Yuan, D. A. Stein, T. Lim, D. Qiu, S. Coughlin, Z. Liu, Y. Wang, R. Blouch, H. M. Moulton, P. L. Iversen and D. Yang, *J. Virol.*, 2006, **80**, 11510.

97. T. S. Deas, I. Binduga-Gajewska, M. Tilgner, P. Ren, D. A. Stein, H. M. Moulton, P. L. Iversen, E. B. Kauffman, L. D. Kramer and P. Y. Shi, *J. Virol.*, 2005, **79**, 4599.

98. T. S. Deas, C. J. Bennett, S. A. Jones, M. Tilgner, P. Ren, M. J. Behr, D. A. Stein, P. L. Iversen, L. D. Kramer, K. A. Bernard and P. Y. Shi, *Antimicrob. Agents Chemother.*, 2007, **51**, 2470.

99. G. McClorey, H. M. Moulton, P. L. Iversen, S. Fletcher and S. D. Wilton, *Gene Ther.*, 2006, **13**, 1373.

100. G. McClorey, A. M. Fall, H. M. Moulton, P. L. Iversen, J. E. Rasko, M. Ryan, S. Fletcher and S. D. Wilton, *Neuromuscul. Disord.*, 2006, **16**, 583.

101. S. Fletcher, K. Honeyman, A. M. Fall, P. L. Harding, R. D. Johnsen, J. P. Steinhaus, H. M. Moulton, P. L. Iversen and S. D. Wilton, *Mol. Ther.*, 2007, **15**, 1587.

102. B. W. Neuman, D. A. Stein, A. D. Kroeker, M. J. Churchill, A. M. Kim, P. Kuhn, P. Dawson, H. M. Moulton, R. K. Bestwick, P. L. Iversen and M. J. Buchmeier, *J. Virol.*, 2005, **79**, 9665.

103. B. W. Neuman, D. A. Stein, A. D. Kroeker, A. D. Paulino, H. M. Moulton, P. L. Iversen and M. J. Buchmeier, *J. Virol.*, 2004, **78**, 5891.

104. E. van den Born, D. A. Stein, P. L. Iversen and E. J. Snijder, *J. Gen. Virol.*, 2005, **86**, 3081.

105. M. Alonso, D. A. Stein, E. Thomann, H. M. Moulton, J. C. Leong, P. Iversen and D. V. Mourich, *J. Fish Dis.*, 2005, **28**, 399.

106. Y. J. Zhang, D. A. Stein, S. M. Fan, K. Y. Wang, A. D. Kroeker, X. J. Meng, P. L. Iversen and D. O. Matson, *Vet. Microbiol.*, 2006, **117**, 117.

107. L. D. Tilley, B. L. Mellbye, S. E. Puckett, P. L. Iversen and B. L. Geller, *J. Antimicrob. Chemother.*, 2007, **59**, 66.

108. N. N. Kipshidze, P. L. Iversen, P. Overlie, T. Dunlap, B. Titus, D. Lee, J. Moses, P. O'Hanley, M. Lauer and M. B. Leon, *Cardiovasc. Revasc. Med.*, 2007, **8**, 230.

109. K. Lorent, S. Y. Yeo, T. Oda, S. Chandrasekharappa, A. Chitnis, R. P. Matthews and M. Pack, *Development*, 2004, **131**, 5753.

110. H. Zetterberg, W. A. Campbell, H. W. Yang and W. Xia, *J. Biol. Chem.*, 2006, **281**, 11933.

111. R. P. Matthews, N. Plumb-Rudewiez, K. Lorent, P. Gissen, C. A. Johnson, F. Lemaigre and M. Pack, *Development*, 2005, **132**, 5295.

112. Y. H. Ching, T. K. Ghosh, S. J. Cross, E. A. Packham, L. Honeyman, S. Loughna, T. E. Robinson, A. M. Dearlove, G. Ribas, A. J. Bonser, N. R. Thomas, A. J. Scotter, L. S. Caves, G. P. Tyrrell, R. A. Newbury-Ecob, A. Munnich, D. Bonnet and J. D. Brook, *Nat. Genet.*, 2005, **37**, 423.

113. N. S. Zinkevich, D. V. Bosenko, B. A. Link and E. V. Semina, *BMC Dev. Biol.*, 2006, **6**, 13.

114. C. Stoetzel, V. Laurier, L. Faivre, A. Megarbane, F. Perrin-Schmitt, A. Verloes, D. Bonneau, J. L. Mandel, M. Cossee and H. Dollfus, *J. Hum. Genet.*, 2006, **51**, 81.

115. Z. Khuchua, Z. Yue, L. Batts and A. W. Strauss, *Circ. Res.*, 2006, **99**, 201.

116. R. M. Nissen, A. Amsterdam and N. Hopkins, *BMC Dev. Biol.*, 2006, **6**, 28.

117. M. Asai-Coakwell, C. R. French, K. M. Berry, M. Ye, R. Koss, M. Somerville, R. Mueller, V. van Heyningen, A. J. Waskiewicz and O. J. Lehmann, *Am. J. Hum. Genet.*, 2007, **80**, 306.

118. V. C. Seitan, P. Banks, S. Laval, N. A. Majid, D. Dorsett, A. Rana, J. Smith, A. Bateman, S. Krpic, A. Hostert, R. A. Rollins, H. Erdjument-Bromage, P. Tempst, C. Y. Benard, S. Hekimi, S. F. Newbury and T. Strachan, *PLoS Biol.*, 2006, **4**, e242.

119. S. A. Boyadjiev, J. C. Fromme, J. Ben, S. S. Chong, C. Nauta, D. J. Hur, G. Zhang, S. Hamamoto, R. Schekman, M. Ravazzola, L. Orci and W. Eyaid, *Nat. Genet.*, 2006, **38**, 1192.

120. M. Arniges, E. Vazquez, J. M. Fernandez-Fernandez and M. A. Valverde, *J. Biol. Chem.*, 2004, **279**, 54062.

121. P. Songhet, D. Adzic, S. Reibe and K. B. Rohr, *Dev. Dyn.*, 2007, **236**, 633.

122. I. Stalmans, D. Lambrechts, F. De Smet, S. Jansen, J. Wang, S. Maity, P. Kneer, M. von der Ohe, A. Swillen, C. Maes, M. Gewillig, D. G. Molin, P. Hellings, T. Boetel, M. Haardt, V. Compernolle, M. Dewerchin, S. Plaisance, R. Vlietinck, B. Emanuel, A. C. Gittenberger-de Groot, P. Scambler, B. Morrow, D. A. Driscol, L. Moons, C. V. Esguerra, G. Carmeliet, A. Behn-Krappa, K. Devriendt, D. Collen, S. J. Conway and P. Carmeliet, *Nat. Med.*, 2003, **9**, 173.

123. D. Yimlamai, L. Konnikova, L. G. Moss and D. G. Jay, *Dev. Biol.*, 2005, **279**, 44.

124. B. Tucker, R. I. Richards and M. Lardelli, *Hum. Mol. Genet.*, 2006, **15**, 3446.

125. P. J. Jensen, J. D. Gitlin and M. O. Carayannopoulos, *J. Biol. Chem.*, 2006, **281**, 13382.

126. T. E. North and L. I. Zon, *Dev. Dyn.*, 2003, **228**, 568.

127. J. L. Jadrich, M. B. O'Connor and E. Coucouvanis, *Development*, 2006, **133**, 1529.

128. P. N. Valdmanis, I. A. Meijer, A. Reynolds, A. Lei, P. MacLeod, D. Schlesinger, M. Zatz, E. Reid, P. A. Dion, P. Drapeau and G. A. Rouleau, *Am. J. Hum. Genet.*, 2007, **80**, 152.

129. D. M. Garrity, S. Childs and M. C. Fishman, *Development*, 2002, **129**, 4635.

130. J. A. Sayer, E. A. Otto, J. F. O'Toole, G. Nurnberg, M. A. Kennedy, C. Becker, H. C. Hennies, J. Helou, M. Attanasio, B. V. Fausett, B. Utsch, H. Khanna, Y. Liu, I. Drummond, I. Kawakami, T. Kusakabe, M. Tsuda, L. Ma, H. Lee, R. G. Larson, S. J. Allen, C. J. Wilkinson, E. A. Nigg, C. Shou, C. Lillo, D. S. Williams, B. Hoppe, M. J. Kemper, T. Neuhaus, M. A. Parisi, I. A. Glass, M. Petry, A. Kispert, J. Gloy,

A. Ganner, G. Walz, X. Zhu, D. Goldman, P. Nurnberg, A. Swaroop, M. R. Leroux and F. Hildebrandt, *Nat. Genet.*, 2006, **38**, 674.

131. D. Ng, N. Thakker, C. M. Corcoran, D. Donnai, R. Perveen, A. Schneider, D. W. Hadley, C. Tifft, L. Zhang, A. O. Wilkie, J. J. van der Smagt, R. J. Gorlin, S. M. Burgess, V. J. Bardwell, G. C. Black and L. G. Biesecker, *Nat. Genet.*, 2004, **36**, 411.

132. E. Chen, J. D. Larson and S. C. Ekker, *Blood*, 2006, **107**, 4364.

133. A. Vaknin-Dembinsky, K. Balashov and H. L. Weiner, *J. Immunol.*, 2006, **176**, 7768.

134. M. Deniziak, C. Thisse, M. Rederstorff, C. Hindelang, B. Thisse and A. Lescure, *Exp. Cell Res.*, 2007, **313**, 156.

135. S. E. Craven, D. French, W. Ye, F. de Sauvage and A. Rosenthal, *Blood*, 2005, **105**, 3528.

136. R. M. Squillace, D. M. Chenault and E. H. Wang, *Dev. Biol.*, 2002, **250**, 218.

137. E. A. Otto, B. Schermer, T. Obara, J. F. O'Toole, K. S. Hiller, A. M. Mueller, R. G. Ruf, J. Hoefele, F. Beekmann, D. Landau, J. W. Foreman, J. A. Goodship, T. Strachan, A. Kispert, M. T. Wolf, M. F. Gagnadoux, H. Nivet, C. Antignac, G. Walz, I. A. Drummond, T. Benzing and F. Hildebrandt, *Nat. Genet.*, 2003, **34**, 413.

138. B. Hinkes, R. C. Wiggins, R. Gbadegesin, C. N. Vlangos, D. Seelow, G. Nurnberg, P. Garg, R. Verma, H. Chaib, B. E. Hoskins, S. Ashraf, C. Becker, H. C. Hennies, M. Goyal, B. L. Wharram, A. D. Schachter, S. Mudumana, I. Drummond, D. Kerjaschki, R. Waldherr, A. Dietrich, F. Ozaltin, A. Bakkaloglu, R. Cleper, L. Basel-Vanagaite, M. Pohl, M. Griebel, A. N. Tsygin, A. Soylu, D. Muller, C. S. Sorli, T. D. Bunney, M. Katan, J. Liu, M. Attanasio, F. O'Toole J, K. Hasselbacher, B. Mucha, E. A. Otto, R. Airik, A. Kispert, G. G. Kelley, A. V. Smrcka, T. Gudermann, L. B. Holzman, P. Nurnberg and F. Hildebrandt, *Nat. Genet.*, 2006, **38**, 1397.

139. S. A. Harvey and M. P. Logan, *Development*, 2006, **133**, 1165.

140. A. Granata, D. Savery, J. Hazan, B. M. Cheung, A. Lumsden and N. A. Quaderi, *Dev. Biol.*, 2005, **277**, 417.

141. O. Ossipova, N. Bardeesy, R. A. DePinho and J. B. Green, *Nat. Cell Biol.*, 2003, **5**, 889.

142. Z. Sun, A. Amsterdam, G. J. Pazour, D. G. Cole, M. S. Miller and N. Hopkins, *Development*, 2004, **131**, 4085.

143. I. Stancheva, A. L. Collins, I. B. Van den Veyver, H. Zoghbi and R. R. Meehan, *Mol. Cell*, 2003, **12**, 425.

144. T. Yoshida, L. A. Phylactou, J. B. Uney, I. Ishikawa, K. Eto and S. Iseki, *J. Anat.*, 2005, **206**, 437.

145. J. Schonberger, L. Wang, J. T. Shin, S. D. Kim, F. F. Depreux, H. Zhu, L. Zon, A. Pizard, J. B. Kim, C. A. Macrae, A. J. Mungall, J. G. Seidman and C. E. Seidman, *Nat. Genet.*, 2005, **37**, 418.

146. B. De Cat, S. Y. Muyldermans, C. Coomans, G. Degeest, B. Vanderschueren, J. Creemers, F. Biemar, B. Peers and G. David, *J. Cell Biol.*, 2003, **163**, 625.

CHAPTER 4

Peptide–Peptide Nucleic Acid Conjugates for Modulation of Gene Expression

MARTIN M. FABANI, GABRIELA D. IVANOVA AND MICHAEL J. GAIT[*]

Medical Research Council, Laboratory of Molecular Biology, Hills Road, Cambridge CB2 0QH, UK

4.1 Steric Block as a Method of Gene Expression Control

When Zamecnik and Stephenson first proposed synthetic oligodeoxyribo-nucleotides as reagents to bind to Rous sarcoma virus RNA to inhibit viral replication, they designed an oligonucleotide (ON) to target the initiation site for protein translation in the expectation that it would form an RNA–DNA duplex and physically block the RNA, thus preventing protein synthesis.[1] Several years later it was found that an alternative mechanism probably operates when DNA oligomers bind to RNA targets inside cells, that of recognition of the hybrid by the cellular enzyme RNase H and subsequent RNA cleavage. This second attribute of DNA ONs, which extends to their phosphorothioate-(PS-) modified counterparts, became established as the predominant 'antisense' mechanism of action and led to the first industrial development of this class of therapeutic ONs. The original concept of the steric block mechanism of inhibition of protein translation continued to be studied by scientists,[2,3] but only recently have steric block ONs been considered seriously as potential therapeutics. In addition, we now know that through duplex

RSC Biomolecular Sciences
Therapeutic Oligonucleotides
Edited by Jens Kurreck
© Royal Society of Chemistry 2008

147. T. Koide, T. Hayata and K. W. Cho, *Development*, 2006, **133**, 2395.
148. T. L. Carrel, M. L. McWhorter, E. Workman, H. Zhang, E. C. Wolstencroft, C. Lorson, G. J. Bassell, A. H. Burghes and C. E. Beattie, *J. Neurosci.*, 2006, **26**, 11014.
149. R. M. Lawn, D. P. Wade, M. R. Garvin, X. Wang, K. Schwartz, J. G. Porter, J. J. Seilhamer, A. M. Vaughan and J. F. Oram, *J. Clin. Invest.*, 1999, **104**, R25.
150. F. Q. Li, R. E. Person, K. Takemaru, K. Williams, K. Meade-White, A. H. Ozsahin, T. Gungor, R. T. Moon and M. Horwitz, *J. Biol. Chem.*, 2004, **279**, 2873.
151. X. Fang, H. Ji, S. W. Kim, J. I. Park, T. G. Vaught, P. Z. Anastasiadis, M. Ciesiolka and P. D. McCrea, *J. Cell Biol.*, 2004, **165**, 87.
152. K. Dutton, J. R. Dutton, A. Pauliny and R. N. Kelsh, *Genesis*, 2001, **30**, 188.
153. C. Ring, S. Ogata, L. Meek, J. Song, T. Ohta, K. Miyazono and K. W. Cho, *Genes Dev.*, 2002, **16**, 820.

formation a number of other RNA-processing events can also be inhibited sterically, for example nuclear splicing, which is required for the processing of most mammalian gene transcripts, and which involves numerous steps of RNA-protein recognition.[4] Another novel example is the inhibition of endogenous microRNAs by complementary synthetic ON constructs, which has recently been used *in vivo*.[5] Note that there is often confusion in the use of the word 'antisense', which is sometimes used interchangeably for both RNase H and steric block mechanisms. In this chapter, the word antisense is used strictly to imply that the oligomer is likely to work predominantly by an RNase H mechanism of action.

An advantage of the steric block approach is the ability to use a considerably wider range of synthetic analogues (see Section 4.2) than is possible with conventional antisense, since there is no requirement for recognition of the RNA–DNA hybrid by a cellular enzyme. Instead, the issue is to obtain tight binding to the RNA target by the analogue while also showing good resistance to nuclease degradation. This allows a greater ability to manipulate ON chemistry and composition to obtain improved pharmacological parameters. In principle, a steric block oligomer might be expected to show a greater specificity (lower off-target effects) compared to antisense, since the binding to a wrong sequence will often not have any biological consequence, because most parts of the RNA (*e.g.* coding) are not affected by such oligomer binding. One disadvantage is the need for at least a stoichiometric amount of the oligomer on the RNA target for full inhibition, whereas antisense and small interfering RNAs (siRNAs) are catalytic in action. In practice however, the amounts of many messenger RNA (mRNA) targets inside cells are small and the issue therefore becomes simply a question of delivering sufficient oligomer into cells to obtain the required biological effect.

4.2 Peptide Nucleic Acids in Comparison with Other Steric Block Oligomer Types

Steric block ONs are designed not to contain any section which, on binding to RNA, may induce cleavage by RNase H. Among a number of ON analogues tested, one of the first to become widely established as a steric block agent was that consisting fully of 2′-*O*-methylated (OMe) nucleosides[6] and usually also containing all PS linkages.[7–9] This type of ON (OMe-PS) recently entered clinical trials in Holland for the treatment of one type of Duchenne muscular dystrophy through an 'exon skipping' steric block mechanism.[10] Unfortunately, OMe ONs in the absence of PS linkages do not enter cells very well in culture and also do not have sufficient stability *in vivo*. The addition of PS linkages helps to counter both of these problems, but also gives rise to poorer binding to RNA, and therefore there is a need to use relatively long ONs (20–30 residues) to observe a strong biological effect. Several other types of negatively charged, phosphate-containing analogues have emerged more recently with better binding to RNA and high resistance to nuclease degradation, for

example 2'-*O*-methoxyethyl (MOE) (with or without PS linkages)[3] and N3' →
P5'-phosphoramidates (NPs).[11] A lipidated phosphorothioamidate ON, targeted to the essential RNA component of telomerase,[12] has recently entered
clinical trials for treatment of chronic lymphocytic leukaemia (GRN163L,
Geron). All of these analogue types simulate conformations similar to that
adopted by an oligoribonucleotide when bound to an RNA target (*i.e.* an
A-like structure).

Another important analogue that also simulates a ribonucleotide conformation is locked nucleic acid (LNA), which binds very tightly to RNA. For
example, LNA ONs have been used in steric block mode for *in vivo* tumour
growth inhibition.[13] Commonly LNA is best used in combination with another
nucleotide derivative as a mixmer, for example with 2'-deoxyribonucleotides[14]
or with OMe residues[15] to modulate the binding characteristics. An alternating
LNA–DNA 16-mer that contained all PS linkages was found to give efficient
splice switching in mice.[16] In our own studies we found that OMe-LNA mixmers bound strongly to the *trans*-activation responsive element (TAR) of
human immunodeficiency virus 1 (HIV-1) viral RNA to block Tat-dependent
transcription.[17] Further, the mixmers were delivered into the nuclei of
Henrietta Lacks (HeLa) cells in culture very efficiently and resulted in
dose-dependent and sequence-dependent inhibition of Tat-dependent *trans*-
activation as judged by a luciferase reporter assay.[17,18]

Two types of charge-neutral analogues, peptide nucleic acids (PNAs;
Figure 4.1) and phosphorodiamidate morpholino ONs (PMOs), demonstrate a
radical departure from the normal ON backbone structure, but nevertheless
retain very strong RNA-binding characteristics. PMOs have been found to
have remarkable steric block properties in cells, especially when conjugated to a
cell-penetrating peptide (CPP), and are also in clinical trials for a range of
indications (AVI Biopharma). Since PMOs are covered in detail in Chapter 3,
they are not mentioned further here, except for comparative purposes. The
other well-known charge-neutral analogue PNA was developed by the group of
Nielsen and colleagues in Denmark in the early 1990s.[19] PNA has an achiral
open-chain backbone that forms strong interactions with both DNA and RNA
ONs, which obey Watson–Crick hydrogen bonding rules.[20] In contrast to OMe
and NP oligomers, a 13-mer PNA was shown to arrest the translation of
protein synthesis within the coding region of an mRNA *in vitro*.[21]

Although PNA quickly became of great value in diagnostic and other *in vitro*
applications, it soon became apparent that, despite being charge neutral, PNA
does not enter cells any more readily than do negatively charged ONs. Various
methods of PNA delivery have been devised, such as electroporation,[22] complexation with a DNA ON together with polyethyleneimine[23] or cationic
liposomes,[24] lipofection of acridine and other polyheteroaromate functionalized PNA,[25] and photochemically induced delivery.[26] In a recent comparison
of a number of transfection methods,[27] it seems clear that there is a significant
variability in the efficiency of delivery, for example depending on the particular
sequence and design of the complementary DNA as well as on the type of
transfection agent used and the amount of cell toxicity resultant. Nevertheless,

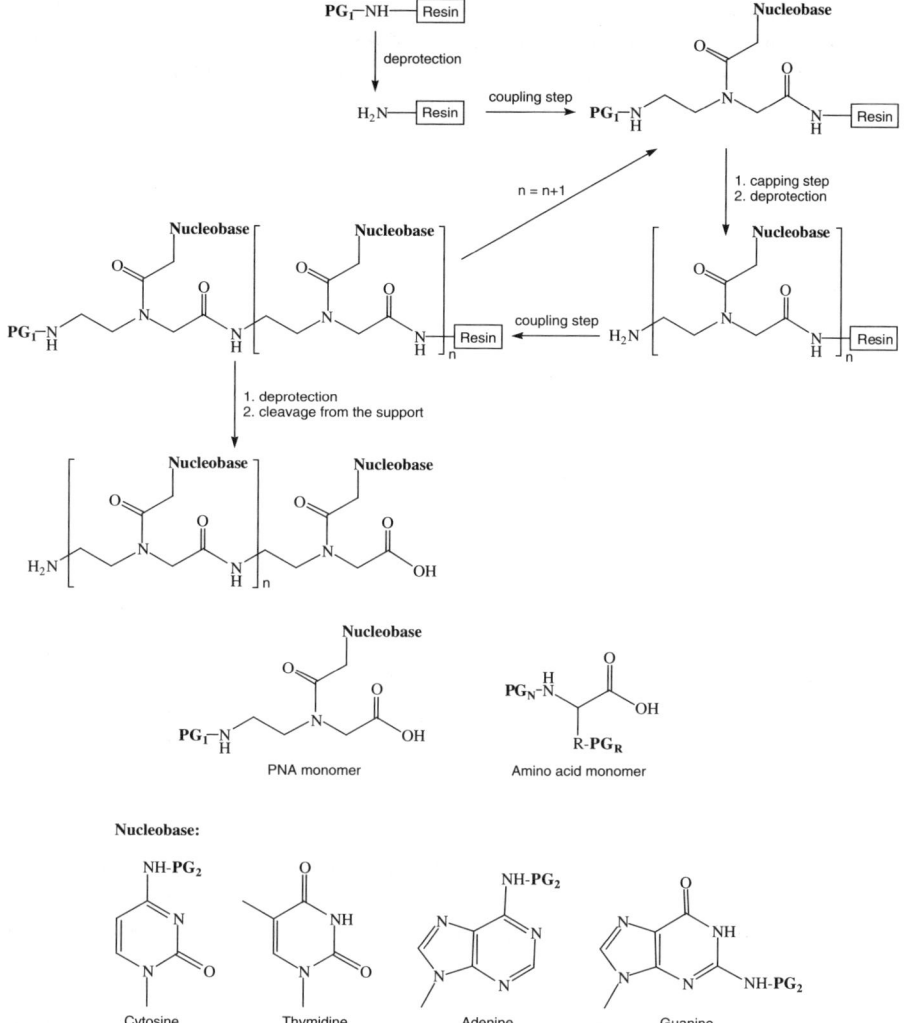

Figure 4.1 Basic steps in solid-phase PNA synthesis and structures of PNA and amino acid monomers.

splice correction activities with IC_{50} values of around 200 nM were obtainable, for example, by such transfection methods.[27]

4.3 Cellular Delivery: The Promise of Cell Penetrating Peptides

The idea of covalent conjugation of ONs to peptides to enhance cell delivery is now at least 20 years old.[28] Particular interest was generated in the mid-1990s when a peptide derived from helix 3 of the *Antennapaedia* homeodomain protein from *Drosophila melanogaster* (residues 43–58, later known as Penetratin) was conjugated to antisense phosphodiester or PS ONs and shown to have biological activity in neurons.[29–31] Similar attributes of cell delivery were also afforded to a short peptide from the basic region of HIV-1 Tat protein (residues 48–60).[32] Since then, very few examples have been published of success in delivering biologically active, negatively charged ONs or siRNA cargoes into cells in culture by covalent attachment of such types of CPP, some of which are also known as 'protein transduction domains' (reviewed recently by Juliano[33] and Turner *et al.*[34,35]). By contrast, many of these peptides have shown remarkable properties of cell penetration, either on their own or as conjugates with a range of other types of molecule (proteins, liposomes, nanoparticles, *etc.*; reviewed by Lindsay[36]). Initially, there were suggestions of an unusual non-endocytotic mechanism of uptake of the CPP Penetratin.[37] But more recently some doubt has been cast on this interpretation and, for example, a fluorescent derivative of Penetratin and the Tat (48–60) peptide were shown to enter cells by an energy-dependent endosomal uptake route when live-cell imaging is used.[38,39]

It has become recognized that negatively charged siRNA, antisense or steric block ONs may be packaged by certain CPP-type peptides which have characteristics that generally include both a region of cationic charge and a region of hydrophobic residues, in a similar way to that achieved by cationic liposomes. Penetratin peptide is in this category of peptide and, for example, it has been shown to enhance the delivery of non-conjugated DNA.[40] Much better results have been achieved by the design of synthetic peptides specifically for packaging and cell delivery, such as MPG for DNA ONs and siRNA[41,42] and pep-2 for uncharged PNA and a negatively charged hydroxypyrrole PNA-like analogue (HyPNA-pPNA).[43] The efficiencies of such delivery peptides are similar to, or even better than in some cases, cationic liposome packaging reagents, and delivery may also extend to primary cells in culture and give rise to reduced toxicity. In a recent example, pep-3 (a 15-mer) was able to deliver a steric block HyPNA-pPNA ON at a 20:1 ratio even better than pep-2 into cancer cell lines, and a PEGylated version of pep-3 was shown to exhibit strong antitumour activity in mouse xenografts.[44]

These sorts of packaging effects have complicated the interpretation of results in the area of covalent peptide conjugates of siRNA and steric block ONs, especially where it is not clear as to the purity of the synthetic conjugate. For example, we showed recently that the addition of Penetratin peptide to a fluorescent steric block ON greatly enhanced the apparent overall cell uptake into endosomes.[45] In general, using model systems we have not been able to find cellular activities for carefully purified covalent conjugates of well-known CPPs with steric block ONs[45] and have found only weak cellular activities with conjugates of CPPs with siRNA.[35] By contrast, peptide conjugates have proved to be extremely promising for the enhancement of mammalian cell delivery of charge-neutral PNA, and recently a number of significant advances have been made.

4.4 Synthesis of Peptide–PNA Conjugates

Chemical synthesis of PNA involves peptide-like coupling reactions to form amide bonds in the same way as standard peptide synthesis. Thus it is not surprising that quite early on in PNA development peptide–PNA conjugates were synthesized by continuous synthesis on a single solid support. However, fragment conjugation of peptide to a PNA in solution, each prepared separately by solid-phase synthesis, is generally more popular and is often more convenient for structure–activity studies. There seems to be no general rule as yet as to which type of chemical linkage is optimal for a particular application, since the factors that affect biological activity tend to be complex. However, it does seem that labile linkages often show higher activity (*e.g.* Bendifallah *et al.*,[46] Abes *et al.*,[47]). Stable linkages can be prepared by either continuous synthesis or fragment conjugation, whereas a labile linkage can only be formed by the latter approach. A comprehensive review of ON–peptide conjugation chemistry and applications has been published recently.[48]

4.4.1 Continuous Synthesis

Several strategies for both manual and automated solid-phase synthesis of PNA have been developed.[49–51] They are based on a cycle of chemical reactions, as shown in Figure 4.1, which require PNA monomeric building blocks protected with two orthogonal groups, one for the N-terminus (PG_1 in Figure 4.1) and one for the amino function on three out of the four nucleobases (PG_2). The two most frequently used compatible combinations of protecting groups PG_1 and PG_2 are 9-fluorenylmethoxycarbonyl (Fmoc) and benzhydryloxycarbonyl (Bhoc) and *tert*-butyloxycarbonyl (Boc) and carbobenzyloxy (Cbz) for the N-terminus and amino function, respectively. In the former case, the Fmoc group is removed at every cycle by mild basic treatment with 20% piperidine and 70% dimethylformamide (DMF), whereas in the latter case the Boc protecting group is removed with 95% trifluoroacetic acid and 5% *meta*-cresol. A wide variety of coupling agents can be used, in principle, to form the amide bond, but the most popular coupling agents are 2-(1*H*-7-azabenzotriazol-1-yl)-1,1,3,3-tetramethyluronium hexafluorophosphate (HATU) and benzotriazol-1-yl-tris-pyrrolidinophosphonium hexafluorophosphate (PyBOP). Recently it has been found that PyBOP can, under some conditions, form O^4-phosphonium compounds of the nucleobase guanine, which can be converted into C^4-modified guanine-derived PNAs by nucleophiles,[52] and thus PyBOP should be used with care. The Bhoc nucleobase-protecting groups are removed at the end of the synthesis by treatment with 95% trifluoroacetic acid, whereas Cbz removal requires harsh acidic treatment with HF or trifluoromethane sulfonic acid. Because of this problem, Fmoc/Bhoc chemistry is more often preferred and we have recently provided protocols for PNA and PNA–peptide syntheses using a robotic Apex 396 synthesizer.[53] Some other alternative combinations of protecting groups are Fmoc/4-methoxytriphenylmethyl (Mmt)[54] and Fmoc/Cbz.[55]

Peptide–PNA conjugates with stable amide linkages can be synthesized by first making the PNA sequence on the support and then elongating the peptide part,[46,50] or *vice versa*,[56] using either Fmoc or Boc chemistry. In some conjugates, the peptide and the PNA moieties are spaced by a short polyether linkage [8-amino-3,5-dioxo-octanoic acid (AEEA), also known as an O-linker] added to the growing chain by coupling with the corresponding Fmoc-protected acid.[57,58] In general there are no restrictions to combining both amino acids and PNA residues in the same assembly.

For many diagnostic applications, a series of PNA and their peptide conjugates must be synthesized on a small scale. Therefore, a semi-automatic procedure for parallel synthesis of up to 96 PNA or peptide–PNA conjugates has been developed using a Boc-protection strategy for PNA synthesis, or a combination of Boc and Fmoc chemistry for conjugate synthesis.[59]

For strand invasion of duplex DNA, rather than a single PNA strand, a bis-PNA[60] is used, because tethering two sequences reduces the entropic penalty paid during the formation of the four-stranded invasion complex (in which one PNA strand binds by Watson–Crick base pairing while a second PNA strand binds to the PNA–DNA hybrid by Hoogsteen pairing). Furthermore, addition of a peptide sequence to the N-terminus of bis-PNA increases the rate of hybridization through non-sequence-specific electrostatic interactions.[61] Such conjugates can be synthesized continuously using automated methods and an Fmoc-protection strategy, or *via* chemoselective oxime ligation.[62]

The synthesis of branched peptide–PNA conjugates requires a more flexible strategy, since the PNA and the peptide should be elongated in a truly orthogonal fashion.[63,64] It is desirable that the N-terminal protecting group on PNA monomers PG_1 (Figure 4.1) is stable under the cleavage conditions for the N-terminal protecting group on amino acid monomers (PG_N; usually the Fmoc group) and their acid-labile side-chain protecting groups PG_R. At the same time the protecting groups on the nucleobases and on the amino acid side-chains should be cleaved together under the final deprotection conditions. Thus, PNA monomers with a 1-(4,4-dimethyl-2,6-dioxacyclohexylidene)ethyl (Dde) protecting group on the N-terminus and Mmt on the nucleobases appears to be fully orthogonal to the commercially available Fmoc amino acid monomers.[63,64] The orthogonality of Dde and Fmoc groups is based on their different mechanisms for deprotection. While Fmoc is cleaved by basic elimination, Dde is removed by treatment with nucleophiles (*trans*-elimination). A further possible application of this dual Dde–Fmoc strategy is the synthesis of PNA molecular beacons (the complementary PNA sequence is constructed with Fmoc protected PNA monomers).

4.4.2   Fragment Conjugation

An alternative strategy for peptide–PNA conjugate synthesis is a solution-phase fragment coupling between a peptide and a PNA, each containing a specific functionality and each synthesized separately on solid phase and purified prior to the conjugation step. This route is more popular for studies that involve synthesis of a range of peptides attached to the same PNA. Although theoretically only a stoichiometric amount of each reagent is needed to obtain the desired conjugate, often the less expensive reagent, usually the peptide, is used in excess to drive the conjugation reaction to completion. In recent years, a number of stable or labile linkage chemistries have been developed to conjugate a peptide to a PNA moiety, and described below are some of the most well-known methods used for gene-modulation applications. Among stably linked conjugates, the thiol–maleimide conjugation method requires addition of a cysteine residue to the peptide and preparation of an *N*-maleimide derivative of the PNA[65] (Figure 4.2a). The latter can be obtained by reacting the terminal primary amino group of purified PNA with the hetero-bifunctional cross-linker succinimidyl 4-(*N*-maleimidomethyl) cyclohexane-1-carboxylate. The solution-phase coupling is very fast when carried out in potassium phosphate buffer (pH 7.0). Similarly, ester-containing conjugates (Figure 4.2b) can be prepared from PNAs bearing a glycyl(hydroxy-methyl)benzoic acid residue at the N-terminus, and this linkage is potentially labile to esterases within cells.[46]

We have recently described the formation of stably linked thioether peptide–PNAs by conjugation of a peptide containing a cysteine at the C-terminus with N-terminally bromoacetylated PNA in BisTris. HBr buffer [pH 7.5; Figure 4.3].[47] This work is based on original thioether conjugate formation between bromoacetyl oligodeoxynucleotides and cysteine-containing peptides.[66]

By far the most popular linkage is one based on a disulfide bond between the peptide and the PNA residue. Such a linkage is expected to be cleaved within the reducing environment of the cell and was first used by Langel and colleagues for biological studies that involved Transportan–PNA and Pene-tratin–PNA.[67] To obtain a disulfide linkage, both the peptide and the PNA should contain an additional cysteine residue, since the disulfide bond is formed between the cysteine thiol groups. The asymmetric disulfide conjugation reaction requires a pre-activation of one of the thiols, which is achieved

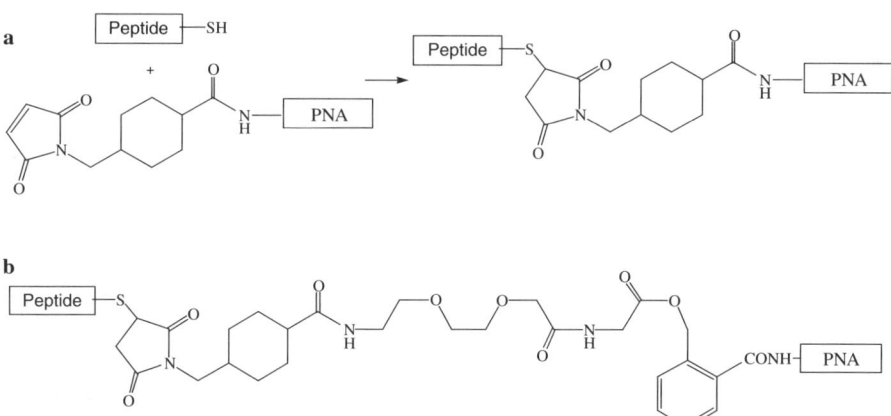

Figure 4.2   Formation of (a) thiol-maleimide and (b) ester-linked peptide–PNA conjugates.

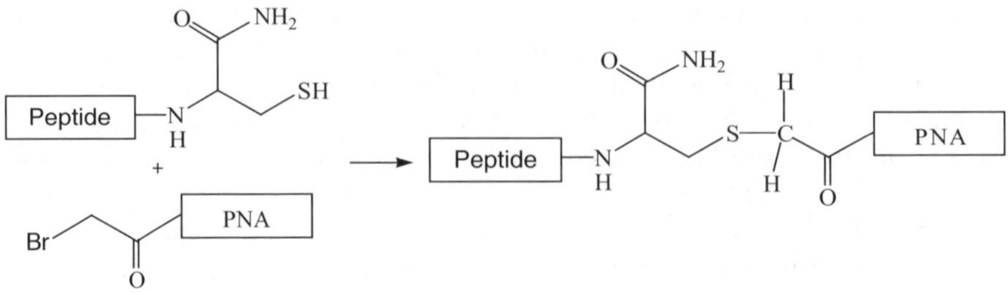

Figure 4.3   Formation of thioether-linked peptide–PNA conjugates.

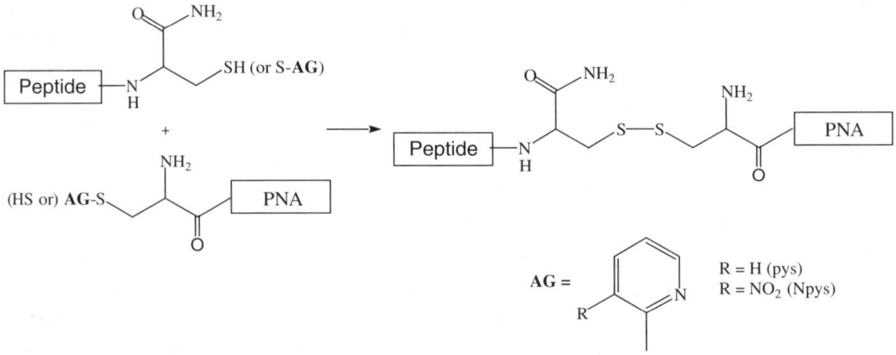

Figure 4.4   Formation of disulfide-linked peptide–PNA conjugates.

typically by use of either the pyridylsulfenyl (pys) or 3-nitropyridylsulfenyl (Npys) group (Figure 4.4). We have used the disulfide linkage extensively in studies of peptide–PNA conjugates targeted to the HIV-1 TAR RNA[58] and very recently in the luciferase splice correction model.[47]

A convergent synthetic route for PNA-N-to-C-peptide conjugate construction involves regioselective chemical ligation of a thioester peptide fragment to an N-terminal cysteine-modified PNA under denaturing conditions (Figure 4.5a).[68] Furthermore, a native ligation approach, but using a bifunctional PNA derivative with a C-terminal thioester and an N-terminal thiaproline (Figure 4.5b), can be applied to the synthesis of trimeric peptide–PNA–peptide conjugates.[69,70]

4.5   Steric Block Biological Activities of Peptide–PNA Conjugates

The field of peptide–ON conjugates was much stimulated by Corey's early studies that showed a 48 000-fold enhancement of binding to RNA of a synthetic oligodeoxyribonucleotide conjugated to a lysine-rich peptide.[71] Soon after, Corey and colleagues described the synthesis of PNA conjugates of the

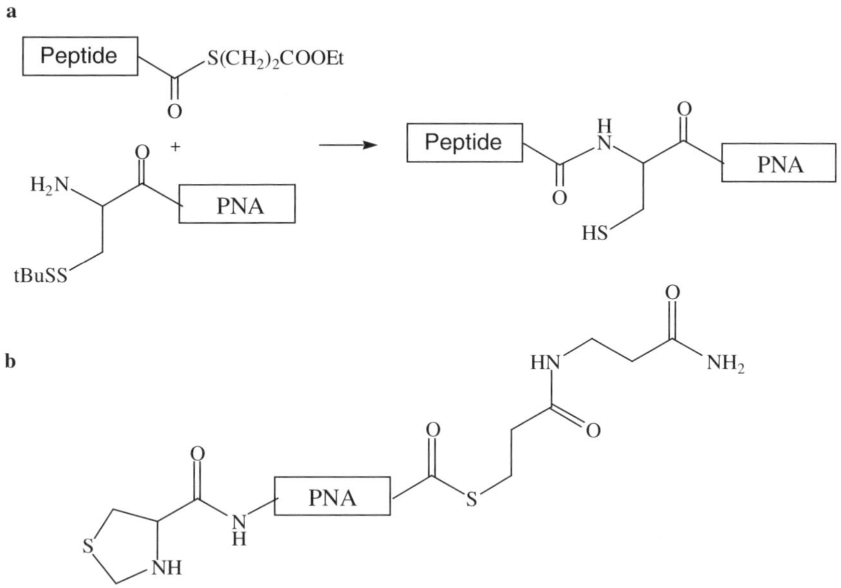

Figure 4.5 (a) Formation of a PNA-*N-to-C*-peptide conjugate by native ligation and (b) a bifunctional PNA used in synthesis of peptide–PNA–peptides.

Penetratin peptide and showed substantially enhanced uptake into DU145 prostate cancer cells compared to that of unconjugated PNA.[72] However, much controversy in this field has centred, in particular, around to what extent improvements in cell uptake afforded by an attached peptide are correlated with steric block biological activities of the PNA within cells.

In principle, RNA-targeting applications of PNA may be divided into those that occur in the cytosol, such as inhibition of protein translation, and those that occur in the cell nucleus. A number of RNA targets on an mRNA may be used to block protein translation by PNA, for example the AUG initiator sequence, 5′-terminus and other regions of the 5′-untranslated region (UTR), or even the 3′-UTR, but in general it has not been possible to inhibit translation significantly in cells by PNA targeting the coding region of a gene,[73] even though it has been shown that it is possible to block translation *in vitro*.[21] Table 4.1 lists a number of PNA–peptides that have been investigated for modulation of gene expression in mammalian cells.

Among the very few model systems studied for protein-expression inhibition by a PNA–peptide, the Langel group, working towards the modification of pain transmission, showed very early on that a 21-mer PNA disulfide-coupled to the CPPs Penetratin or Transportan (a hybrid of a section of the neuropeptide galanin and the wasp venom peptide mastoparan) blocked expression of the galanin receptor mRNA in human Bowes cells.[67] More recently, this group has obtained remarkably potent inhibition (down to 70–80 nM EC$_{50}$) in the same biological model with shorter PNAs targeting different regions of

Table 4.1 Some recent PNA–peptide conjugates used in gene modulation applications in cells.

PN–peptide conjugate (peptide sequence)	PNA	Conjugation type	Target/host	Reference
Penetratin–PNA (RQIKIWFQNRRMKWKK) Transportan–PNA (GWTLNSAGYLLGKINLKALAALAKKIL)	21-mer	Disulfide	Galanin mRNA/human Bowes cells and rats	67
TP10–PNA (CAGYLLGKINLKALAALAKKIL)	12-, 15- and 21-mer (+1 K)	Disulfide	Galanin mRNA/human Bowes cells	74
MAP–PNA (KLALKLALKAALKLA)	12-mer (+1 K)	Straight-through synthesis	Nociceptin/orphanin FQ mRNA/neonatal rat cardiomyocytes and CHO cells	77
D(AAKK)$_4$–PNA	19-mer (+1 K)	Straight-through synthesis	Caveolin I mRNA/HeLa and primary endothelial cells	78
Penetratin–PNA (QIKIWFQNRRMKWKK)	13-mer	Disulfide	RNA component of telomerase/JR8 cells	75
(FL)$_3$–PNA	14-mer (+1 K)	Straight-through synthesis	Inducible nitric oxide synthase mRNA/RAW 264.7 cells	76
Penetratin–PNA	18-mer (+1 K)	Disulfide		90

Peptide–PNA	Length	Synthesis	Target	Ref.
(CRQIKIWFQNRRMKWKK) Tat–PNA (CYGRKKRRQRRR) Transportan–PNA (GWTLNSAGYLLGK$_C$INLKALAALAKKIL)			Luciferase/β-globin aberrant splice site/HeLa pLuc reporter system	
Transportan–PNA (GWTLNSAGYLLGK$_C$INLKALAALAKKIL) D-Arg$_{(n)}$–PNA Tat–PNA (GRKKRRQRRPPQ) Penetratin–PNA (CRQIKIWFQNRRMKWK)	18-mer	Various: Maleimide/Ester/ Disulfide	Luciferase/β-globin aberrant splice site/HeLa pLuc reporter system	46
K$_8$–PNA	15-mer (+1 K)	Straight-through synthesis	CD40 mRNA/BCL$_1$ cells and primary macrophages	85
R$_6$Penetratin–PNA (RRRRRR-RQIKIWFQNRRMKWKKC)	18-mer (+4 K)	Stable/disulfide	Luciferase/β-globin aberrant splice site/HeLa pLuc reporter system	92
R$_6$Penetratin–PNA Penetratin–PNA TransportanPNA Tat–PNA	16-mer (+4 K)	Stable/disulfide	TAR RNA/HeLa TAR-Luc reporter system	58
K$_7$–PNA NLS–PNA (PKKKRKV)	15-mer (+1 K)	Straight-through synthesis	Encapsidation signal duck hepatitis virus/duck primary hepatocytes	96

galanin-receptor mRNA, conjugated to a truncated Transportan version TP10.[74] PNA–peptide conjugates have also been used to inhibit telomerase activity, by binding to the RNA component of the enzyme complex. For example, a 13-mer PNA disulfide-linked to Penetratin was able to inhibit telomerase activity in JR8 melanoma cells when incubated for 144 hours ($IC_{50} \sim 7\,\mu M$).[75] PNAs stably linked to short hydrophobic peptides rich in Leu and Phe have shown increased cell internalization and activity *in vitro*.[76] Stimulated macrophages incubated with anti-inducible nitric oxide synthase (iNOS) PNA–peptide showed up to 44% decrease in nitric oxide synthase activity compared to non-targeted controls.

In a different model system, an amphipathic peptide, MAP, disulfide-conjugated to a PNA complementary to the nociceptin–orphanin FQ receptor mRNA was shown by the Bienert research group to impart improved cell uptake and steric block biological effects in both CHO cells and neonatal rat cardiomyocytes.[77] A 19-mer PNA complementary to human caveolin 1 mRNA, when linked to a D(AAKK)$_4$ peptide, was able to inhibit expression in HeLa cells ($IC_{50} = 2\,\mu M$) and even in primary endothelial cells, but only when added at high ($>15\,\mu M$) concentrations, whereas the same PNA delivered by liposomal packaging and a complementary DNA sequence had an IC_{50} of 25 nM.[78] A nuclear localization sequence (NLS) and PNA conjugate was completely inactive in this assay.

For inhibition of translation in cells, steric block can rarely match quantitatively the efficacy of RNase H inducing antisense or siRNA, where the RNA is degraded. Most studies, therefore, that involve PNA–peptide conjugates have been directed to nuclear RNA targets, such as the redirection of splicing.[4,79] PNA and other steric blocking ONs are particularly well-suited to this application, since siRNA is believed to be active primarily in the cytosol, and conventional PS-containing antisense ONs are generally poorer competitors for the nuclear RNA regions targeted by protein complexes, such as the spliceosome.

Although the cellular uptake pathway of PNA conjugates involving hydrophobic peptides, such as Transportan, is not yet clear, there is agreement that PNAs conjugated to cationic peptides are taken up primarily by endocytosis into cytosolic vesicles and become sequestered there.[39,78,80,81] The key issue for biological activity is thus whether the attached peptide can bring about sufficient release of the PNA from such vesicles, such as can be obtained artificially through co-administration of a lysosomotropic agent (*e.g.* chloroquine or calcium ions).[58,82,83] Peptides that have been classed as 'cell-penetrating' vary substantially in their ability to influence endosomal release and thus allow the PNA to enter the nucleus, as judged by splicing redirection assays.

In a HeLa cell culture assay, the addition of just four lysines on the C-terminus of a PNA 18-mer (PNA-4K) complementary to an aberrant splice site (654) within a DNA construct containing a human β-globin intron within the gene for enhanced green fluorescent protein (EGFP; Figure 4.6) was enough to result in the observation of some splice correction and up-regulation of EGFP.[9] The activities of the PNA-4K and PMO oligomers were found to be much higher that those for negatively charged OMe-PS oligomers with free

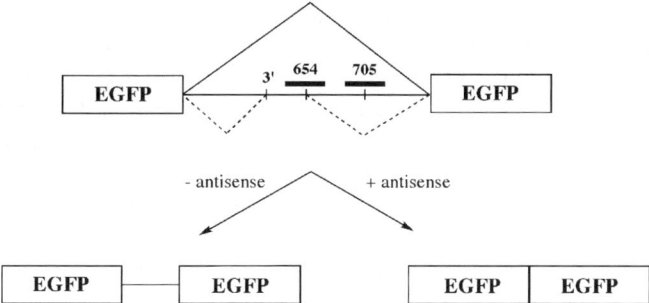

Figure 4.6 The EGFP splice-correction model[9] in which oligonucleotides or PNA complementary to the 654 site can bring about up-regulation of EGFP. A similar construction has been used with the 705 site to up-regulate luciferase production.[87]

uptake, but nevertheless 3–10 µM concentrations were required to see any significant level of splice correction. However, in transgenic mouse studies with the same EGFP 654 splice-correction model, PNA-4K 654 showed activity in several tissues (heart, kidney, lung, liver, muscles and small intestine) when delivered intraperitoneally (i.p.) at $50\,mg\,kg^{-1}$ for up to 4 days, and generally showed higher splicing correction (up to 40%) than the corresponding MOE and PMO oligomers.[84] No visible signs of pathology were seen for the mice at this dose.

In a different model involving the downregulation of murine CD40 pre-mRNA *via* splice redirection, it was shown that 8 Lys residues on the N-terminus of a PNA complementary to the splice site increased activity significantly with free uptake, compared to 4 Lys residues, but once again high concentrations of conjugate (3–10 µM) were required to see strong downregulation of this target in either B-cell lymphoma 1 (BCL$_1$) cells or primary murine macrophages.[85] Albertshofer *et al.* examined stably linked PNA conjugates of variously spaced oligo-Lys peptides by free incubation with cells and showed some small enhancements of splicing redirection in certain cases compared to Lys$_8$.[86] Preliminary pharmacokinetic studies of Lys$_8$-PNA in mice showed a broad tissue distribution and only modest elimination *via* excretion within 24 hours.[86]

The studies were continued with a series of 28 cationic peptides based on Lys or Arg that had amphipathic character.[87] This work led to improved peptide–PNA conjugates that showed stronger splicing redirection activity at 1–6 µM concentrations without toxicity, but half-lives of the peptide parts of the conjugates in 25% mouse serum were found to be just a few hours or even less. The PNA itself was found to be stable to nucleases and proteases in serum. Two such peptide–PNA conjugates, in which the peptide part utilized metabolically more stable D-amino acids, were radiolabelled, and studies in mice showed remarkably different biodistributions, demonstrating that peptide sequence alteration may be used to modulate *in vivo* pharmacology of a PNA. *In vivo* efficacies for these constructs have not yet been reported, however.

Many of the recent results on the use of CPPs to improve PNA cell delivery have been observed in a splice redirection model developed by the group of Kole.

This model is based on an HeLa cell integrated plasmid test system containing a gene that codes for firefly luciferase but interrupted by an aberrant β-globin intron,[88] similar to that already mentioned for EGFP (Figure 4.6),[9,84] but where aberrant splicing occurs at a different site (705). Upregulation of luciferase by redirection of splicing is obtained by the introduction into the cells of an 18-mer ON complementary to the aberrant 705 splice site. In a collaboration between our group and that of Bernard Lebleu and colleagues in Montpellier, we found that $(Lys)_8$–PNA stably linked conjugate was not very effective at 1 μM concentration and splicing correction was observed only when the lysosomotropic agent chloroquine was added.[83] There was also evidence of some cell cytotoxicity when 2.5 μM or higher concentrations were used. We found also that stably or disulfide-linked conjugates of the Tat (48–60) peptide with the 18-mer PNA had similar sequestration properties in endosomes and that the conjugates were inactive in the splicing assay at 1 μM concentration unless chloroquine was added.[89] Thus we concluded that neither $(Lys)_8$ nor Tat constructs are good starting points for CPP–PNA conjugate design.

Bendifallah *et al.*, evaluated a range of peptides, such as Tat, Penetratin, R_9 and Transportan, conjugated to a PNA 18-mer in several alternative ways.[46] The only conjugates to show significant splicing correction when delivered to the HeLa 705 cells at 2–5 μM were those that contained Transportan, but no simple relationship could be determined between the type or placement of the linker between peptide and PNA and the splicing correction activity, although labile linkages were in general of higher activity. Addition of serum to the test media had a dramatic negative effect on splicing-correction activity in all cases. Very similar results were obtained by El-Andaloussi and colleagues for disulfide-linked conjugates of Tat, Penetratin and Transportan, and only at concentrations of 5–10 μM was significant splice-correction activity seen, and only in the absence of serum.[90] Serum suppression of splicing correction activity seems to be a general phenomenon common to most PNA-peptides and PMO–peptides (Chapter 3). Partly, this may result from proteolysis of the peptide part in serum, whereas the PNA is known to be stable, but cationic peptides may also bind to serum proteins.

We have recently evaluated more Arg-rich peptides for conjugation with PNA. Substantial splicing correction was obtained at 2 μM for $(R-Ahx-R)_4$ stably conjugated to PNA 18-mer (where Ahx stands for aminohexanoyl) in the absence of chloroquine,[89] at a similar level to that obtained for the $(R-Ahx-R)_4$-Ahx–βAla–PMO construct,[91] a lead peptide conjugate for *in vivo* studies which showed very little *in vitro* cytotoxicity (see Chapter 3). However, most impressive of all has proved to be a conjugate of the 18-mer PNA disulfide-linked or linked through a stable thioether bridge to an R_6–Penetratin peptide. We found that such conjugates gave rise to extremely good luciferase upregulation, corresponding to about 50–60% splicing correction at 1 μM concentration, as judged by a reverse transcription polymerase chain reaction (RT-PCR) assay for transcript length.[47] A mutant in which a W residue was replaced by an L residue at position 6 of the Penetratin part of the R_6–Penetratin–PNA conjugate showed slightly higher activity, even though such an amino acid substitution had previously been shown to inhibit Penetratin peptide from crossing membranes.[92] This suggests that the sequence requirements of the R_6–Penetratin peptide needed to impart cell uptake and endosomal release of the PNA are considerably different to those of Penetratin peptide alone. The level of activity seen with R_6–Penetratin–PNA is higher than that for other CPP–PNA constructs published to date. By further peptide-sequence manipulation, hopefully it may be possible to bring the splicing-correction activity levels for PNA–peptides in this model splice-correction system down to the nM range.

PNAs conjugated to peptides have potential also as antiviral or virucidal agents. We have described the chemical synthesis of a range of CPPs conjugated to a 16-mer PNA that targets the HIV-1 TAR RNA apical stem-loop, which is situated at the 5′-end of all HIV-1 transcripts and which is the site of action of the *trans*-activator protein Tat.[58] These were tested in an HeLa cell assay of inhibition of Tat-dependent *trans*-activation involving stably integrated luciferase-expressing plasmids. The activity depends on nuclear entry of the PNA. Stably linked conjugates (*via* a simple polyether linker) were unable to elicit significant inhibition of Tat-dependent *trans*-activation upon incubation up to 2.5 µM tested. However, two types of disulfide-linked PNA conjugate (Transportan or R_6–Penetratin), when incubated for 24 hours, showed significant inhibitory activity (IC_{50} values of 0.5–1.0 µM), whereas conjugates of PNA with Tat (48–60) or Penetratin peptides failed to show activity unless chloroquine was co-administered.[58]

In parallel studies from the Pandey group, a similar disulfide-linked conjugate of a 16-mer PNA with Transportan elicited anti-HIV activity in chronically infected H9 cells by free delivery at 1–5 µM concentrations.[93] Further, this Transportan–PNA conjugate showed high activity ($IC_{50} \sim 50$ nM) as a virucidal agent when pseudotyped vesicular stomatitis virus (VSV)-HIV virions were pre-treated with conjugate before CEM-cell infection.[94] Similar levels of virucidal activity were found for several other CPP conjugates of the 16-mer PNA, but their corresponding antiviral activities were less good and IC_{50} values varied from 400 to 1100 nM.[95] It is not known currently why virucidal activity is so high compared to antiviral activity, but the authors suggested that the PNA–peptide conjugates may enter HIV-1 virions and inhibit reverse transcription of HIV-1 viral RNA. CPP–PNA conjugates may, therefore, have most potential for development as virucidal agents. Perhaps HIV-1 RNA undergoing reverse transcription within an infected cell cytosol or RNA being transcribed Tat-dependently from the HIV-1 pro-virus in the cell nucleus may present a less readily accessible TAR target as occurs in virion RNA.

Robaczewska *et al.*,[96] have shown that an unconjugated 15-mer PNA directed to the viral encapsidation signal of duck hepatitis B virus was able to decrease the viral genomic DNA by 30% in infected duck primary hepatocytes, compared to mismatch controls. Upon conjugation to an oligoarginine (R_7) peptide the construct was able to decrease the viral DNA genome by 65% under the same experimental conditions. However, an NLS–PNA conjugate did not improve antiviral activity compared to PNA alone.

It was recognized from the very early days that pyrimidine-rich PNAs can strand-invade double-stranded DNA containing complementary purine-rich sequences to form double strands or triple strands with two PNA strands and one DNA strand.[97] The subsequent development of this approach to the targeting of double-stranded DNA and chromosomal DNA within cells has been reviewed recently.[98] For example, a mixed sequence 17-mer *anti* c-myc PNA conjugated to the NLS peptide PKKKRKV by stable linkage exerted steric block antigene activity through downregulation of c-myc oncogene expression by interfering with transcription, but only at a relatively high concentration (10 μM).[99] Perhaps more effective as a strand-invasion method is the use of bis-PNA linked *via* polyether linkage and attached to a cationic peptide D–(AAKK)$_4$ with rates of invasion 100 times faster than bis-PNA not attached to the peptide.[61]

Another important study area is the bactericidal activity of PNA conjugated stably to a membrane-active peptide KFFKFFKFFK.[57,58] In some cases, PNA targeted to ribosomal RNA or mRNA was able to cure HeLa cell cultures of bacterial infections. Although toxicity of the peptide in *in vivo* studies has precluded industrial development of this PNA–peptide conjugate, modified and less toxic peptides are now being used in the commercial development of PMO conjugates.[100]

4.6   Towards *in vivo* Therapeutic Applications

In contrast to PMOs (Chapter 3), PNAs and their peptide conjugates are much less well-developed so far for *in vivo* therapeutic use. First, a number of commercial and licensing factors appear to have affected the industrial development of PNA in general as therapeutic agents. In addition, the cost of commercial custom PNA synthesis and PNA monomers has been fairly high, perhaps partly because of more modest synthetic yields compared to some other analogues. No significant efforts to manufacture PNA on large scale seem to have been reported. Thus, as yet, no PNA or PNA–peptide has been taken to clinical trials. There are good signs that this situation may change soon with the recent advent of a number of new PNA supply companies.

Some interesting preliminary conclusions can be drawn from the very few published animal studies of PNA and PNA–peptides directed towards their use as possible therapeutics. Firstly, PNA–4K showed some activity in several tissues (heart, kidney, lung, liver, muscles and small intestine) when delivered i.p. at 50 mg kg^{-1} for up to four days in transgenic mice with the EGFP 654 splice-correction model without showing acute toxicity.[84] Secondly, an 18-mer complementary to the enhancer Eμ intronic sequence, stably linked to a 7-mer NLS peptide by continuous synthesis, that is able to inhibit c-myc oncogene expression in Burkitt's Lymphoma (BL) cells[99] was shown to have little or no toxicity in a human tumour severe combined immunodeficiency disease (SCID) mice models.[101] Very recently this has shown inhibition of BL cell-induced tumour growth in SCID mice when inoculated subcutaneously.[102] Thus PNA–peptide conjugates have shown efficacy in therapeutic mouse models.

In results from Isis Pharmaceuticals, a PNA attached to K_8 was reported to have broad tissue distribution in mice after intravenous injection and a relatively slow excretion level.[86] Two different Lys/Arg-rich amphipathic D-peptides stably conjugated to a PNA were shown to have distributed substantially differently in mice.[87] These results show that the pharmacology of PNA can be manipulated significantly by alteration of the attached peptide sequence. As we await further reports of biodistribution and efficacy for PNA–peptides, the following issues become clear. Firstly, some peptide–PNA conjugates have shown in cell lines a significant cytotoxicity that is peptide sequence-dependent and, secondly, there is a wide variation in stability of the peptide component of PNA conjugates in the presence of serum.[86,87] Thus, future *in vivo* studies will rely on optimizing these parameters, while at the same time obtaining high biological activity in cellular models. For example, we are currently modifying and improving the peptide component of the R_6–Penetratin–PNA paradigm to attain such features required for *in vivo* testing.

Another promising alternative to peptide conjugation of PNA towards *in vivo* studies is to utilize a PNA analogue that has been engineered to reintroduce some negative charges, such as a steric block HyPNA-pPNA ON. This has been packaged recently with a specifically designed PEGylated 15-mer CPP, pep-3, and shown to have *in vivo* activity in a mouse tumour model.[44] Whether peptide packaging or peptide conjugation of PNA offers the best hope for PNA therapeutics at the moment is an open question. Nevertheless, ample opportunities for therapeutic development now seem to be at hand, and the next few years should hopefully bring PNA and their peptide conjugates or complexes closer to the clinic.

Acknowledgements

We thank Andrey Arzumanov, Donna Williams and David Owen for their work on the PNA–peptide projects and Bernard Lebleu and Said Abes (University of Montpellier 2) for productive collaboration on the splice correction model system.

References

1. P. C. Zamecnik and M. L. Stephenson, *Proc. Natl. Acad. Sci. U. S. A.*, 1978, **75**, 280.
2. C. Boiziau, R. Kurfurst, C. Cazenave, V. Roig, N. T. Thuong and J. J. Toulmé, *Nucleic Acids Res.*, 1991, **39**, 1113.
3. B. F. Baker, S. S. Lot, T. P. Condon, S. Cheng-Flourney, E. A. Lesnik, H. M. Sasmor and C. F. Bennett, *J. Biol. Chem.*, 1997, **272**, 11994.
4. R. Kole, M. Vacek and T. Williams, *Oligonucleotides*, 2004, **14**, 65.
5. J. Krützfeldt, N. Rajewsky, R. Braich, K. G. Rajeev, T. Tuschl, M. Manoharan and M. Stoffel, *Nature*, 2005, **438**, 685.
6. F. Morvan, H. Porumb, G. Degols, I. Lefebvre, A. Pompon, B. S. Sproat, B. Rayner, C. Malvy, B. Lebleu and J.-L. Imbach, *J. Med. Chem.*, 1993, **36**, 280.

7. G. Schmajuk, H. Sierakowska and R. Kole, *J. Biol. Chem.*, 1999, **274**, 21783.

8. J. C. Schmitz, D. Yu, S. Agrawal and E. Chu, *Nucleic Acids Res.*, 2001, **29**, 415.

9. P. Sazani, S.-H. Kang, M. A. Maier, C. Wei, J. Dillman, J. Summerton, M. Manoharan and R. Kole, *Nucleic Acids Res.*, 2001, **29**, 3965.

10. M. Bremmer-Bout, A. Aartsma-Rus, E. J. de Meijer, W. E. Kaman, A. A. M. Janson, R. H. A. M. Vossen, G. B. van Ommen, J. T. den Dunnen and J. C. T. van Deutekom, *Mol. Ther.*, 2004, **10**, 232.

11. M. Faria, D. G. Spiller, C. Dubertret, J. S. Nelson, M. R. H. White, D. Scherman, C. Hélène and C. Giovannangeli, *Nat. Biotechnol.*, 2001, **19**, 40.

12. A. Asai, Y. Oshima, Y. Yamamoto, T. Uochi, H. Kusaka, S. Akinaga, Y. Yamashita, K. Pongracz, R. Pruzan, E. Wunder, M. Piatyszek, S. Li, A. C. Chin, C. B. Harley and S. Gryaznov, *Cancer Res.*, 2003, **63**, 3931.

13. K. Fluiter, A. L. M. A. ten Asbroek, M. B. de Wissel, M. E. Jakobs, M. Wissenbach, H. Olsson, O. Olsen, H. Oerum and F. Baas, *Nucleic Acids Res.*, 2003, **31**, 953.

14. D. A. Braasch, Y. Liu and D. R. Corey, *Nucleic Acids Res.*, 2002, **30**, 5160.

15. J. L. Childs, M. D. Disney and D. H. Turner, *Proc. Natl. Acad. Sci. U.S.A.*, 2002, **99**, 11091.

16. J. Roberts, E. Palma, P. Sazani, H. Ørum, M. Cho and R. Kole, *Mol. Ther.*, 2006, **14**, 471.

17. A. Arzumanov, A. P. Walsh, V. K. Rajwanshi, R. Kumar, J. Wengel and M. J. Gait, *Biochemistry*, 2001, **40**, 14645.

18. A. Arzumanov, D. A. Stetsenko, A. D. Malakhov, S. Reichelt, M. D. Sørensen, B. R. Babu, J. Wengel and M. J. Gait, *Oligonucleotides*, 2003, **13**, 435.

19. M. Egholm, O. Buchardt, P. E. Nielsen and R. H. Berg, *J. Am. Chem. Soc.*, 1992, **114**, 1895.

20. M. Egholm, O. Buchardt, L. Christensen, C. Behrens, S. M. Freier, D. A. Driver, R. H. Berg, S. K. Kim, B. Norden and P. Nielsen, *Nature*, 1993, **365**, 566.

21. N. Dias, S. Dheur, P. E. Nielsen, S. Gryaznov, A. Van Aerschot, P. Herdewijn, C. Hélène and T. E. Saison-Behmoaras, *J. Mol. Biol.*, 1999, **294**, 403.

22. M. A. Shammas, C. G. Simmons, D. R. Corey and R. J. Shmookler Reis, *Oncogene*, 1999, **18**, 6191.

23. L. J. Brandén, A. J. Mohamed and C. I. Smith, *Nat. Biotechnol.*, 1999, **17**, 784.

24. B.-S. Herbert, A. E. Pitts, S. I. Baker, S. E. Hamilton, W. E. Wright, J. W. Shay and D. R. Corey, *Proc. Natl. Acad. Sci. U. S. A.*, 1999, **96**, 14276.

25. T. Shiraishi, N. Bendifallah and P. E. Nielsen, *Bioconjugate Chem.*, 2006, **17**, 189.

26. M. Folini, K. Berg, E. Millo, R. Villa, L. Prasmickaite, M. G. Daidone, U. Benatti and N. Zaffaroni, *Cancer Res.*, 2003, **63**, 3490.

27. F. W. Rasmussen, N. Bendifallah, V. Zachar, T. Shiraishi, T. Fink, P. Ebbesen, P. E. Nielsen and U. Koppelhus, *Oligonucleotides*, 2006, **16**, 43.
28. M. Lemaitre, B. Bayard and B. Lebleu, *Proc. Natl. Acad. Sci. U. S. A.*, 1987, **84**, 648.
29. B. Allinquant, P. Hantraye, P. Mailleux, K. Moya, C. Bouillot and A. Prochiantz, *J. Cell Biol.*, 1995, **128**, 919.
30. D. Derossi, S. Calvet, A. Trembleau, A. Brunissen, G. Chassaing and A. Prochiantz, *J. Biol. Chem.*, 1996, **271**, 18188.
31. C. M. Troy, D. Derossi, A. Prochiantz, L. A. Greene and M. L. Shelanski, *J. Neurosci.*, 1996, **16**, 253.
32. E. Vivès, P. Brodin and B. Lebleu, *J. Biol. Chem.*, 1997, **272**, 16010.
33. R. L. Juliano, *Curr. Opin. Mol. Ther.*, 2005, **7**, 132.
34. J. J. Turner, A. Arzumanov, G. Ivanova, M. Fabani and M. J. Gait, in *Peptide conjugates of oligonucleotide analogs and siRNA for gene modulation*, ed. U. Langel, Boca Raton, 2006.
35. J. J. Turner, S. Jones, M. Fabani, G. Ivanova, A. Arzumanov and M. J. Gait, *Blood, Cells, Mol. Diseases*, 2007, **38**, 1.
36. M. A. Lindsay, *Curr. Opin. Pharmacol.*, 2002, **2**, 587.
37. A. Prochiantz, *Curr. Opin. Neurobiol.*, 1996, **6**, 629.
38. P. E. G. Thorén, D. Persson, P. Isakson, M. Goksör, A. Onfelt and B. Nordén, *Biochem. Biophys. Res. Commun.*, 2003, **307**, 100.
39. J.-P. Richard, K. Melikov, E. Vivès, C. Ramos, B. Verbeure, M. J. Gait, L. V. Chernomordik and B. Lebleu, *J. Biol. Chem.*, 2003, **278**, 585.
40. G. Dom, C. Shaw-Jackson, C. Matis, O. Bouffioux, J. J. Picard, A. Prochiantz, M.-P. Mingeot-Leclercq, R. Brasseur and R. Rezsohazy, *Nucleic Acids Res.*, 2003, **31**, 556.
41. L. Chaloin, P. Vidal, P. Lory, J. Méry, N. Lautredou, G. Divita and F. Heitz, *Biochem. Biophys. Res. Comm.*, 1998, **243**, 601.
42. F. Simeoni, M. C. Morris, F. Heitz and G. Divita, *Nucleic Acids Res.*, 2003, **31**, 2717.
43. M. C. Morris, L. Chaloin, M. Choob, J. Archdeacon, F. Heitz and G. Divita, *Gene Ther.*, 2004, **11**, 757.
44. M. C. Morris, E. Gros, G. Aldrian-Herrada, M. Choob, J. Archdeacon, F. Heitz and G. Divita, *Nucleic Acids Res.*, 2007, **35**, e49.
45. J. J. Turner, A. A. Arzumanov and M. J. Gait, *Nucleic Acids Res.*, 2005, **33**, 27.
46. N. Bendifallah, F. W. Rasmussen, V. Zachar, P. Ebbesen, P. E. Nielsen and U. Koppelhus, *Bioconjugate Chem.*, 2006, **17**, 750.
47. S. Abes, J. J. Turner, G. D. Ivanova, D. Owen, D. Williams, A. Arzumanov, P. Clair, M. J. Gait and B. Lebleu, *Nucleic Acids Res.*, 2007, **35**, 4495.
48. N. Venkatesan and B. H. Kim, *Chem. Rev.*, 2006, **106**, 3712.
49. J. C. Norton, J. J. Waggenspack, E. Varnum and D. R. Corey, *Bioorg. Med. Chem.*, 1995, **3**, 437.
50. L. D. Mayfield and D. R. Corey, *Anal. Biochem.*, 1999, **268**, 410.
51. D. A. Braasch, C. J. Nulf and D. R. Corey, *Curr. Prot. Nucl. Acid Chem.*, 2002, 4.11.1.

52. S. Pritz, Y. Wolf, C. Klemm and M. Bienert, *Tetrahedron Lett.*, 2006, **47**, 5893.
53. J. J. Turner, D. Williams, D. Owen and M. J. Gait, *Curr. Prot. Nucl. Acid Chem.*, 2005, 4.28.1.
54. G. Breipohl, J. Knolle, D. Langner, G. O'Malley and E. Uhlmann, *Bioorg. Med. Chem. Lett.*, 1996, **6**, 665.
55. S. A. Thomson, J. A. Josey, R. Cadilla, M. D. Gaul, C. F. Hassman, M. J. Luzzio, A. J. Pipe, K. L. Reed, D. J. Ricca, R. W. Wiethe and S. A. Noble, *Tetrahedron*, 1995, **51**, 6179.
56. S. Basu and E. Wickstrom, *Bioconjugate Chem.*, 1997, **8**, 481.
57. L. Good, S. K. Awasthi, R. Dryselius, O. Larsson and P. E. Nielsen, *Nat. Biotechnol.*, 2001, **19**, 360.
58. J. J. Turner, G. D. Ivanova, B. Verbeure, D. Williams, A. Arzumanov, S. Abes, B. Lebleu and M. J. Gait, *Nucleic Acids Res.*, 2005, **33**, 6837.
59. S. K. Awasthi and P. E. Nielsen, *Comb. Chem. High Throughput Screening*, 2002, **5**, 253.
60. M. Egholm, L. Christensen, K. L. Dueholm, O. Buchardt, J. Coull and P. E. Nielsen, *Nucleic Acids Res.*, 1995, **23**, 217.
61. K. Kaihatsu, D. A. Braasch, A. Cansizolglu and D. R. Corey, *Biochemistry*, 2002, **41**, 11118.
62. P. Neuner, P. Gallo, L. Orsatti, L. Fontana and P. Monaci, *Bioconjugate Chem.*, 2003, **14**, 276.
63. J. J. Diaz-Mochon, L. Bialy and M. Bradley, *Organic Lett.*, 2004, **6**, 1127.
64. L. Bialy, J. J. Diaz-Mochon, E. Specker, L. Keinicke and M. Bradley, *Tetrahedron*, 2005, **61**, 8295.
65. J. G. Harrison, C. Frier, R. Laurant, R. Dennis, K. D. Raney and S. Balasubramanian, *Bioorg. Med. Chem. Lett.*, 1999, **9**, 1273.
66. K. Arar, A.-M. Aubertin, A.-C. Roche, M. Monsigny and M. Mayer, *Bioconjugate Chem.*, 1995, **6**, 573.
67. M. Pooga, U. Soomets, M. Hällbrink, A. Valkna, K. Saar, K. Rezaei, U. Kahl, J.-X. Hao, X.-J. Xu, Z. Wiesenfeld-Hallin, T. Hökfelt, T. Bartfai and Ü. Langel, *Nat. Biotechnol.*, 1998, **16**, 857.
68. M. C. de Koning, D. V. Fillipov, N. Meuuwenoord, M. Overhand, G. A. van der Marel and J. H. van Boom, *Syn. Lett.*, 2001, 1516.
69. M. C. de Koning, D. V. Filippov, N. Meeuwenoord, M. Overhand, G. A. van der Marel and J. H. van Boom, *Tetrahedron Lett.*, 2002, **43**, 8173.
70. M. C. de Koning, D. V. Filippov, G. A. van der Marel, J. H. van Boom and M. Overhand, *Eur. J. Org. Chem.*, 2004, **850**.
71. C. R. Corey, *J. Am. Chem. Soc.*, 1995, **117**, 9373.
72. C. G. Simmons, A. E. Pitts, L. D. Mayfield, J. W. Shay and D. R. Corey, *Bioorg. Med. Chem. Lett.*, 1997, **7**, 3001.
73. D. F. Doyle, D. A. Braasch, C. G. Simmons, B. A. Janowski and D. R. Corey, *Biochemistry*, 2001, **40**, 53.
74. K. Kilk, A. Elmquist, K. Saar, M. Pooga, T. Land, T. Bartfai, U. Soomets and U. Langel, *Neuropeptides*, 2004, **38**, 316.

75. R. Villa, M. Folini, S. Lualdi, S. Veronese, M. G. Daidone and N. Zaffaroni, *FEBS Lett.*, 2000, **473**, 241.
76. S. Scarfi, M. Giovine, A. Gasparini, G. Damonte, E. Millo, M. Pozzolini and U. Benatti, *FEBS Lett.*, 1999, **451**, 264.
77. J. Oehlke, G. Wallukat, Y. Wolf, A. Ehrlich, B. Wiesner, H. Berger and M. Bienert, *Eur. J. Biochem.*, 2004, **271**, 3043.
78. K. Kaihatsu, K. E. Huffman and D. R. Corey, *Biochemistry*, 2004, **43**, 14340.
79. D. R. Mercatante and R. Kole, *Biochim. Biophys. Acta*, 2002, **1587**, 126.
80. K. Braun, P. Peschke, R. Pipkorn, S. Lampel, M. Wachsmuth, W. Waldeck, E. Friedrich and J. Debus, *J. Mol. Biol.*, 2002, **318**, 237.
81. U. Koppelhus, S. K. Awasthi, V. Zachar, H. U. Holst, P. Ebbeson and P. E. Nielsen, *Antisense Nucleic Acid Drug Dev.*, 2002, **12**, 51.
82. T. Shiraishi, S. Pankratova and P. E. Nielsen, *Chem. Biol.*, 2005, **12**, 923.
83. S. Abes, D. Williams, P. Prevot, A. R. Thierry, M. J. Gait and B. Lebleu, *J. Controlled Release*, 2006, **110**, 595.
84. P. Sazani, F. Gemignani, S.-H. Kang, M. A. Maier, M. Manoharan, M. Persmark, D. Bortner and R. Kole, *Nat. Biotechnol.*, 2002, **20**, 1228.
85. A. M. Siwkowski, L. Malik, C. C. Esau, M. A. Maier, E. V. Wancewicz, K. Albertshofer, B. P. Monia, C. F. Bennett and A. B. Eldrup, *Nucleic Acids Res.*, 2004, **32**, 2695.
86. K. Albertshofer, A. M. Siwkowski, E. V. Wancewicz, C. C. Esau, T. Watanabe, K. C. Nishihara, G. A. Kinberger, L. Malik, A. B. Eldrup, M. Manoharan, R. S. Geary, B. P. Monia, E. E. Swayze, R. H. Griffey, C. F. Bennett and M. A. Maier, *J. Med. Chem.*, 2005, **48**, 6741.
87. M. A. Maier, C. C. Esau, A. M. Siwkowski, E. V. Wancewicz, K. Albertshofer, G. A. Kinberger, N. S. Kadaba, T. Watanabe, M. Manoharan, C. F. Bennett, R. H. Griffey and E. E. Swayze, *J. Med. Chem.*, 2006, **49**, 2534.
88. S.-H. Kang, M.-J. Cho and R. Kole, *Biochemistry*, 1998, **37**, 6235.
89. S. Abes, H. M. Moulton, J. J. Turner, P. Clair, J.-P. Richard, P. L. Iversen, M. J. Gait and B. Lebleu, *Biochem. Soc. Trans.*, 2007, **35**, 53.
90. S. El-Andaloussi, H. J. Johansson, P. Lundberg and U. Langel, *J. Gene Med.*, 2006, **8**, 1262.
91. S. Abes, H. M. Moulton, P. Clair, P. Prevot, D. S. Youngblood, R. P. Wu, P. L. Iversen and B. Lebleu, *J. Controlled Release*, 2006, **116**, 304.
92. M. Lindgren, X. Gallet, U. Soomets, M. Hällbrink, E. Bråkenhielm, M. Pooga, R. Brasseur and U. Langel, *Bioconjugate Chem.*, 2000, **11**, 619.
93. N. Kaushik, A. Basu, P. Palumbo, R. L. Nyers and V. N. Pandey, *J. Virol.*, 2002, **76**, 3881.
94. B. Chaubey, S. Tripathi, S. Ganguly, D. Harris, R. A. Casale and V. N. Pandey, *Virology*, 2005, **331**, 418.
95. S. Tripathi, B. Chaubey, S. Ganguly, D. Harris, R. A. Casale and P. K. Pandey, *Nucleic Acids Res.*, 2005, **33**, 4345.
96. M. Robaczewska, R. Narayan, B. Siegneres, O. Schorr, A. Thermetr, A. J. Podhajska, F. Trepo, P. E. Nielsen and L. Cova, *J. Hepatol.*, 2005, **42**, 180.

97. P. E. Nielsen, M. Egholm, R. H. Berg and O. Buchardt, *Science*, 1991, **254**, 1497.
98. K. Kaihatsu, B. A. Janowski and D. R. Corey, *Chem. Biol.*, 2004, **11**, 749.
99. G. Cutrona, E. M. Carpaneto, M. Ulivi, S. Roncella, O. Landt, M. Ferrarini and L. C. Boffa, *Nat. Biotechnol.*, 2000, **18**, 300.
100. L. D. Tilley, B. L. Melbye, S. E. Puckett, P. L. Iversen and B. L. Geller, *J. Antimicrob. Chemother.*, 2007, **59**, 66.
101. L. C. Boffa, G. Cutrona, M. Cilli, M. R. Mariani, S. Matis, M. Pastorini, G. Damonte, E. Millo, S. Roncella and M. Ferrarini, *Oligonucleotides*, 2005, **15**, 85.
102. L. C. Boffa, G. Cutrona, M. Cilli, S. Matis, G. Damonte, M. R. Mariani, E. Millo, M. Moroni, S. Roncella, F. Fedeli and M. Ferrarini, *Cancer Gene Ther.*, 2007, **14**, 220.

CHAPTER 5

Locked Nucleic Acid: Properties and Therapeutic Aspects

TROELS KOCH,* CHRISTOPH ROSENBOHM, HENRIK F. HANSEN, BO HANSEN, ELLEN MARIE STRAARUP AND SAKARI KAUPPINEN

Santaris Pharma A/S, Bøge Allé 3, DK-2970, Hørsholm Denmark

5.1 Introduction

In 1978 Zamecnik and Stephenson[1] showed for the first time that messenger RNA (mRNA) repression could be achieved by single-stranded oligonucleotides (ONs). This mechanistic approach for gene inhibition was later called the antisense (AS) principle. The simplicity of this new principle captivated many scientists at that time, as did the therapeutic potential that directly followed from it.

Effective disease intervention by AS is based on a thorough understanding of the underlying causative genetic mechanisms of the disease, and it was not until 20 years later that this fundamental understanding matured to a deep enough level. Furthermore, it was clear from the very beginning that native ONs were useless as AS therapeutics, and therefore Zamecnik and Stevenson's pioneering work stimulated a world-wide medicinal chemistry effort aimed at synthesizing ON analogues with the diverse set of needed properties for AS to achieve better nuclease stability and higher RNA affinity.

The first generation of chemically modified AS drugs were the phosphorothioates (PS) and, although capable of repressing mRNA targets *in vitro*, their *in vivo* potency was more questionable. The PS modification improved nuclease

RSC Biomolecular Sciences
Therapeutic Oligonucleotides
Edited by Jens Kurreck
© Royal Society of Chemistry 2008

stability, but at the expense of *lower* RNA affinity. This feature, combined with rather high systemic toxicities,[2] led to a narrow therapeutic window. This is the main reason for the many failures in the clinic for the first-generation AS drugs, and unfortunately these failures were frequently related to the AS concept itself.

Since the PS showed improved nuclease stability, but lower RNA affinity, the second-generation AS drugs were developed to improve the binding affinity. Examples of second-generation AS drugs are ONs comprising 2'-substitution chemical modifications, such as 2'-fluoro (2'-F), 2'-*O*-methylated (2'-*O*-Me), 2'-*O*-methoxyethyl (2'-MOE) and 2'-fluoroarabinonucleic acid (2'-FANA). Depending on sequence motifs these modifications typically increased the thermal stability of the RNA–hybrid duplexes with 0.5–1.5 °C per modification.[3,4] Today, many of these second-generation drugs are at advanced phases in clinical trials, and they do show prospects compared to the first-generation drugs. However, to obtain sufficient potency and efficacy, the length of the second-generation AS drugs was unchanged compared to the first-generation (18–20-mers) drugs, indicating that the affinity was still not adequately addressed.

The discovery of locked nucleic acid [LNA – defined as an ON comprising one or more 2'-*O*,4'-*C*-methylene-β-D-ribofuranosyl nucleotide building blocks (=LNA monomeric unit)] in 1997 has introduced a new member of the third generation AS drugs. When LNA nucleotides (nt's) were incorporated into ONs they turned out to induce the highest binding affinity ever reported in the field for both complementary RNA and DNA.[5,6] Dependent on sequence motifs, LNA nt's typically increase the thermal stability by 4–8 °C per modification. Combined with high biostability *in vitro* and *in vivo*, LNA was quickly positioned as a key player in the field of single-stranded RNA inhibition. In this chapter we present the therapeutic aspects of LNA to illustrate the LNA platform for single-stranded ON drugs. We describe the basic chemical and biophysical properties of LNA, and show that LNA can be produced in bulk amounts at competitive pricing, which is a market prerequisite for a broad pharmaceutical use. We also show how LNA can lead to the development of shorter-than-usual ONs ('shortmers') that exhibit dramatically improved pharmacological activity compared to ONs based on other principles and chemistries. The pharmacological properties of LNAs will be illustrated for targeting both mRNAs and microRNAs (miRNAs). Finally, we show the interim data from the first clinical trial of SPC2996, a 16-mer directed against the B-cell lymphoma 2 (Bcl-2) mRNA for the treatment of CLL.

5.2 Basic Properties of Locked Nucleic Acids

The first ONs comprising the 2'-*O*,4'-*C*-methylene-β-D-ribofuranosyl nucleotide building block were synthesized in 1997 by the Wengel group at the University of Copenhagen. Such ONs were designated locked nucleic acids (LNAs; Figure 5.1A)[6] to depict that the bicyclic structure of LNA nucleosides

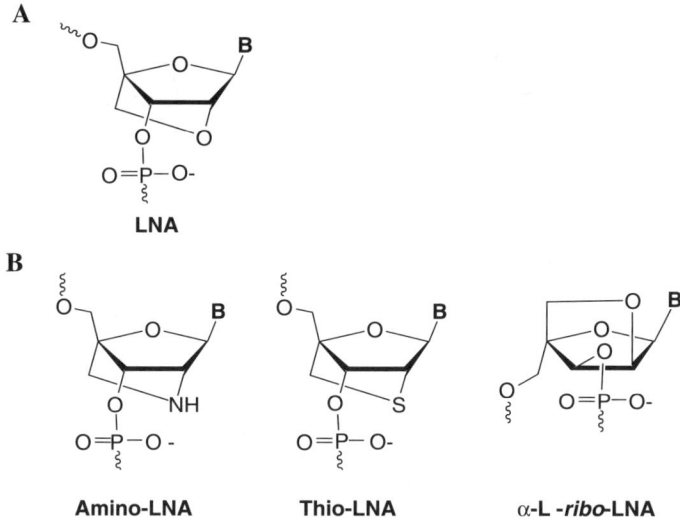

Figure 5.1 A. Molecular structure of LNA. B. Molecular structures of selected LNA-analogues.

locks the conformational flexibility of the ribose ring. In this pioneering paper the remarkable increase in the melting temperature (T_m) followed by the incorporation of the bicyclic unit(s) was reported for the first time. Prior to this work many medicinal chemists had designed and synthesized numerous nucleic acid analogues that also displayed 'conformational restriction' aimed at inducing increased affinity, but the hybridization properties of these were inferior to those of LNA.[4,7–10] The bicyclic structure of the LNA nucleoside served as a general template for analogues of LNA with similar favourable hybridization properties. Prominent LNA-analogue members are amino-, thio- and α-L-LNA (Figure 5.1B).[11–32] The synthesis and basic properties valuable for therapeutic applications have recently been extensively reviewed.[33]

In the remaining part of Section 5.2 the fundamental properties of LNA are described. Section 5.2 focuses on the properties that are of particular therapeutic relevance for the use of LNA as a single-stranded RNA inhibitor. LNA and LNA analogues have numerous other applications as probes for diagnostics and other fields of life science, but these applications are outside the scope of this chapter and the interested reader can consult the cited literature for further information.[33–40]

5.2.1 LNA–RNA Hybrid Structure

The furanose ring of deoxyribose exists at room temperature in an equilibrium between C2′-*endo* (*S*-type) and C3′-*endo* (*N*-type) conformation with an energy barrier of only ~2 kcal mol^{-1} (Figure 5.2).[41] The C2′-*endo* (*S*-type) conformation

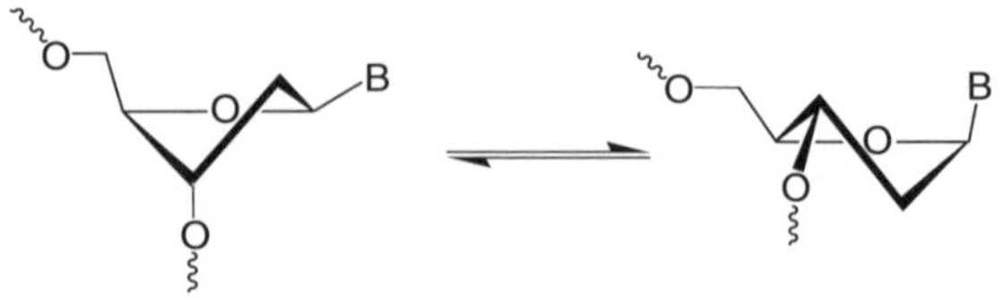

C2′-*endo* (*S*-type)   C3′-*endo* (*N*-type)

Figure 5.2   Conformations of the natural deoxyribose.

gives rise to the B-form helix, whereas the C3′-*endo* (*N*-type) conformation gives rise to the A-form helix. For deoxyribose the *S*-type conformation is slightly lowered in energy ($\sim 0.6\,\mathrm{kcal\,mol^{-1}}$) compared to the *N*-type and explains why DNA is found in the B-form helix. For ribose the preference is for the *N*-type, and thus RNA adopts the A-form helix associated with higher hybridization stability.

The LNA nucleoside has a fixed C3′-*endo* sugar pucker (*N*-type, $_3$E),[42–44] and LNA ONs adopt right-handed helices and hybridize according to the antiparallel Watson–Crick (WC) base-pairing pattern.[43,45–48] The glycosidic angle of included LNA nt's is in the *anti* range and with the 2′-*O*,4′-*C* methylene bridge positioned at the brim of the minor groove of the duplex. When an increasing number of LNA nucleosides are included in an ON the LNA–RNA duplex is progressively altered towards the A-type. When only three LNA nt's are included in a 9-mer ON the hybrid adopts an almost canonical A-type duplex geometry (Figure 5.3). This correlates with the observed helical thermostability *per LNA-nucleoside* that reaches a maximum with less than 50% LNA nt's.[5,6,43,49] This phenomenon, called 'structural saturation', is explained by the fact that LNA nt's induce conformational change of nearby deoxy-nt's to attain a higher degree of *N*-type. This induction of altered conformation is most pronounced for deoxy-nt's towards the 3′-end. Structural saturation is a very important property of LNAs since only a few inclusions can transform the entire ON into a high-affinity structure. Thus, the high-affinity conformation of a single LNA inclusion is amplified by its induction of a conformational change of juxtaposed DNA nt's. The consequence of 'structural saturation' is that *few* LNA inclusions give rise to a higher affinity increase *per residue* than many inclusions. Many classical 2′-modifications (2′-*O*-Me and 2′-MOE) show the exact opposite trend in which the largest effect *per residue* is reached if the modifications are grouped in consecutive blocks or if the strands are fully modified.[3,21] Structural saturation and the fact that LNA fits perfectly within the WC base-paring framework is one of the reasons for the design freedom of LNA, and also explains why the LNA mixmer design provides such a high duplex stability.[20]

5.2.2   Hybridization Properties

The inherent property of LNA nucleotide substitutions is the high binding affinity in almost any possible design variant.[5,6,13,21,44,50] LNA hybrids

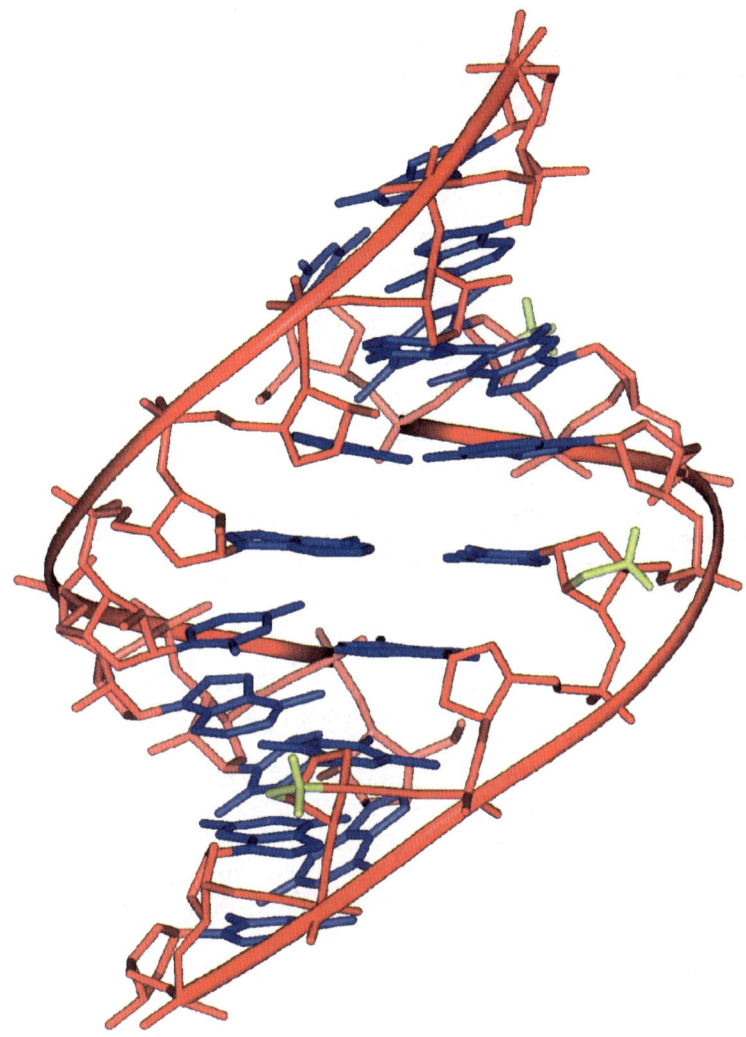

Figure 5.3 Structure of a 9-mer LNA–RNA hetero duplex. The LNA oligonucleotide comprises three LNA nucleotides, shown in yellow.

obey the WC hydrogen bonding rules and form right-handed helices and classical hypo-chromicity during hybridization. Thermal denaturation of the duplexes is associated with hyper-chromicity and the T_{m} is determined classically.

LNA nt's can be combined with DNA, RNA or analogues such as 2'-*O*-Me (Table 5.1). The affinity increase is largest towards RNA and ranges typically between 5 and 7 °C per modification. The largest increase is seen when LNA nt's are included in PS, which provides a record high increase of 9–10 °C per modification (Table 5.1, entry 6). This is a unique feature of LNA since the PS

Table 5.1 Hybridization studies with 9-mer LNA against complementary DNA and RNA.[a]

Entry	Sequence	DNA Compl. T_m (°C)	RNA Compl. T_m (°C)
1: Reference DNA	5′-d(GTGATATGC)	28	28
2: LNA–DNA chimera	5′-d(GT^{L}GAT^{L}AT^{L}GC)	44	50
3: LNA–RNA chimera	5′-r(GT^{L}GAT^{L}AT^{L}GC)	55	63
4: Fully modified LNA	5′-(G^{L}T^{L}G^{L}A^{L}T^{L}A^{L}T^{L}G^L ^{Me}C^L)	64	74
5: Fully PS	5′-d(G$_S$T$_S$G$_S$A$_S$T$_S$A$_S$T$_S$G$_S$C)	21	17
6: PS-LNA–DNA chimera	5′-d(G$_S$T^L$_S$G$_S$A$_S$T^L$_S$A$_S$T^L$_S$G$_S$C)	41	47
7: 2′-O-Me-RNA	5′-(GTGATATGC)	33	49
8: 2′-O-Me-RNA–DNA chimera	5′-(GTGATATGC)	22	34
9: LNA–2′-O-Me-RNA chimera	5′-(GT^{L}GAT^{L}AT^{L}GC)	53	64

[a]A, nucleotide monomer with an adenin-9-yl base; C, nucleotide monomer with a cytosin-1-yl base; G, nucleotide monomer with a guanin-9-yl base; T, nucleotide monomer with a thymin-1-yl base. ^{Me}C, nucleotide monomer with a 5-methylcytosin-1-yl base. Oligo-2′-deoxyribonucleotide sequences are depicted as d(sequence); oligoribonucleotide sequences as r(sequence); and 2′-O-Me-oligoribonucleotide residues are underlined. LNA monomers are shown in boldface with superscript 'L'. PS designates phosphorothioates and subscript 'S' denotes a phosphorothioate linkage.

modification is known to reduce the affinity of DNA but also for 2′-modified ON analogues (*e.g.* 2′-O-Me). However, when PS are used in LNA the affinity is almost not affected.

When six 2′-O-Me RNA modifications are included in a DNA ON the T_m decreases against the DNA complement, while the T_m increase is only 1 °C per modification against the RNA complement (Table 5.1, entry 8). When the ON becomes fully 2′-O-Me RNA-modified the T_m increases approximately by 0.5 °C and 2.2 °C against DNA and RNA, respectively (Table 5.1, entry 7). This demonstrates clearly the difference in the 'mixing power' and thus the design freedom between LNA and 2′-analogues.[3,21,51] However, when three LNA residues are included in the ON together with the six 2′-O-Me RNA residues the affinity-lowering effect by spiking 2′-O-Me RNA residues into ONs can be restored entirely (Table 5.1, entry 9). Recently, LNA has also been included in 2′-O-MOE RNA[52] and, as in the example with 2′-O-Me RNA, a few LNA-nucleotide substitutions increased significantly the affinity of 2′-O-MOE RNA. These ONs were found to be the most effective when tested in miRNA inhibition *in vitro*. Despite the fact that LNA hybridizes with high affinity to its nucleic acid target the binding is highly specific in terms of WC base pairing. Mismatch studies have shown that a ΔT_m of −21 °C is found for T/C mismatches and ΔT_m of −14 °C is found for T/G mismatches.[5,53] However, the ΔT_m can vary from a few degrees to more than 20 °C, depending on the nucleotide mismatch and sequence context.[53]

From a therapeutic point of view the interplay between PS-internucleoside linkages and LNA nucleosides is of particular interest. Compared to phosphordiesters, PS are known to provide improved pharmacokinetic and cellular uptake properties, whereas a clear disadvantage is their decreased affinity to complementary RNA sequences. This is an important explanation for the many failures of the first- and second-generation AS drugs. By including LNA-nt's and PS in one ON the improved pharmacokinetic and cellular uptake properties are *combined* with much higher affinity and potency.[33]

5.2.3 Thermodynamic and Kinetic Considerations

The design rationale for making the $2'-O,4'-C$-oxymethylene link was to make a high-affinity RNA analogue by pre-organizing, *i.e.* locking the furanose in an *N*-type conformation and the benefit was intended to be twofold:

(1) a smaller hybridization entropy penalty;
(2) pre-organization of the molecule in a high-affinity conformation.[6,50]

Early reports supported a major role of reduced entropy penalty,[44,54] but later more comprehensive studies proved that a more favourable hybridization enthalpy and base stacking were the major components contributing to the high duplex stability.[45,49,55-57] In general, enthalpy–entropy compensation is observed and both entropy (pre-organization) and enthalpy (stacking) are contributing, but not simultaneously for a given sequence.[53] For most sequences hybridization is driven by enthalpy, but in a few cases by entropy.[58] LNA pyrimidines are more stabilizing than the purines, especially A^L; however, the contributions here are also, to some extent, sequence dependent.[53]

The hybridization kinetics of LNA follow closely the mechanisms known for nucleic acids: initial formation of a nucleation complex followed by fast annealing of the duplex.[55] Christensen *et al.* showed that the overall K_D values – calculated from the T_m curves at various temperatures – decreased from 20 to 0.3 nM as one to three LNA nt's were included.[55] This indicates that LNA hybridizes with high *on*-rates ($2 \times 10^7 \, M^{-1} s^{-1}$) and the increased K_D values between DNA and LNA were reflected by markedly lower *off*-rates of LNA.

5.2.4 Biochemical Properties

5.2.4.1 General Design of LNA

LNA can be mixed in any combination within other nucleic acids and analogues, thereby increasing their affinity. As described above this is a unique property of LNA compared to classical $2'$-modifications and offers the largest range of high affinity design possibilities seen within the field of ONs. Depending on the intended target for LNA and the mechanism whereby inhibition of RNA expression is most effective, the LNA ON design is divided into two main categories: mixmers or gapmers (Figure 5.4). In a mixmer LNA residues are dispersed along the sequence of the ONs (Figure 5.4 line 1), while in a

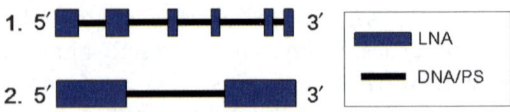

Figure 5.4   Mixmer and gapmer oligonucleotide designs.

gapmer two continuous LNA segments at both ends of the ON are separated by a central nucleic acid segment (Figure 5.4 line 2). For inhibiting mRNA expression ON designs of the gapmer design are the most potent, but for miRNA inhibition where RNase H recruitment (see below) is not feasible, the most effective LNA ONs are mixmers (*vide infra*). The slow *off*-rate of LNA is especially important for antagonizing miRNAs, since this is thought to occur by sequestration of the miRNA by complementary LNA ON.

5.2.4.2   Nuclease Resistance

By virtue of its synthetic origin incorporation of LNA nucleosides increase nuclease resistance.[59] The degree of protection is dependent on the number of LNA residues, their position and incorporation of other stabilizing entities (*e.g.* PS internucleoside linkages). To secure high protection against 3′-exonuclease [*e.g.* snake venom phosphordiesterase (SVPD)], two LNA nt's have to be incorporated at the 3′-end. The nuclease resistance is independent of whether the LNA dimer is placed in the ultimate or penultimate position.[6,22,59–61] Increasing the number of LNA nucleotide substitutions to five in an 11-mer ON of a mixmer design provides resistance to SVPD digestion.[23] Resistance to SVPD digestion is also seen for 20-mer LNA mixmers and gapmers that contain 8 to 11 residues.[62]

The central DNA segment in the gapmer design is rather sensitive to endonucleolytic activity, especially if the segment gets longer that four nt's. However, it has been shown that the LNA flanks also protect the DNA segment against endonucleolytic degradation (*e.g.* S1-endonuclease) since degradation of a 4LNA-7DNA-5LNA diester gapmer is slower compared to the iso-sequential PS.[22] Degradation of 18-mer LNA phosphordiester gapmers in human serum as a function of LNA-nucleotide incorporation was nicely shown by Kurreck *et al.*[63] Increasing the LNA segment from one to four at either end increased the half-life from 4 to 15 hours compared to 1.5 hours for the native DNA. The stability of a LNA that contains three or four modifications at either end of the molecule is also better than the corresponding 2′-*O*-Me gapmer and PS with respective half-lives of 12 and 10 hours.

Since longer central DNA segments will be degraded – albeit rather slowly – they have to be protected further and the most convenient way to do this is by PS internucleoside linkages. The PS linkage is easy to introduce under LNA synthesis and serves as a substrate for RNase H. We have observed that gap sizes ranging from 9 to 11 PS linkages are entirely stable in rat serum over 24 hours (unpublished data).

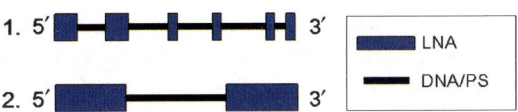

Figure 5.4   Mixmer and gapmer oligonucleotide designs.

gapmer two continuous LNA segments at both ends of the ON are separated by a central nucleic acid segment (Figure 5.4 line 2). For inhibiting mRNA expression ON designs of the gapmer design are the most potent, but for miRNA inhibition where RNase H recruitment (see below) is not feasible, the most effective LNA ONs are mixmers (*vide infra*). The slow *off*-rate of LNA is especially important for antagonizing miRNAs, since this is thought to occur by sequestration of the miRNA by complementary LNA ON.

5.2.4.2   *Nuclease Resistance*

By virtue of its synthetic origin incorporation of LNA nucleosides increase nuclease resistance.[59] The degree of protection is dependent on the number of LNA residues, their position and incorporation of other stabilizing entities (*e.g.* PS internucleoside linkages). To secure high protection against 3′-exonuclease [*e.g.* snake venom phosphordiesterase (SVPD)], two LNA nt's have to be incorporated at the 3′-end. The nuclease resistance is independent of whether the LNA dimer is placed in the ultimate or penultimate position.[6,22,59–61] Increasing the number of LNA nucleotide substitutions to five in an 11-mer ON of a mixmer design provides resistance to SVPD digestion.[23] Resistance to SVPD digestion is also seen for 20-mer LNA mixmers and gapmers that contain 8 to 11 residues.[62]

The central DNA segment in the gapmer design is rather sensitive to endonucleolytic activity, especially if the segment gets longer that four nt's. However, it has been shown that the LNA flanks also protect the DNA segment against endonucleolytic degradation (*e.g.* S1-endonuclease) since degradation of a 4LNA-7DNA-5LNA diester gapmer is slower compared to the iso-sequential PS.[22] Degradation of 18-mer LNA phosphordiester gapmers in human serum as a function of LNA-nucleotide incorporation was nicely shown by Kurreck *et al.*[63] Increasing the LNA segment from one to four at either end increased the half-life from 4 to 15 hours compared to 1.5 hours for the native DNA. The stability of a LNA that contains three or four modifications at either end of the molecule is also better than the corresponding 2′-*O*-Me gapmer and PS with respective half-lives of 12 and 10 hours.

Since longer central DNA segments will be degraded – albeit rather slowly – they have to be protected further and the most convenient way to do this is by PS internucleoside linkages. The PS linkage is easy to introduce under LNA synthesis and serves as a substrate for RNase H. We have observed that gap sizes ranging from 9 to 11 PS linkages are entirely stable in rat serum over 24 hours (unpublished data).

phosphordiesters, PS are known to provide improved pharmacokinetic and cellular uptake properties, whereas a clear disadvantage is their decreased affinity to complementary RNA sequences. This is an important explanation for the many failures of the first- and second-generation AS drugs. By including LNA-nt's and PS in one ON the improved pharmacokinetic and cellular uptake properties are *combined* with much higher affinity and potency.[33]

5.2.3 Thermodynamic and Kinetic Considerations

The design rationale for making the $2'-O,4'-C$-oxymethylene link was to make a high-affinity RNA analogue by pre-organizing, *i.e.* locking the furanose in an *N*-type conformation and the benefit was intended to be twofold:

(1) a smaller hybridization entropy penalty;
(2) pre-organization of the molecule in a high-affinity conformation.[6,50]

Early reports supported a major role of reduced entropy penalty,[44,54] but later more comprehensive studies proved that a more favourable hybridization enthalpy and base stacking were the major components contributing to the high duplex stability.[45,49,55-57] In general, enthalpy–entropy compensation is observed and both entropy (pre-organization) and enthalpy (stacking) are contributing, but not simultaneously for a given sequence.[53] For most sequences hybridization is driven by enthalpy, but in a few cases by entropy.[58] LNA pyrimidines are more stabilizing than the purines, especially A^L; however, the contributions here are also, to some extent, sequence dependent.[53]

The hybridization kinetics of LNA follow closely the mechanisms known for nucleic acids: initial formation of a nucleation complex followed by fast annealing of the duplex.[55] Christensen *et al.* showed that the overall K_D values – calculated from the T_m curves at various temperatures – decreased from 20 to 0.3 nM as one to three LNA nt's were included.[55] This indicates that LNA hybridizes with high *on*-rates ($2 \times 10^7 \, M^{-1} \, s^{-1}$) and the increased K_D values between DNA and LNA were reflected by markedly lower *off*-rates of LNA.

5.2.4 Biochemical Properties

5.2.4.1 General Design of LNA

LNA can be mixed in any combination within other nucleic acids and analogues, thereby increasing their affinity. As described above this is a unique property of LNA compared to classical $2'$-modifications and offers the largest range of high affinity design possibilities seen within the field of ONs. Depending on the intended target for LNA and the mechanism whereby inhibition of RNA expression is most effective, the LNA ON design is divided into two main categories: mixmers or gapmers (Figure 5.4). In a mixmer LNA residues are dispersed along the sequence of the ONs (Figure 5.4 line 1), while in a

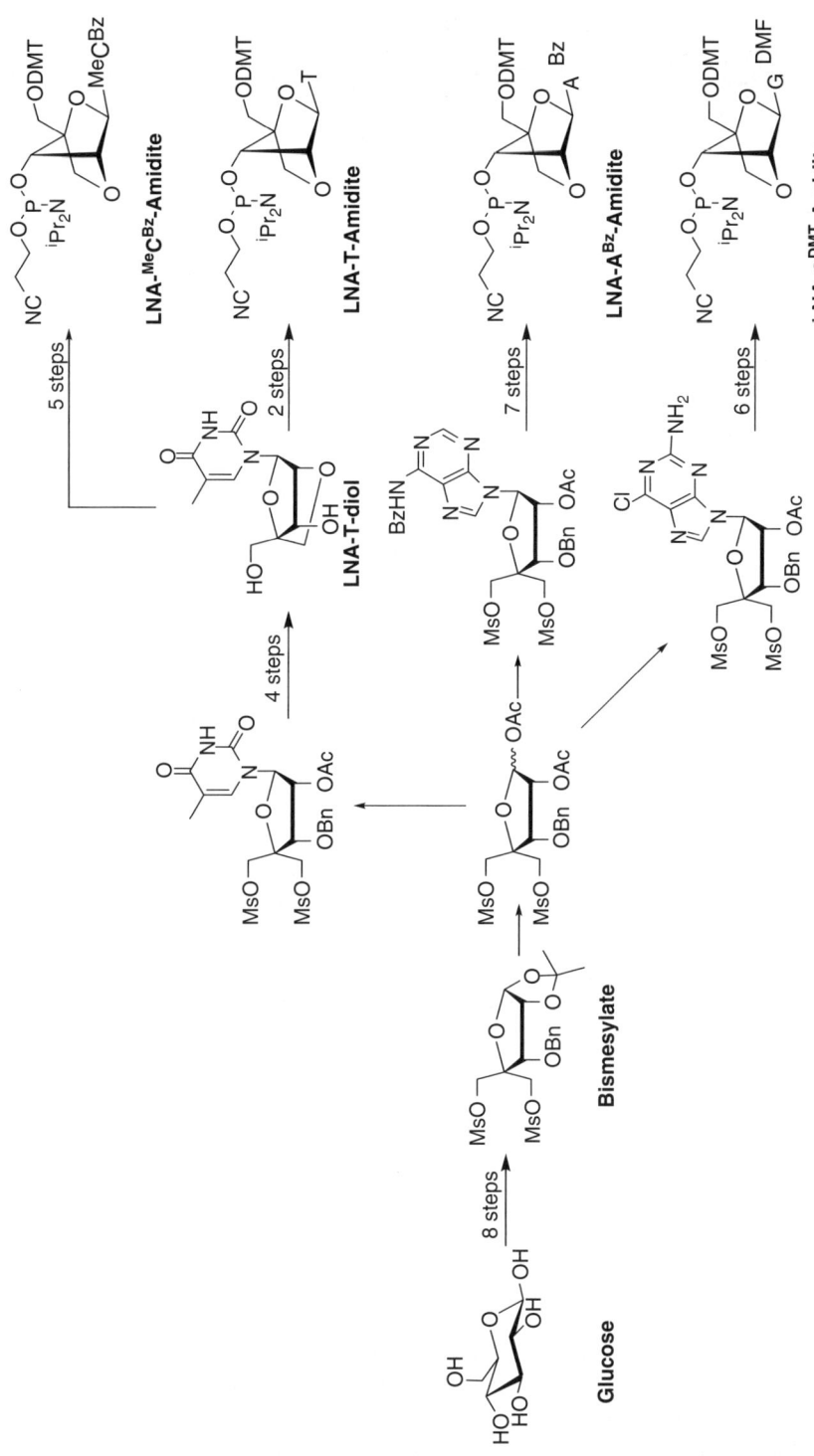

Figure 5.5 The synthesis of LNA phosphoroamidites.

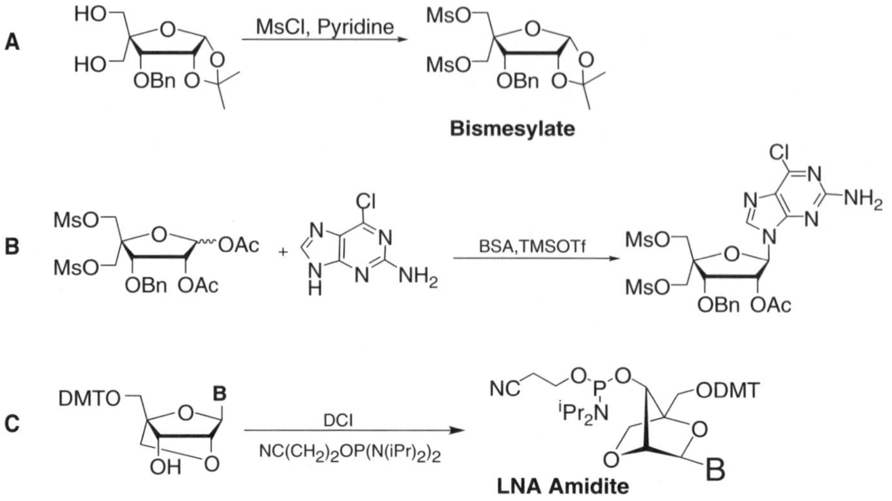

Figure 5.6   Key synthetic improvements.

A major improvement was achieved with development of the bismesylate because it is based on the symmetrical protection of the sugar intermediate and
furthermore provides a crystalline material – the bismesylate (Figure 5.6A).[5]
The needed selectivity in the ring formation step is provided by the stereochemistry inherent in the modified glucose. The alcohol group at the 2′-position
that facilitates the subsequent ring formation is only able to react with the
mesylate on the α-face of the compound, thus giving solely the desired bicyclic
structure.

The original synthesis of the LNA guanine monomer was based on the
coupling between the coupling sugar and *iso*-butyryl-protected guanine. This
approach has several drawbacks of which the regioselectivity of approximately
9:1 between the desired N9 isomer and the N7 isomer was the main concern. It
is critical for the quality of the final product to avoid the formation of the N7
isomer because, on a large scale, it is very hard to separate from the N9 isomer.
The use of 2-amino-6-chloropurine as a synthon for guanine (Figure 5.6B)
solved this problem, providing N9:N7 selectivity in excess of 99:1 – very often
the formation of the N7 isomer is not even detectable.[27] Furthermore, this
synthon is more lipophilic and stable under the reaction conditions employed
later in the synthesis and therefore provides a more robust synthesis than the
original process.

The phosphoramidite synthesis is the final step in the synthesis of the four
LNA monomers (Figure 5.6C) and is therefore also of great importance because the quality of the amidites has direct influence on the coupling efficiencies
in the synthesis of ONs. Just a few percent of impurities have a great influence
on the coupling yield (*e.g.* hydrolysed phosphitylation reagent). The original
phosphitylation protocol for the LNA amidites was based on the use of a
phosphoramidochloridite and a base and often gave yields in the 70–90% range

and typically required silica gel column purification.[5,6] To achieve both higher yields and possibly omit the column purification, a different protocol based on a phosphorodiamidite and 4,5-dicyanoimidazole as an activator was developed.[64] This method routinely provides LNA monomers at yields >90% after work-up and, interestingly, often of so high purity that chromatography is not essential.

It is now a well-established fact that regulatory authorities regard the phosphoramidites as API starting materials for drugs based on nucleic acids. This defines the regulatory level at which GMP begins. Although there are no GMP requirements to the processes leading up to the API starting material level, the processes have to be strictly controlled according to traceability and documentation. Also, the impurities in the amidite batches have to be identified. The Chemical Manufacturing Control (CMC) documentation of LNA amidites is constantly expanding and will be developed as the LNA ON clinical candidates proceed through the clinical phases towards registration.

The significant optimization of the entire synthesis sequence combined with the economy of scale has reduced the costs of LNA amidites to a very competitive level. For manufactures larger than 100 kg scale that we today have realised the average cost per step is now less than $1, and with the knowledge of the LNA amidite synthesis available today we predict that this level will be reached for the entire monomer synthesis at that scale.

5.3.2 Synthesis of Locked Nucleic Acids Oligonucleotides

LNA ONs are composed of LNA and DNA nt's, of which the LNA and DNA amidites are the key starting materials. LNA amidites are produced according to the procedures described above while DNA amidites are commercially available for API synthesis in all scales. Although solution-phase procedures have been developed for large-scale API productions, virtually all ONs for therapeutic applications are manufactured according to solid-phase procedures, where the ONs are synthesized by sequential coupling of the single A-, T-, C/^{Me}C-, or G-amidites until the desired length is reached.

LNA-amidites are adapted to the solid-phase procedures essentially as DNA-amidites, but initially the coupling and oxidation steps resulted in unsatisfactory yields. These issues have now been solved, and today LNA-amidites couple almost as DNA-amidites. The average yield per cycle is typically 98–99%, leaving only 1–2% of non-full length products per cycle.

The fact that LNA ON synthesis is readily adapted from DNA ON synthesis provides the inherent possibilities to mix LNA nt's in any combination with DNA nt's, and to make LNA PS in the same way as DNA PS. From a manufacturing point of view the synthetic similarities between DNA and LNA are very attractive, because the knowledge that is constantly developed in the field of DNA synthesis is easily adapted for LNA synthesis.

The similarities between DNA and LNA synthesis also allow for an efficient outsourcing process because there are several custom manufacturing organizations that specialize in the cGMP production of ONs.

LNA ONs have been synthesized on most small- and middle-scale ON synthesizers. Multiple small-scale synthesis batches from 50 nmole to 1 µmole have been synthesized on Expedites, Mermades, ABI3900 and Äkta instruments. Batch sizes between 1 and 200 µmole are routinely made on Mermade2 and Äkta100 instruments, while batch sizes ranging from 200 µmole to 60 mmole are made on Äkta100 and Oligopilot400 instruments. The largest batch of LNA ON that has been produced is 385 g of the LNA compound SPC2996. The purity of this material was 93% based on ion exchange high-performance liquid chromatography (IEX-HPLC).

LNA ONs are white, hygroscopic solid materials. Despite the hygroscopicity we have not experienced cintering of LNA ONs. Also, solid material LNA ONs are stable as virtually no degradation has been observed for up to two years at 40 °C. In aqueous solution LNA is also highly stable, although a 1–2% hydrolysis is observed after one year at room temperature.

The status today is that there are no obstacles for producing and controlling LNA ONs and, hence, LNA synthesis can easily be adapted to the large-scale instrumentation platforms designed for commercial use. The fact that LNA ONs are significantly shorter than second-generation AS drugs offers many CMC advantages. For instance, a 13-mer LNA needs only 48 synthetic steps where a 20-mer second-generation ON requires 76 steps, and since the complexity increases significantly with ON length this is a marked difference. Also the purity measured, as the full-length vs. $(n-1)$ ratio, will be improved since it is much easier to remove a 12-mer from a 13-mer than to remove a 19-mer from a 20-mer. We believe, therefore, that commercial LNA manufacturing will show an attractive cost and compliance relation.

5.4 Pharmacological Aspects of Locked Nucleic Acid Oligonucleotides Targeting Messenger RNA and MicroRNA

The pharmacological aspects of LNA ONs targeting mRNA are extensively reported in the literature and have recently been reviewed.[33] In Section 5.4.2 dealing with mRNA inhibition we focus on recent data from our own, mainly unpublished, work to illustrate new aspects of mRNA targeting. In Section 5.4.3 on miRNAs we take a slightly different approach and present this novel class of targets more broadly and then, finally, focus on the pharmacological work that we have generated.

First, in Section 5.4.1 we briefly illustrate the general biodistribution and cellular uptake of saline-formulated fully phosphorylated LNA ONs.

5.4.1 Biodistribution of Locked Nucleic Acid Oligonucleotides

Fluiter et al. examined the tissue distribution of tritium-labelled LNA and the LNA analogues amino-LNA, thio-LNA and α-L-LNA.[24] The ONs were

16-mers of the same gapmer design targeting H-ras, and they were administered by subcutaneously implanted mini-pumps. LNA and α-L-LNA had similar biodistributions, except in kidneys, where the uptake of α-L-LNA was higher than that of LNA. Amino- and thio-LNA had increased uptake in liver, while the amino-LNA had the broadest distribution, showing significantly higher uptake in heart, lung, muscle and bone compared to the other analogues.[24]

Tritium labelling was also used to image the biodistribution of the LNA compounds SPC2968 and SPC2996 in mice. The LNA ONs were 16-mer gapmers and designed to target Hypoxia-inducible Factor 1 Alpha (Hif-1α) and Bcl-2, respectively. The biodistribution was measured as whole-body autoradiography in sagittal sections between five minutes and 18 days post dosing. The highest uptake of radioactivity was registered in the kidney cortex, uterus, uvea of the eye, skin, liver, bone marrow, lymph nodes and spleen (Figure 5.7; unpublished data). The tissue half-lives were estimated on the basis of disappearance of tritium-label over 18 days and were found to be remarkably similar across tissues, estimated at approximately 8 days. This is much longer than reported for first-generation PS and can be related to the improved nuclease resistance of LNA.[65] Both ONs showed a very high blood clearance of radioactivity with elimination half-lives of about one hour.

The biodistribution to the haematological compartment in female Naval Medical Research Institute (NMRI) mice one hour after intravenous (i.v.) administration of $50\,mg\,kg^{-1}$ FAM-labelled LNA ON (targeting Hif-1α) showed that systemic delivery of LNA ON led to rapid association with virtually all cell types in the haematological compartments, even in the bone marrow. Tissue resident cells, such as dendritic cells and macrophages in the spleen, were, however, also exposed to LNA ONs rapidly.

Once taken up by the tissues, LNA ONs appear to be taken up by the cells by internalization, and there is a good relation between tissue uptake and endogenous effect, with the kidney being the most pronounced exception.[33] Uptake in haematopoietic cells has been studied by FACS analysis using fluorescently labelled LNA. Following i.v. injection of FITC-SPC2968 the label was detected in virtually all cell types in the bone marrow, spleen and in the

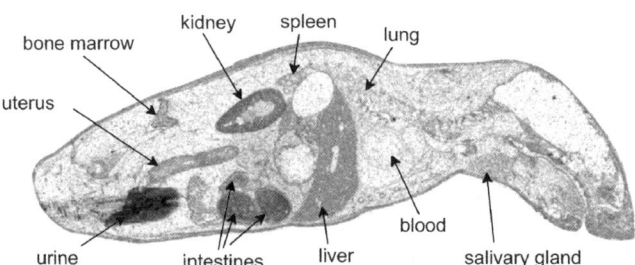

Figure 5.7 Whole body radioluminograms showing the intensity and distribution of tritium labelled SPC2996 (LNA oligonucleotide targeting Bcl-2) in mice four hours after intravenous injection.

circulation. The data indicated that the LNA ON was internalized rather than just adhering to the cell surface, whereas significant differences were observed in the ability of the individual cells to internalize LNA, with macrophages, granulocytes and dendritic cells being the most active.[33]

5.4.2 Inhibition of Messenger RNA Targets

The first *in vivo* study with LNA was published by Wahlestedt and co-workers in 2000.[59] This study used 15-mer phosphorodiester (PO) gapmers and mixmers to repress the transcript coding for the delta opiod receptor in rat brain, and reported that LNA was pharmacologically responsive. In addition, LNA was tolerated with low toxicity in the rat model. Since this first study many improvements have been made, of which a better understanding of the LNA ON design and how LNA ONs interact with the intracellular molecular machinery are the major achievements.

5.4.2.1 Targeting Apolipoprotein B-100

Cardiovascular diseases are a leading cause of death in the industrialized countries. Elevated plasma cholesterol levels are often observed in metabolic disorders, including diabetes and obesity, and are associated with increased risk of atherosclerosis. Cholesterol is transported in the circulation by lipoproteins, primarily the low-density lipoprotein (LDL). Very low-density lipoprotein (VLDL), which is the metabolic precursor of LDL, is assembled in the liver by a process that is dependent on apolipoprotein B-100 (ApoB-100; the main structural apolipoprotein in VLDL and LDL) and the microsomal transfer protein (MTP). High secretion of ApoB-100 to the circulation has a similar phenotype as an increased level of atherogenic particles and thereby is an increased risk for plaque formation. By lowering the formation of ApoB-100 the formation, and therefore the secretion of VLDL to the circulation, is reduced.

In 2004 Soutschek *et al.* reported for the first time the downregulation of ApoB-100 in the liver after systemic delivery of a small interfering RNA (siRNA) duplex.[66] The passenger strand of the duplex was conjugated with cholesterol (chol-siRNA) in order to mediate adequate biodistribution to the liver and hepatocyte uptake. The ApoB-100 message was downregulated by about 50% in both liver and jejunum after three doses of 50 mg kg^{-1}. We designed a 16-mer LNA gapmer with the sequence representing a truncated version of the 21-mer guide strand from the duplex (designated as SPC3197), and a second 16-mer gapmer targeting another recognition sequence in the ApoB mRNA (designated as SPC3302). Downregulation of the ApoB message was observed for both LNA ONs in a dose-dependent manner and a more than 50% downregulation in both liver and jejunum was obtained after three doses of 6.5 mg kg^{-1} (Figure 5.8). Thus, a total dose of about 20 mg adminis-tered over three doses of unconjugated SPC3197 and SPC3302 LNA ONs,

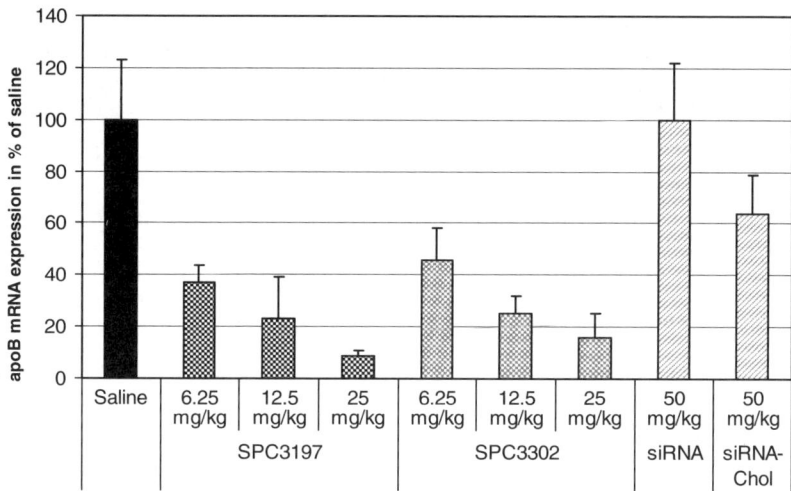

Figure 5.8 Expression of apoB mRNA in C57BL/6 mice treated with 6.25, 12.5 or 25 mg kg^{-1} of SPC3197 or SPC3302 (LNA oligonucleotides targeting apoB) and compared with mice treated with 50 mg kg^{-1} siRNA or cholesterol conjugated siRNA at days 0, 1 and 2. At sacrifice (24 hours after last dosing) liver was sampled and analysed for apoB mRNA expression by quantitative real-time quantitave polymerase chain reaction (RT-qPCR). Data are normalized to glyceraldehyde 3-phosphate dehydrogenase (GAPDH) and presented relative to the saline control group ($n = 7$).

respectively, resulted in the same level of inhibition as 150 mg of the chol-siRNA duplex. Since the *in vitro* potencies of the three tested compounds were almost equal, the marked difference in the *in vivo* effect was most likely a direct consequence of improved cellular uptake (or intracellular distribution) and/or improved biostability of the two LNA ONs compared to the chol-siRNA. The concentration of total cholesterol was followed during the experiment. As a direct consequence of ApoB-100 downregulation, SPC3197 and SPC3302 reduced total cholesterol in plasma, and at a total dose of about 25 mg, total cholesterol was reduced about 50%. Significant reduction of plasma cholesterol by the chol-siRNA duplex was neither observed in our study nor reported by Soutschek and co-workers.[66]

The LNA 16-mers reported above are size reduced compared to the first- and second-generation AS drugs. To further assess the potency of shortened LNA ONs, we used SPC3302 as the sequence model and designed truncated 14-mer and 12-mer versions, called SPC3714 and SPC3716. For clarity, we designated LNA ONs shorter than 16 nt's in length as 'shortmers'.

C57BL/6 mice were dosed at 5 mg kg^{-1} per dose for three consecutive days with SPC3302, SPC3714 and SPC3716. ApoB-100 mRNA expression levels in jejunum were slightly lower in SPC3302-dosed animals compared to those dosed with SPC3714 and SPC3716, with respective reduction of 65%, 60% and 50%. However, in the liver the effect was reversed with the shortmers being

significantly more potent. The ApoB-100 mRNA expression was reduced with
more than 70% with the shortmers compared to about 30% with SPC3302.
As a direct correlation of lowering ApoB-100 mRNA expression in the liver,
total plasma cholesterol levels were reduced by approximately 70%, 50% and
20% following administration with the 12-, 14-, and 16-mer, respectively
(Figure 5.9). In a subsequent dose–response study a 50% reduction in ApoB-100
expression was found after three doses at $1\,\mathrm{mg\,kg^{-1}}$ leading to a similar (50%)
reduction in total plasma cholesterol.

The duration of action of the ONs was also measured. Eight days after one
dose at $5\,\mathrm{mg\,kg^{-1}}$ of SPC3716 and SPC3714 and one dose at $25\,\mathrm{mg\,kg^{-1}}$ of
SPC3302, the reduction in total plasma cholesterol was, respectively, 55%,
50% and 40%, and after a 16-day follow-up the cholesterol levels were still
reduced by 45% in SPC3714 and SPC3716 dosed mice.

The marked size effect observed *in vivo* was inversely related *in vitro*. Thus,
the 16-mer was more potent than the shortmers with IC_{50} values of 1 nM and
5 nM, respectively. This can be explained by the fact that transfection *in vitro*
ensures efficient intracellular delivery of the LNA ONs and the potencies are
therefore driven by the thermodynamics of the LNA–RNA duplexes, *i.e.* the
16-mer has higher T_m than the shortmers, leading to a lower IC_{50} value. By
contrast, *in vivo* the slightly lower T_m of the shortmers is compensated by their
much better cellular uptake (or better intracellular distribution), leading to
their improved pharmacological efficacies.

Increased intake of fat is correlated with increased circulating fat and increased plasma cholesterol, which, in turn, is correlated with increased risk of
atherosclerosis (*vide supra*). Hence, we set out to examine the effect of the
shortmers in hypercholesterolaemic mice. C57BL/6 female mice were fed with a

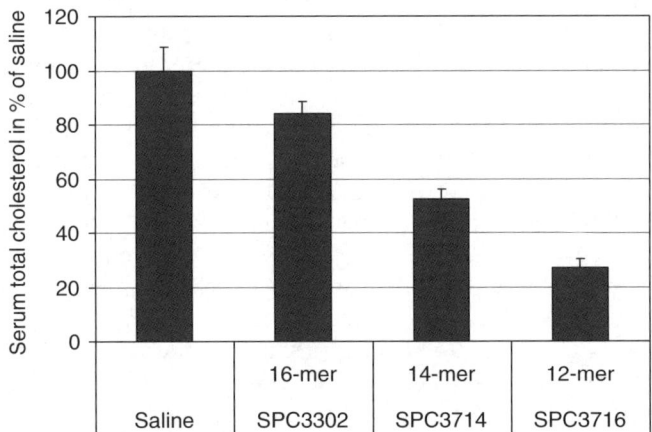

Figure 5.9   Serum total cholesterol in C57BL/6 mice dosed with LNA oligonucleotides
targeting apoB; SPC3302 (16-mer), 3714 (14-mer) or 3716 (12-mer) at
$5\,\mathrm{mg\,kg^{-1}}$ at days 0, 1 and 2. Total cholesterol was measured in
serum 24 hours after last dosing and presented relative to the saline control
($n = 5$).

high-fat diet (60% energy) for 13 weeks, leading to increased plasma cholesterol of about 75% (Figure 5.10). SPC3716 and SPC3714 were dosed at $5\,\mathrm{mg\,kg^{-1}}$ twice weekly for four weeks, resulting in effective silencing of the ApoB-100 mRNA expression levels (>90%) and significantly reduced plasma cholesterol levels (Figure 5.10).

This demonstrates the high cholesterol-lowering potential of SPC3716 and SPC3714, and shows that a hypercholesterolaemic phenotype can revert into normal at doses of less than the $5\,\mathrm{mg\,kg^{-1}}$ LNA ON used in this study.

The above-mentioned data clearly show that shortmers targeting ApoB-100 are more potent than the 16-mers in reducing plasma cholesterol and suggest that the use of shortmers could become a new approach for systemic gene targeting. The most recent approach in siRNA drug research is the design of novel formulations to improve biodistribution and cellular uptake, which would lead to more complex and expensive pharmaceuticals. For example, stable nucleic acid lipid particle (SNALP) formulated siRNA duplexes targeting ApoB-100 have been reported to have improved efficacies compared to chol-siRNA.[33,67,68] These lipid formulations are complex and expensive to manufacture and they are often associated with increased regulatory and safety considerations. The shortmer approach presented here enables us to shortcut these issues by improving RNA inhibition *in vivo* simply by reducing the size of the LNA ONs.

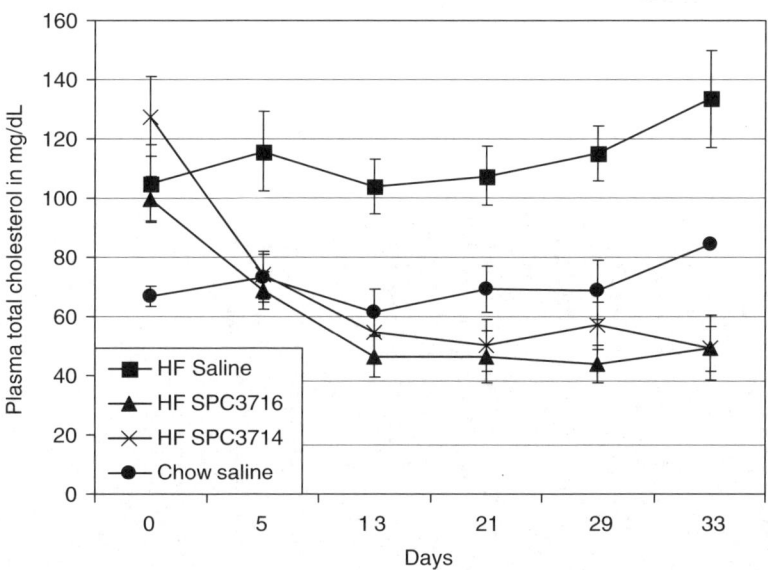

Figure 5.10 Plasma total cholesterol in C57BL/6 mice fed a high fat diet (60% energy) for 10 weeks before start of LNA ON dosing (day 0). LNA ONs were dosed at $2.5\,\mathrm{mg\,kg^{-1}}$ twice weekly for 31 days. Total cholesterol was measured once weekly in plasma ($n = 5$).

5.4.2.2   Targeting Survivin

Survivin is a 17 kDa protein and a key regulator of apoptosis and cell-cycle progression.[69,70] It plays a dual biological function as it protects cells from apoptosis, downstream of the intrinsic apoptotic pathway, and is involved in the regulation of mitosis where the protein selectively accumulates and localizes to various components of the mitotic apparatus during the mitotic spindle formation.[69,70] Survivin belongs to the family of inhibitors of apoptosis proteins and interferes with apoptosis regulation through a direct or indirect interaction with caspases. It is almost exclusively expressed in cancer and is regarded as one of the most cancer-specific targets. Inhibition of survivin expression has, therefore, for many years been of interest for the anticipated valuable therapeutic outcome.

In vitro LNA gapmers have been used to inhibit survivin expression in a range of cell lines – including 15PC3 (prostate), A549 (lung adenocarcinoma) and SW480 (colon carcinoma). Survivin inhibition has effectively led to growth arrest (cell-cycle arrest) of the cells and programmed cell death. The IC_{50} values for survivin mRNA downregulation by 16-mer LNA gapmers have been shown to be in the 1–5 nanomolar range. Strong survivin inhibition clearly leads to polyploidy of the treated cells, as shown in Figure 5.11 where the cells are treated with the lead survivin LNA compound SPC3042.

The antitumour potential of SPC3042 was tested in a xenografted model. SPC3042 was delivered systemically to Balb/C nude mice with subcutaneously inoculated PC3 cells (human prostate). SPC3042 and Taxol were tested alone and in combination. Neither of the drugs alone inhibited the tumour growth, but treatment with SPC3042 sensitized the tumours to treatment with Taxol. Administration with SPC3042 (10 mg kg^{-1} per dose, 12 doses over 28 days) in combination with Taxol (10 mg kg^{-1} per dose, six doses over 28 days) reduced tumour weights by 40% compared to saline treatment. This suggests that

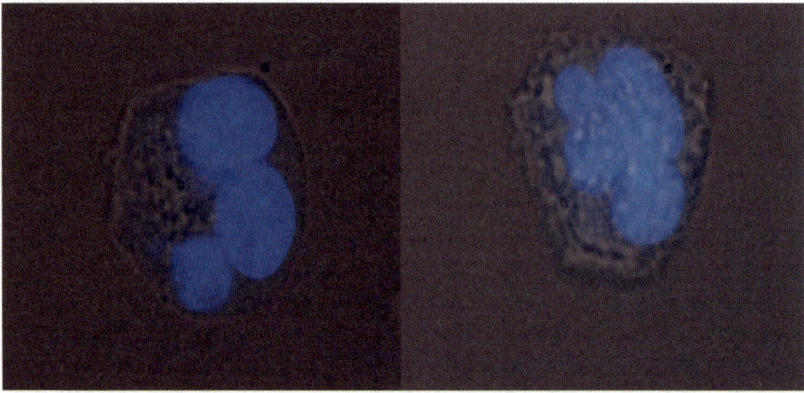

Figure 5.11   Survivin AS treatment induces polyploidy in 15PC3 cells. Cells were transfected with 5 nM AS oligo for 48 hours and visualised by DAPI staining.

SPC3042 can reduce the tumour growth of prostate tumours in combination with Taxol treatment. There is a clear synergistic effect, which indicates that the combination is an improvement of the individual therapeutic potential of both drugs. Combined with a good toxicology profile, as revealed by extensive good laboratory practice (GLP) toxicology studies, SPC3042 holds promise for further development.

5.4.2.3 Targeting Hypoxia-inducible Factor 1 Alpha

Hif-1 is another interesting target for cancer.[71,72] Hif-1 is a heterodimer consisting of Hif-1α and Hif-1β (ARNT) which are constitutively expressed in the cell. Hif-1 is a major transcriptional factor in the cellular response to hypoxia and regulates a large group of genes, among others vascular endothelial growth factor (VEGF), glucose transporter1 (Glut1) and multi-drug resistance1 (MDR1). HIF-1α is continually degraded at normoxic conditions ($t_{1/2} =$ 5 minutes), but is stabilized under hypoxic conditions and translocated to the nucleus, where it dimerizes with HIF-1β to form Hif-1. Blocking Hif-1α effectively inhibits the activity of Hif-1, and this mRNA may therefore serve as an interesting target for the inhibition of cancer growth. The mechanistic effect is likely to be a 'starvation' of the tumour for oxygen and nutrient supply by inhibiting vascularization and metabolism.[73]

SPC2968 is a 16-mer gapmer designed to downregulate the Hif-1α message and was, in fact, the first AS drug reported to inhibit this target. SPC2968 is highly potent against Hif-1α *in vitro* and has an IC_{50} around 1 nM in hypoxic cells. Hif-1α is also expressed under benign conditions with different levels of activities found in many normal tissues. SPC2968 has sequence complementarity to the Hif-1α mRNA in both human and mouse, which makes the mouse a good pharmacological model for obtaining relevant information about the systemic effects related to target inhibition (*e.g.* toxicology).

In a two-week study, SPC2968 showed a clear dose–response relationship in the inhibition of Hif-1α mRNA in both liver and kidney. The wild-type mice used in the study were dosed daily by intraperitoneal injection for 14 days.[74] Of the two organs the activity was most pronounced in the liver, where a 50% reduction in Hif-1α mRNA levels was achieved after a daily dose of only 3.6 mg kg^{-1}. Consistent with the role of Hif-1 as a transcription factor, treatment with SPC2968 also caused a substantial reduction in the mRNA from the Hif-1 regulated gene, VEGF.[33]

Yu and co-workers reported in 2004 an siRNA duplex with high *in vitro* potency against Hif-1α.[75] We thus compared the pharmacological effects of this siRNA with those of SPC2968 in the liver. The molecules were tested in wild-type mice as saline formulations. The mice were treated on three consecutive days (i.v.) with saline, siRNA or SPC2968 at doses ranging from 6.25 mg kg^{-1} to 50 mg kg^{-1}. The animals were sacrificed at one day or four days after the last dose. At day one after the last dose SPC2968 was found to downregulate the target dose dependently by 40%, 60% and 90% after dosing at 6.25, 12.5 and

$25 \, \text{mg kg}^{-1}$, respectively. Four days after the last dose the SPC2968 treatment still resulted in downregulation of Hif-1α by 35%, 40% and 50%, respectively. In contrast, the siRNA did not lead to any observable target downregulation even at a high dose of $50 \, \text{mg kg}^{-1}$, suggesting that further stabilization, conjugation and/or formulation of siRNA duplexes are needed for any practical pharmacological use.

Biodistribution studies have implied that relatively little LNA ON is taken up by xenografted tumours, which is likely to be caused by the poor vascularization of these artificial tumour models. In accordance with this we observed only little tumour inhibition of SPC2968 in Balb/C nude mice bearing DU145 prostate cancer xenografts. To assess uptake in xenografts we transfected the cells with SPC2968 (Lipofectamine 2000 was used to transfect the cells) prior to the subcutaneous inoculation of the cancer cells to secure *ex-vivo* delivery of SPC2968. In the control group the cells were pre-transfected with a scrambled control LNA ON with no effect on Hif-1α expression levels. Following cell inoculation the mice from both groups were dosed twice weekly ($50 \, \text{mg kg}^{-1}$ per dose) to maintain constant *in vivo* concentrations of SPC2968, along with saline control.

The group bearing tumours pre-transfected with the scrambled control and subsequently dosed with SPC2968 did not show any tumour inhibition. The same result was seen in the group bearing tumours pre-transfected with SPC2968 and subsequently dosed with saline. In contrast, the combination of pre-transfection and dosing with SPC2968 resulted in an approximately 35% reduction in tumour weight ($P < 0.001$) compared to that in the controls (Figure 5.12).

This is interesting, given the significant downregulation of Hif-1α even in these lowly vascularized xenografted tumours. The biology in human tumours is highly different to that in the artificial xenograft models, as is the vascularization of human tumours. Thus, the above-mentioned data indicate that SPC2968 is a promising candidate for the treatment of human cancer, where it is expected that inhibition of vascularization and nutrient supply is more important in the prevention of tumour survival.

5.4.2.4 Targeting B-cell Lymphoma 2

B-cell lymphoma 2 (Bcl-2) is a key inhibitor of apoptosis. Over-expression of Bcl-2 is a key event in the pro-survival of a number of cancers, ranging from solid tumours to leukaemias,[76] and several studies have described that over-expression of Bcl-2 is linked with chemoresistance.[77] This coupling has been found for, *e.g.*, prostate cancer[78] and malignant melanoma.[79] High levels of Bcl-2 expression are collectively found to be associated with malignant phenotypes, bad prognosis and resistance to therapy, and Bcl-2 is therefore an interesting target for therapeutic intervention. [For a detailed discussion of the function of Bcl-2 as well as of the development of a PS ON targeting Bcl-2 (Genasense) see Chapter 2]. Here, we present the main findings of our pre-clinical pharmacology studies that formed the basis for the clinical trial with SPC2996 (*vide infra*).

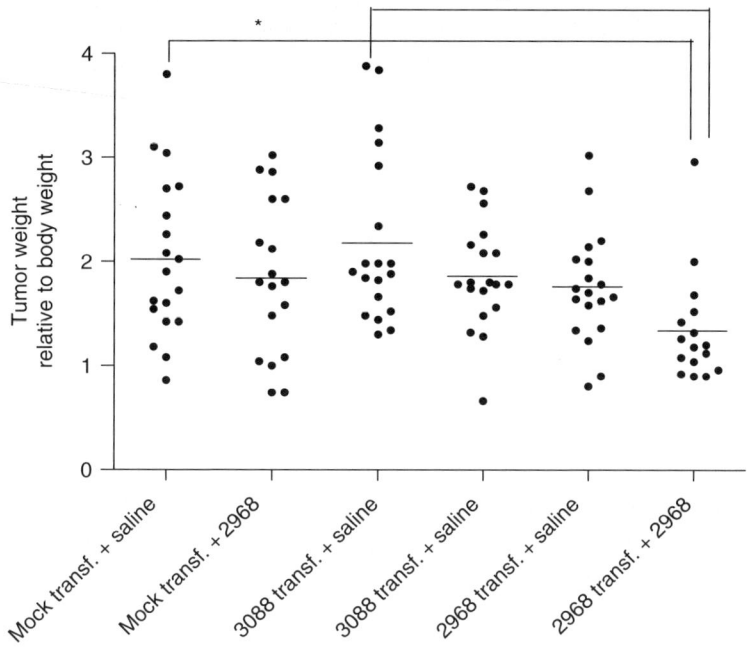

Figure 5.12 Relative tumour weights as a function of saline, SPC3088 (scrambled control) or SPC2968 treatments.

SPC2996 is a 16-mer gapmer directed against the first six codons of the Bcl-2 transcript. SPC2996 downregulated the Bcl-2 mRNA in 518A2 melanoma cells with an IC_{50} value of 5 nM, leading to efficient protein reduction. The protein downregulation was long-lasting as shown by the fact that at least 50% downregulation was observed after 72 hours.

The pro-apoptotic potential of SPC2996 in 518A2 and 15PC3 cells was demonstrated by a marked activation of caspases 3 and 7, and it was further underlined by significant Annexin V binding to the surface of Henrietta Lacks (HeLa) cells subsequent to SPC2996 treatment. After 48 hours pronounced cell death was observed.

The antitumour effect of SPC2996 was studied in two xenograft models: severe combined immunodeficiency disease (SCID) mice bearing human melanoma 518A2 xenografts, and Balb/c nude mice bearing human prostate PC3 xenografts. In the prostate model the mice were dosed at $10\,\mathrm{mg\,kg^{-1}}$ SPC2996 daily for 14 days, leading to a 40% reduction in tumour growth compared to saline-treated animals.

In the melanoma model the mice were dosed with SPC2996 for 14 days at doses of 7, 3.5 and $1.8\,\mathrm{mg\,kg^{-1}}$ per day. The Bcl-2 protein levels were reduced relative to saline control by 62%, 24% and 11%, corresponding to the 7, 3.5

and 1.8 mg kg^{-1} dose levels, while 30% tumour growth inhibition was observed at the 7 mg kg^{-1} per day dose. When SPC2996 (at 7 mg kg^{-1} per day) was combined with the chemotherapeutic drug dacarbazine the tumour growth was reduced approximately 30% more than with treatment of dacarbazine alone.

These data confirm the pharmacological potential of SPC2996 as an anticancer drug both as a single agent and in combination with other chemotherapeutics.

5.4.3 MicroRNA Targeting with Locked Nucleic Acids

MiRNAs are an abundant class of short endogenous RNAs that post-transcriptionally repress expression of protein-coding genes by base-pairing with the 3'-untranslated regions of the target mRNAs.[80] The ~22 nt mature miRNAs are processed sequentially from longer hairpin transcripts by the RNAse III ribonucleases Drosha and Dicer.[81–83] To date more than 4500 miRNAs have been annotated in vertebrates, invertebrates and plants according to the miRBase miRNA database release 9.2 in May 2007, and many miRNAs that correspond to putative genes have also been identified.[84] The identified miRNAs to date most likely represent the tip of the iceberg, and the number of miRNAs might turn out to be very large. Bioinformatic predictions combined with array analyses, small RNA cloning and deep sequencing, and Northern blot validation indicate that the total number of miRNAs in vertebrate genomes is significantly higher than previously estimated and may be as many as 1000.[85–87]

An increasing body of research shows that animal miRNAs play fundamental biological roles in cell growth and apoptosis,[88] haematopoietic lineage differentiation,[89] life-span regulation,[90] photoreceptor differentiation,[91] homeobox gene regulation,[92,93] neuronal asymmetry,[94] insulin secretion,[95] brain morphogenesis,[96] muscle proliferation and differentiation,[97–99] cardiogenesis,[100] homeostasis and function of the immune system[101,102] and late embryonic development in vertebrates.[103] Several studies have identified subclasses of miRNAs directly implicated in the regulation of mammalian brain development and neuronal differentiation.[104–107] Interestingly, many neural miRNAs appear to be temporally regulated in cortical cultures co-purifying with polyribosomes, which suggests that they may control localized translation of dendrite-specific mRNAs, thereby possibly contributing to the molecular basis of learning and memory.[108] This hypothesis has recently gained support from a study which showed that miR-134, a brain-specific miRNA, is present in dendrites, where it represses the local synthesis of the protein kinase LimK1 to regulate spine size.[109]

The number of regulatory mRNA targets of vertebrate miRNAs has been estimated by identifying conserved complementarity to the miRNA seed sequences (nucleotide 2–7 of the miRNA), suggesting that ~30% of the human genes may be miRNA targets.[110] Computational predictions in *Drosophila* provide evidence that a given miRNA has on average ~100 mRNA target sites in the fly, while another recent study reported that vertebrate miRNAs target ~200 mRNAs each, further supporting the notion that miRNAs can regulate the expression of a large fraction of the protein-coding genes in multicellular eukaryotes.[111,112] Most recent reports indicate that miRNAs may not function as developmental switches, but rather play a role in maintaining tissue identity by conferring accuracy to gene-expression programs.[96,103,113–115]

5.4.3.1 MicroRNAs in Human Disease

The expanding inventory of human miRNAs along with their highly diverse expression patterns and high number of potential target mRNAs suggest that miRNAs are involved in a wide variety of human diseases. One is spinal muscular atrophy (SMA), a paediatric neurodegenerative disease caused by reduced protein levels or loss-of-function mutations of the survival of motor neurons (SMN) gene.[116] A mutation in the target site of miR-189 in the human *SLITRK1* gene was recently shown to be associated with Tourette's syndrome,[117] while another recent study reported that the hepatitis C virus (HCV) RNA genome interacts with a host-cell miRNA, the liver-specific miR-122a, to facilitate its replication in the host.[118] Other diseases in which miRNAs or their processing machinery have been implicated include fragile X mental retardation (FXMR) caused by absence of the fragile X mental retardation protein (FMRP),[119,120] DiGeorge syndrome,[121] human immunodeficiency virus (HIV) replication[122] and coronary artery disease.[123] In addition, perturbed miRNA expression patterns have been reported in many human cancers. For example, the human miRNA genes miR15a and miR16-1 are deleted or downregulated in the majority of B-cell CLL cases, where a unique signature of 13 miRNA genes was shown to associate with prognosis and progression.[124,125] The role of miRNAs in cancer is further supported by the fact that more than 50% of the human miRNA genes are located in cancer-associated genomic regions or at fragile sites.[126] Systematic expression analysis of a diversity of human cancers revealed a general downregulation of miRNAs in tumours compared to normal tissues.[127] Interestingly, miRNA-based classification of poorly differentiated tumours was successful, whereas mRNA profiles were highly inaccurate when applied to the same samples. miRNAs have also been shown to be deregulated in breast cancer,[128] lung cancer[129] and colon cancer,[130] while the miR-17-92 cluster, which is amplified in human B-cell lymphomas, and miR-155, which is upregulated in Burkitt's lymphoma, have been identified as the first human miRNA oncogenes.[131,132] Hence, disease-associated human miRNAs could represent a novel group of viable targets for therapeutic intervention.

5.4.3.2 MicroRNA Targeting Using Locked Nucleic Acid

Detection and analysis of the miRNAs is not trivial, and the small size and sometimes low level of expression of different miRNAs require the use of sensitive and specific research tools. Valoczi *et al.*[133] described in 2004 a highly efficient detection of miRNAs by Northern blot analysis using LNA-modified

ON probes and demonstrated their significantly improved sensitivity by detecting different miRNAs in animals and plants. Besides being efficient as Northern probes, LNA-modified ONs have proved highly useful for *in situ* localization of miRNAs in cells and tissues. The spatial and temporal expression patterns of 115 conserved vertebrate miRNAs has been determined directly by whole-mount *in situ* hybridizations with LNA-modified DNA probes similar to those used for Northern blots[133] on embryos and larvae of different developmental stages.[103] The *in situ* patterns correlated well with zebrafish miRNA array data and with the available miRNA expression data for mammals. Temporal accumulation of miRNAs in developing zebrafish embryos suggested that miRNAs may not directly act as developmental switches, but are involved in the maintenance of tissue identity.[103]

In a subsequent study, Kloosterman *et al.* (2006)[134] describe a detailed analysis on the conditions for LNA probe-based *in situ* detection of miRNAs in zebrafish embryos, and demonstrate the utility of the method in Xenopus and mouse embryos. Using LNA-modified probes for miR-206, miR-124a and miR-122a in 72-hour-old zebrafish embryos revealed the expected patterns for all three miRNAs, whereas no *in situ* signals could be detected with the corresponding DNA probes, in accordance with the significantly increased affinity of LNA ONs towards complementary RNA molecules. For miR-122a and miR-206, specific *in situ* staining was lost upon introduction of a single central mismatch in the LNA probe, whereas two central mismatches were needed for adequate discrimination for miR-124a. Notably, specific detection of miR-206 and miR-124a could be achieved using shortened LNA versions complementary to a 14-nt region at the 5'-end of the miRNA, thereby improving their specificity.[134] The successful use of very short LNA ONs has significant implications for therapeutic applications, such as inhibition of disease-related miRNAs by LNA-antimiRs.

5.4.3.3   *In vitro Inhibition of MicroRNA Function Using LNA–AntimiRs*

The small size of miRNA genes makes it difficult to create loss-of-function mutants for functional analysis. Another potential problem is that many miRNA genes are present in several copies per genome occurring in different loci, which makes it even more difficult to obtain mutant phenotypes. An alternative approach to creating miRNA gene knock-outs has been reported by Hutvagner *et al.* (2004)[135] and Leaman *et al.* (2005),[136] in which 2'-*O*-Me AS ONs were used as potent and irreversible inhibitors of siRNA and miRNA function *in vitro* and *in vivo* in *D. melanogaster* and *Caenorhabditis elegans*, thereby inducing a loss-of-function phenotype. Recent studies have reported that LNA-modified ONs can also mediate specific inhibition of miRNA function *in vitro*.[137–139]

Antagonism of miR-21 using LNA–DNA-mixed AS ONs in cultured glioblastoma cells triggered activation of caspases and lead to increased apoptotic cell death.[137] In this study, LNA AS inhibitors were used in parallel with the 2'-*O*-Me-ONs to inhibit miR-21 and showed similar efficacy and specificity in the same range of concentrations when transfected with the same type of cationic liposomes.[137] Using the well-characterized interaction between the *D. melanogaster* bantam miRNA and its target gene hid as a model, Ørom *et al.* (2006)[139] have described the efficacy and specificity of the LNA-based silencing method. LNA-antimiRs could readily inhibit exogenously introduced miRNAs with high specificity and, furthermore, inhibit endogenous bantam in *D. melanogaster* cells, leading to upregulation of its cognate target protein hid. The method showed stoichiometric and reliable inhibition of the targeted miRNA and would thus be applicable to functional analysis of miRNAs and validation of putative target genes.[139]

Lecellier *et al.* (2005)[138] used LNA-antimiRs to inhibit the human miR-32 that was shown to effectively limit primate foamy virus type 1 (PFV-1) replication. To assess the antiviral effect of miR-32 they designed LNA-antimiRs against miR-32, and when the LNA-antimiR was co-transfected with PFV-1 in HeLa and in baby hamster kidney 21 (BHK-21) cells the translational inhibition by miR-32 was significantly reduced.[138] At 10 nM LNA-antimiR concentrations the antiviral effect of miR-32 was lost, leading to accumulation of PFV-1.

5.4.3.4 Antagonizing MicroRNAs for Therapeutics

The aforementioned studies reporting that LNA-modified ONs can mediate potent and specific inhibition of miRNA function *in vitro*, along with the improved miRNA recognition properties of LNA in Northern blot analyses and *in situ* hybridization, suggest that LNA ONs might be useful in antagonizing miRNAs *in vivo*.[103,133,134] A major challenge in understanding the biology of miRNAs is to identify their target mRNAs. Although computational analyses suggest that a given miRNA may have hundreds of targets,[110,112,140] only a limited number has been experimentally confirmed.[141] Microarray expression profiling has been used to detect genes downregulated in response to exogenous miRNAs.[114] However, the introduction of an exogenous miRNA into cells that do not normally express it may lead to the identification of non-physiological targets. In contrast, specific inhibition of an endogenous miRNA *in vivo* has the potential to pinpoint the physiological targets and their sequence determinants.

In a recent study, Elmén *et al.* found that a systemically administered short, unconjugated LNA-antimiR ON complementary to the liver-specific miRNA-122 is effective in the silencing of miR-122 in mice.[142] Antagonism of miR-122 is potent, dose-dependent and due to the formation of stable heteroduplexes between the LNA-antimiR and miR-122. Functional antagonism of miR-122 was inferred from a low cholesterol phenotype and de-repression within 24 hours of 199 predicted target genes having at least one six-nucleotide miR-122 seed match in their 3′-untranslated regions (3′-UTRs). Interestingly, within the different variants of seeds, significant enrichment of 3.9-fold for de-repressed genes harbouring seven nt seeds with M8 matches was observed.[142] Expression profiling extended to three weeks after the last LNA-antimiR dose revealed that most of the changes in liver-gene expression were normalized to control levels coinciding with normalized miR-122 and plasma cholesterol levels.

Previously, cholesterol conjugated 2′-*O*-Me AS ONs (antago-mirs) have been used to silence miRNAs, including miR-122.[143] In a similar study, unconjugated 2′-MOE AS ONs were used to inhibit miR-122 in both normal and diet-induced obese mice.[144] Both studies reported a low plasma cholesterol loss-of-function phenotype.[143,144] Antagonism of miR-122 by LNA-antimiR confirmed this phenotype, resulting in reduction of plasma cholesterol by 40%.[142] The decrease in plasma cholesterol levels was maximal one week post-treatment and then returned slowly to baseline. Thus, the observed changes in gene expression appear to precede reduction in cholesterol, which, in turn, appears longer lasting than changes in miR-122 target gene expression. This suggests that coordinated changes in miR-122-associated gene networks are responsible for the control of cholesterol and lipid metabolism in the liver. It is therefore tempting to speculate that, in human therapeutics, to affect an entire metabolic network in this way may have advantages over the inhibition of a single metabolic enzyme.

In conclusion, the recent findings by Elmén *et al.*[142] show that the high affinity antagonism of miR-122 in mice by steric hindrance using an LNA-antimiR ON can result in potent functional inhibition suggesting that ONs that comprise LNA may be valuable tools for identifying miRNA targets *in vivo* and for studying the biological role of miRNAs and miRNA-associated gene-regulatory networks in a physiological context. In addition, the high metabolic stability of LNA-antimiRs, due in part to increased nuclease resistance, their small size and apparent lack of acute toxicity or changes in liver morphology[142] imply that LNA-antimiRs may be well-suited as a novel class of potential therapeutics for disease-associated miRNAs.

5.5   Clinical Phase I/II Study of SPC2996

Data from the initial Phase I/II clinical trial of the Bcl-2 mRNA-directed ON, SPC2996, in patients with advanced CLL was presented at the 2007 Annual Meeting of the American Society of Clinical Oncology. Prior to this first clinical study of SPC2996, comprehensive GLP toxicology studies were conducted. A detailed presentation of the preclinical clinical trial application (CTA) and investigational new drug (IND) supporting studies is beyond the scope of this chapter, but a brief outline of the results is included below. For a more detailed account the interested reader can consult Koch and Ørum.[33]

SPC2996 was well-tolerated in rodents and Cynomolgus monkeys at concentrations substantially above clinically relevant doses. In the key organs for SPC2996 uptake – kidney and liver – the compound showed relatively low toxicological potential. As an illustration of this, 10 out of 10 monkeys showed no increase in alanine or serine aminotransferase activities in plasma after repeated doses of $20\,mg\,kg^{-1}$. Even after repeated administration of $60\,mg\,kg^{-1}$, 9 out of 10 animals showed rises of no more than twice the upper limit of normal. Notably, even very high plasma concentrations of SPC2996 did not produce clinical signs of the acute toxicities (complement activation and prolongation of bleeding times) previously described for other classes of ONs. Low doses of SPC2996 (3 and $6\,mg\,kg^{-1}$) showed an effective and long-lived reduction in Bcl-2 mRNA and protein in Cynomolgus monkey livers. Furthermore, SPC2996 exhibited plasma half-lives in primates of between 1.2 and 13 hours, depending on level and frequency of dosing. However, once taken up by tissues, the compound was eliminated much more slowly with tissue half-lives of approximately eight days (mice) to several weeks (monkeys; unpublished data).

5.5.1 Clinical Phase I/II Data

The Phase I/II study was designed as an open label, international, multi-centre, dose escalation trial in Denmark, France, the UK and the US, in patients with refractory or relapsed CLL. The main objectives for the Phase I/II study of SPC2996 – targeting Bcl-2 – were to investigate the safety and to define the biological active dose. The primary endpoints were to define the maximum tolerated repeat dose (MTD) when given approximately every other day for two weeks, and to measure the drug-induced changes in the biological surrogate marker, mBcl-2. Clinical assessment parameters, including measurement of the height of the leukaemia and changes in lymph-node swelling, were regarded as secondary endpoints. Patients received SPC2996 by intravenous infusion in saline over a two hour period on six occasions within two weeks. The patients were studied in five cohorts (A to E) administered escalating doses, respectively, of 0.2, 0.5, 1.0, 2.0 and $4.0\,mg\,kg^{-1}$ on each occasion. Here only the data from the top dose are presented.

Bcl-2 expression levels were determined by quantitative polymerase chain reaction (qPCR) analysis on RNA-purified whole blood samples. The pooled patient data at each time point for Bcl-2 expression levels are presented relative to 18S rRNA (total RNA). Regression analysis of the data in Group E ($4\,mg\,kg^{-1}$) is shown in Figure 5.13. A statistically significant downward trend in Bcl-2 mRNA levels was observed over the course of the two week treatment period.

The effect of the drug on the total lymphocyte counts was examined in the same patient group (Figure 5.14). The normal lymphocyte level is approximately $5 \times 10^9\,L^{-1}$, whereas the CLL patients in this study had elevated lymphocyte counts on entry ranging from 10 to about $60 \times 10^9\,L^{-1}$. Remarkably, six out of six patients receiving the drug at $4\,mg\,kg^{-1}$ showed a rapid reduction in lymphocyte levels within 24 hours after the first dose and five out of six patients went on to achieve more than 50% reduction. Interestingly, the patients with the highest lymphocyte counts on commencement of treatment showed the strongest response.

It was furthermore observed in Group E that two patients had partial tumour response (PR) as defined as $>50\%$ reduction in lymph-node diameter, and three patients had stable disease. One of the patients with PR was ongoing

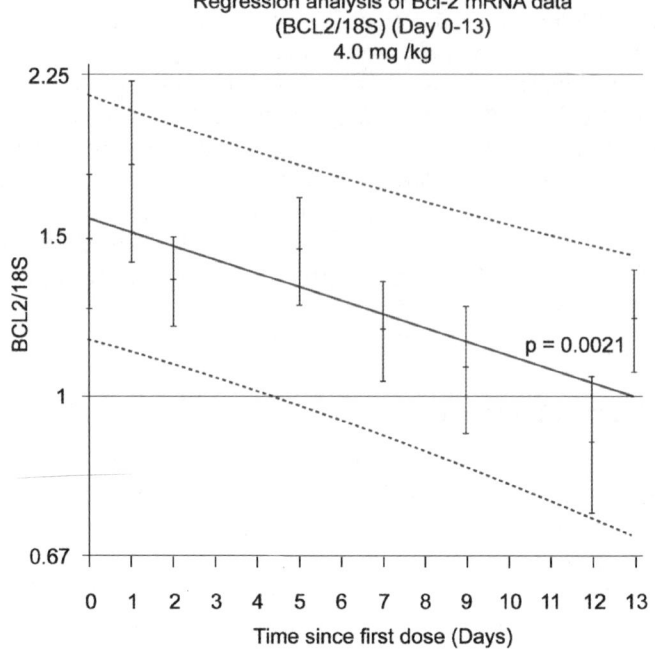

Figure 5.13   Regression analyses of Bcl-2 expression for dosis group 4 mg/kg show downregulation of Bcl-2 with statistical significance.

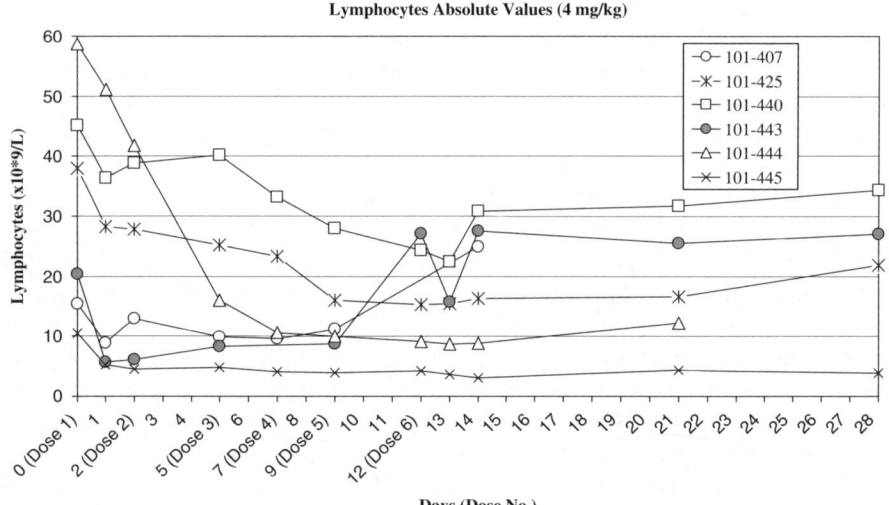

Figure 5.14   Absolute lymphocyte counts for dosis group 4 mg/kg.

after the end of the study, and the other had PR between 28 and 42 days, continuing with stable disease for four months. Overall, the time to progression in Group E was determined to be 122 days.

These clinical data are the first to indicate efficacy of a LNA ON – SPC2996 – in humans. It is emphasized that the data are early and based on a small number of patients. It is interesting to note that statistically significant down-regulation of Bcl-2 mRNA was measured during the dosing period which indicates that mRNA inhibition is causally associated with the clinical observations.

For RNA inhibitors it is important to establish the link between target downregulation and clinical efficacy. In the past such a connection has been difficult to prove with earlier generations of AS drugs. The combination of target downregulation and clinical efficacy observed in this trial is likely to be at least in part attributable to the improved properties offered by LNA.

5.6 Concluding Remarks

In this chapter we have shown that the basic properties of LNA are well suited to produce potent single-stranded RNA inhibitors. The high affinity of LNA offers, for the first time, a real opportunity to combine size reduction with high potency and efficacy. A result of this is shown here where the highest efficacy in targeting ApoB-100 is obtained with LNA 12- and 14-mers. Such short, highly efficacious ONs open a new perspective for hybridization-based drugs. The data are, furthermore, a confirmation that the combination of LNA nt's and PS in chimeric shortmers combine the 'the best of the two worlds': High potency, biostability and size reduction offered by LNA substitutions – with improved bio-availability and cellular uptake offered by the PS.

The marginal cellular uptake experienced with siRNA is sought by introducing complex and expensive conjugations or lipid formulations.[33,67,68] By using single-stranded LNA ONs an improved efficacy is obtained by the exact opposite strategy: use less complex – and less expensive molecules – shortmers.

Lately, the recognition of miRNA as a new important class of targets during pathogenesis has added new perspectives to ON-based drugs and further underlined their need. In contrast to siRNA, which is only useful for mRNA silencing, LNA ONs can equally well be used to inhibit miRNA expression. We have shown here how the design flexibility of LNA can be transferred to produce highly potent miRNA inhibitors and how the shortmer concept can also be used to make LNA-antimiRs with *in vivo* efficacies that are unrivalled in the field.

The clinical data obtained by SPC2996 illustrate that the use of LNA in humans can downregulate a disease-associated mRNA target and elicit a clinical response. Besides being a proof of the concept for the future therapeutic potential of LNA, the clinical results represent a step forward in the development of better treatments against Bcl-2 mediated malignancies.

Finally, we hope that this chapter illustrates that the properties of LNA – and the technologies developed around it – are positioning LNA as a unique single-stranded RNA inhibitor. Taken together, all of the data presented confirm that LNA in our eyes represents a transformational new platform in the field of RNA inhibition and may thus realise a broader therapeutic potential of single stranded ONs.

Acknowledgements

We are grateful to Keith McCullagh and Zenia M. Størling for their valuable suggestions and comments during the review process of the manuscript. Part of the work described in this chapter was performed by our colleagues at Santaris Pharma and has yet to be published. We thank them for their contributions and assistance, all of which are greatly appreciated. The structure of the 9-mer LNA–RNA heteroduplex (Figure 5.3) was kindly provided by Associate Professor Michael Petersen at the University of Southern Denmark.

References

1. P. C. Zamecnik and M. L. Stephenson, *Proc. Natl. Acad. Sci. U. S. A.*, 1978, **75**, 280.
2. A. A. Levin, *Biochim. Biophys. Acta*, 1999, **1489**, 69.
3. P. Martin, *Helv. Chim. Acta*, 1995, **78**, 486.
4. Susan M. Freir and Karl-Heinz Altmann, *Nucleic Acid Res.*, 1997, **25**, 4429.
5. A. Koshkin, S. K. Singh, P. Nielsen, V. K. Rajwanshi, R. Kumar, M. Meldgaard, C. E. Olsen and J. Wengel, *Tetrahedron*, 1998, **54**, 3607.
6. S. K. Singh, P. Nielsen, A. Koshkin and J. Wengel, *Chem. Commun.*, 1998, 455.
7. J. Wengel, *Acc. Chem. Res.*, 1999, **32**, 301.
8. M. Bolli, J. C. Litten, R. Schutz and C. J. Leumann, *Chem. Biol.*, 1996, **3**, 197.
9. P. Herdewijn, *Liebigs Ann.*, 1996, **1**, 1337.
10. J. Wang and M. D. Matteucci, *Bioorg. Med. Chem. Lett.*, 1997, **7**, 229.
11. R. Kumar, S. K. Singh, A. A. Koshkin, V. K. Rajwanshi, M. Meldgaard and J. Wengel, *Bioorg. Med. Chem. Lett.*, 1998, **8**, 2219.
12. S. K. Singh, R. Kumar and J. Wengel, *J. Org. Chem.*, 1998, **63**, 10035.
13. J. Wengel, A. Koshkin, S. K. Singh, P. Nielsen, M. Meldgaard, V. K. Rajwanshi, R. Kumar, J. Skouv, C. B. Nielsen, N. Jakobsen and C. E. Olsen, *Nucleosides Nucleotides*, 1999, **18**, 1365.
14. M. D. Sørensen, M. Petersen and J. Wengel, *Chem. Commun.*, 2003, 2130.
15. P. A. Hralicka, B. R. Babu, M. D. Sørensen, N. Harrit and J. Wengel, *J. Am. Chem. Soc.*, 2005, **127**, 13293.
16. P. J. Hrdlicka, B. R. Babu, M. D. Sørensen and J. Wengel, *Chem. Commun.*, 2004, 1478.

17. B. R. Babu, P. J. Hrdlicka, C. J. McKenzie and J. Wengel, *Chem. Commun.*, 2005, 1705.
18. V. K. Rajwanshi, A. E. Haakansson, R. Kumar and J. Wengel, *Chem. Commun.*, 1999, 2073.
19. V. K. Rajwanshi, A. E. Haakansson, B. M. Dahl and J. Wengel, *Chem. Commun.*, 1999, 1395.
20. V. K. Rajwanshi, A. E. S. M. D. Håkansson, S. Pitsch, S. K. Singh, R. Kumar, P. Nielsen and J. Wengel, *Angew. Chem. Int. Ed.*, 2000, **39**, 1656.
21. J. Wengel, M. Petersen, K. E. Nielsen, G. A. Jensen, A. E. Hakansson, R. Kumar, M. D. Sørensen, V. K. Rajwanshi, T. Bryld and J. P. Jacobsen, *Nucleosides Nucleotides Nucleic Acids*, 2001, **20**, 389.
22. M. Frieden, S. M. Christensen, N. D. Mikkelsen, C. Rosenbohm, C. A. Thrue, M. Westergaard, H. F. Hansen, H. Ørum and T. Koch, *Nucleic Acids Res.*, 2003, **31**, 6365.
23. M. D. Sørensen, L. Kvaerno, T. Bryld, A. E. Håkansson, B. Verbeure, G. Gaubert, P. Herdewijn and J. Wengel, *J. Am. Chem. Soc.*, 2002, **124**, 2164.
24. K. Fluiter, M. Frieden, J. Vreijling, C. Rosenbohm, M. B. De Wissel, S. M. Christensen, T. Koch, H. Ørum and F. Baas, *Chembiochem*, 2005 **6**, 1104.
25. N. Kumar, K. E. Nielsen, S. Maiti and M. Petersen, *J. Am. Chem. Soc.*, 2006, **128**, 14.
26. P. J. Hrdlicka, T. S. Kumar and J. Wengel, *Chem. Commun.*, 2005, 4279.
27. C. Rosenbohm, D. S. Pedersen, M. Frieden, F. R. Jensen, S. Arent, S. Larsen and T. Koch, *Bioorg. Med. Chem.*, 2004, **12**, 2385.
28. A. A. Koshkin, *J. Org. Chem.*, 2004, **69**, 3711.
29. B. R. Babu, Raunak, N. E. Poopeiko, M. Juhl, A. D. Bond, V. S. Parmar and J. Wengel, *Eur. J. Org. Chem.*, 2005, **11**, 2297.
30. B. R. Babu and J. Wengel, *Chem. Commun.*, 2001, **20**, 2114.
31. B. R. Babu, A. K. Prasad, S. Trikha, N. Thorup, V. S. Parmar and J. Wengel, *J. Chem. Soc., Perkin Trans. 1*, 2002, 2509.
32. Raunak, B. B. Ravindra, M. D. Sørensen, V. S. Parmar, N. H. Harrit and J. Wengel, *Org. Biomol. Chem.*, 2004, **2**, 80.
33. T. Koch and H. Ørum, in *Antisense Drug Technologies; Principles, Strategies and Applications*, ed. S. T. Crooke, Taylor & Francis Group, Oxford, 2007, Vol. 19, p. 519.
34. D. Latorra, K. Campbell, A. Wolter and J. M. Hurley, *Hum. Mutat.*, 2003, **22**, 79.
35. A. Valoczi, C. Hornyik, N. Varga, J. Burgyan, S. Kauppinen and Z. Havelda, *Nucleic Acids Res.*, 2004, **32**, e175.
36. W. P. Kloosterman, E. Wienholds, E. de Bruijn, S. Kauppinen and R. H. Plasterk, *Nat. Methods*, 2006, **3**, 27.
37. U. A. Ørum, S. Kauppinen and A. H. Lund, *Gene*, 2006, **372**, 137.
38. N. Jacobsen, J. Bentzen, M. Meldgaard, M. H. Jakobsen, M. Fenger, S. Kauppinen and J. Skouv, *Nucleic Acids Res.*, 2002, **30**, e100.

39. H. Ørum, M. H. Jacobsen, T. Koch, J. Vuust and M. B. Borre, *Clin. Chem.*, 1999, **45**, 1898.

40. S. Obika, *Chem. Pharm. Bull. (Tokyo)*, 2004, **52**, 1399.

41. K. A. Brameld and W. A. Goddard, *J. Am. Chem. Soc.*, 1999, **121**, 985.

42. S. Obika, D. Nanbu, Y. Hari, J. A. K. Morio, Y. In, T. Ishida and T. Imanishi, *Tetrahedron Lett.*, 1997, **38**, 8735.

43. M. Egli, G. Minasov, M. Teplova, R. Kumar and J. Wengel, *Chem. Commun.*, 651.

44. S. Obika, D. Nanbu, Y. Hari, J. A. K. Morio, T. Doi and T. Imanishi, *Tetrahedron Lett.*, 1998, **39**, 5401.

45. K. E. Nielsen, S. K. Singh, J. Wengel and J. P. Jacobsen, *Bioconjugate Chem.*, 2000, **11**, 228.

46. K. E. Nielsen, J. Rasmussen, R. Kumar, J. Wengel, J. P. Jacobsen and M. Petersen, *Bioconjugate Chem.*, 2004, **15**, 449.

47. M. Petersen, C. B. Nielsen, K. E. Nielsen, G. A. Jensen, K. Bondensgaard, S. K. Singh, V. K. Rajwanshi, A. A. Koshkin, B. M. Dahl, J. Wengel and J. P. Jacobsen, *J. Mol. Recognit.*, 2000, **13**, 44.

48. M. Petersen, K. Bondensgaard, J. Wengel and J. P. Jacobsen, *J. Am. Chem. Soc.*, 2002, **124**, 5974.

49. A. Koshkin, P. Nielsen, M. Meldgaard, V. K. Rajwanshi, S. K. Singh and J. Wengel, *J. Am. Chem. Soc.*, 1998, **120**, 13252.

50. S. K. Singh and J. Wengel, *Chem. Commun.*, 1998, 1247.

51. E. Kierzek, A. Ciesielska, K. Pasternak, D. H. Mathews, D. H. Turner and R. Kierzek, *Nucleic Acids Res.*, 2005, **33**, 5082.

52. S. Davis, B. Lollo, S. Freier and C. Esau, *Nucleic Acids Res.*, 2006, **34**, 2294.

53. P. M. McTigue, R. J. Peterson and J. D. Kahn, *Biochemistry*, 2004, **43**, 5388.

54. G. A. Jensen, S. K. Singh, R. Kumar, J. Wengel and J. P. Jacobsen, *J. Chem Soc., Perkin Trans. 2*, 2001, 1224.

55. U. Christensen, N. Jacobsen, V. K. Rajwanshi, J. Wengel and T. Koch, *Biochem. J.*, 2001, **354**, 481.

56. L. Kvaerno and J. Wengel, *Chem. Commun.*, 1999, 657.

57. L. Kvaerno, R. Kumar, B. M. Dahl, C. E. Olsen and J. Wengel, *J. Org. Chem.*, 2000, **65**, 5167.

58. H. Kaur, A. Arora, J. Wengel and S. Maiti, *Biochemistry*, 2006, **45**, 7347.

59. C. Wahlestedt, P. Salmi, L. Good, J. Kela, T. Johnsson, T. Hokfelt, C. Broberger, F. Porreca, J. Lai, K. Ren, M. Ossipov, A. Koshkin, N. Jakobsen, J. Skouv, H. Ørum, M. H. Jacobsen and J. Wengel, *Proc. Natl. Acad. Sci. U. S. A.*, 2000, **97**, 5633.

60. M. Frieden, H. F. Hansen and T. Koch, *Nucleosides Nucleotides Nucleic Acids*, 2003, **22**, 1041.

61. K. Morita, C. Hasegawa, M. Kaneko, S. Tsutsumi, J. Sone, T. Ishikawa, T. Imanishi and M. Koizumi, *Bioorg. Med. Chem. Lett.*, 2001, **12**, 73.

62. S. Obika, R. Hemamayi, T. Masuda, T. Sugimoto, S. Nakagawa, T. Mayumi and T. Imanishi, *Nucleic Acids Res. Suppl.*, 2001, **1**, 145.
63. J. Kurreck, E. Wyszko, C. Gillen and V. A. Erdmann, *Nucleic Acids Res.*, 2002, **30**, 1911.
64. D. S. Pedersen, C. Rosenbohm and T. Koch, *Synthesis*, 2002, 802.
65. R. S. Geary, T. A. Watanabe, L. Truong, S. M. Freier, E. A. Lesnik, N. B. Sioufi, H. Sasmot, M. Manoharan and A. A. Levin, *J. Pharmacol. Exp. Ther.*, 2001, **296**, 890.
66. A. Soutschek, A. Aakinc and K. C. R. Bramlage, *Nature*, 2004, **432**, 173.
67. J. Soutschek, A. Akinc, B. Bramlage, K. Charisse, R. Constien, M. Donoghue, S. Elbashir, A. Geick, P. Hadwiger, J. Harborth, M. John, V. Kesavan, G. Lavine, R. K. Pandey, T. Racie, K. G. Rajeev, I. Rohl, I. Toudjarska, G. Wang, S. Wuschko, D. Bumcrot, V. Koteliansky, S. Limmer, M. Manoharan and H. P. Vornlocher, *Nature*, 2004, **432**, 173.
68. T. S. Zimmermann, A. C. Lee, A. Akinc, B. Bramlage, D. Bumcrot, M. N. Fedoruk, J. Harborth, J. A. Heyes, L. B. Jeffs, M. John, A. D. Judge, K. Lam, K. McClintock, L. V. Nechev, L. R. Palmer, T. Racie, I. Rohl, S. Seiffert, S. Shanmugam, V. Sood, J. Soutschek, I. Toudjarska, A. J. Wheat, E. Yaworski, W. Zedalis, V. Koteliansky, M. Manoharan, H. P. Vornlocher and I. MacLachlan, *Nature*, 2006, **441**, 111.
69. F. Li, G. Ambrosini, E. Y. Chu, J. Plescia, S. Tognin, P. C. Marchisio and D. C. Altieri, *Nature*, 1998, **396**, 580.
70. J. Chen, W. Wu, S. K. Tahir, P. E. Kroeger, S. H. Rosenberg, L. M. Cowsert, F. Bennett, S. Krajewski, M. Krajewska, K. Welsh, J. C. Reed and S. C. Ng, *Neoplasia*, 2000, **2**, 235.
71. K. L. Talks, H. Turley, K. C. Gatter, P. H. Maxwell, C. W. Pugh, P. J. Ratcliffe and A. L. Harris, *Am. J. Pathol.*, 2000, **157**, 411.
72. H. Zhong, A. M. De Marzo, E. Laughner, M. Lim, D. A. Hilton, D. Zagzag, P. Buechler, W. B. Isaacs, G. L. Semenza and J. W. Simons, *Cancer Res.*, 1999, **59**, 5830.
73. D. Lando, J. J. Gorman, M. L. Whitelaw and D. J. Peet, *Eur. J Biochem.*, 2003, **270**, 781.
74. H. Ørum, *IEEE Eng. Med. Biol. Mag.*, 2005, **24**, 81.
75. E. Z. Yu, Y. Y. Li, X. H. Liu, E. Kagan and R. M. McCarron, *Lab. Invest.*, 2004, **84**, 553.
76. J. C. Reed, *Oncology (Williston. Park)*, 2004, **18**, 11.
77. J. C. Reed, *Curr. Opin. Oncol.*, 1995, **7**, 541.
78. K. N. Chi, M. E. Gleave, R. Klasa, N. Murray, C. Bryce, D. E. Lopes de Menezes, S. D'Aloisio and A. W. Tolcher, *Clin. Cancer Res.*, 2001, **7**, 3920.
79. B. Jansen, V. Wacheck, E. Heere-Ress, H. Schlagbauer-Wadl, C. Hoeller, T. Lucas, M. Hoermann, U. Hollenstein, K. Wolff and H. Pehamberger, *Lancet*, 2000, **356**, 1728.

80. D. P. Bartel, *Cell*, 2004, **116**, 281.
81. G. Hutvagner, J. McLachlan, A. E. Pasquinelli, E. Balint, T. Tuschl and P. D. Zamore, *Science*, 2001, **293**, 834.
82. R. F. Ketting, S. E. Fischer, E. Bernstein, T. Sijen, G. J. Hannon and R. H. Plasterk, *Genes Dev.*, 2001, **15**, 2654.
83. Y. Lee, C. Ahn, J. Han, H. Choi, J. Kim, J. Yim, J. Lee, P. Provost, O. Radmark, S. Kim and V. N. Kim, *Nature*, 2003, **425**, 415.
84. S. Griffiths-Jones, R. J. Grocock, S. van Dongen, A. Bateman and A. J. Enright, *Nucleic Acids Res.*, 2006, **34**, D140.
85. I. Bentwich, A. Avniel, Y. Karov, R. Aharonov, S. Gilad, O. Barad, A. Barzilai, P. Einat, U. Einav, E. Meiri, E. Sharon, Y. Spector and Z. Bentwich, *Nat. Genet.*, 2005, **37**, 766.
86. E. Berezikov, V. Guryev, B. J. van de, E. Wienholds, R. H. Plasterk and E. Cuppen, *Cell*, 2005, **120**, 21.
87. X. Xie, J. Lu, E. J. Kulbokas, T. R. Golub, V. Mootha, K. Lindblad-Toh, E. S. Lander and M. Kellis, *Nature*, 2005, **434**, 338.
88. J. Brennecke, D. R. Hipfner, A. Stark, R. B. Russell and S. M. Cohen, *Cell*, 2003, **113**, 25.
89. C. Z. Chen, L. Li, H. F. Lodish and D. P. Bartel, *Science*, 2004, **303**, 83.
90. M. Boehm and F. Slack, *Science*, 2005, **310**, 1954.
91. X. Li and R. W. Carthew, *Cell*, 2005, **123**, 1267.
92. E. Hornstein, J. H. Mansfield, S. Yekta, J. K. Hu, B. D. Harfe, M. T. McManus, S. Baskerville, D. P. Bartel and C. J. Tabin, *Nature*, 2005, **438**, 671.
93. S. Yekta, I. H. Shih and D. P. Bartel, *Science*, 2004, **304**, 594.
94. R. J. Johnston and O. Hobert, *Nature*, 2003, **426**, 845.
95. M. N. Poy, L. Eliasson, J. Krutzfeldt, S. Kuwajima, X. Ma, P. E. Macdonald, S. Pfeffer, T. Tuschl, N. Rajewsky, P. Rorsman and M. Stoffel, *Nature*, 2004, **432**, 226.
96. A. J. Giraldez, R. M. Cinalli, M. E. Glasner, A. J. Enright, J. M. Thomson, S. Baskerville, S. M. Hammond, D. P. Bartel and A. F. Schier, *Science*, 2005, **308**, 833.
97. J. F. Chen, E. M. Mandel, J. M. Thomson, Q. Wu, T. E. Callis, S. M. Hammond, F. L. Conlon and D. Z. Wang, *Nat. Genet.*, 2006, **38**, 228.
98. C. Kwon, Z. Han, E. N. Olson and D. Srivastava, *Proc. Natl. Acad. Sci. U. S. A.*, 2005, **102**, 18986.
99. N. S. Sokol and V. Ambros, *Genes Dev.*, 2005, **19**, 2343.
100. Y. Zhao, E. Samal and D. Srivastava, *Nature*, 2005, **436**, 214.
101. A. Rodriguez, E. Vigorito, S. Clare, M. V. Warren, P. Couttet, D. R. Soond, S. van Dongen, R. J. Grocock, P. P. Das, E. A. Miska, D. Vetrie, K. Okkenhaug, A. J. Enright, G. Dougan, M. Turner and A. Bradley, *Science*, 2007, **316**, 608.
102. T. H. Thai, D. P. Calado, S. Casola, K. M. Ansel, C. Xiao, Y. Xue, A. Murphy, D. Frendewey, D. Valenzuela, J. L. Kutok, M. Schmidt-Supprian, N. Rajewsky, G. Yancopoulos, A. Rao and K. Rajewsky, *Science*, 2007, **316**, 604.

103. E. Wienholds, W. P. Kloosterman, E. Miska, E. Alvarez-Saavedra, E. Berezikov, E. de Bruijn, H. R. Horvitz, S. Kauppinen and R. H. Plasterk, *Science*, 2005, **309**, 310.

104. E. A. Miska, E. Alvarez-Saavedra, M. Townsend, A. Yoshii, N. Sestan, P. Rakic, M. Constantine-Paton and H. R. Horvitz, *Genome Biol.*, 2004, **5**, R68.

105. A. M. Krichevsky, K. S. King, C. P. Donahue, K. Khrapko and K. S. Kosik, *RNA*, 2003, **9**, 1274.

106. L. F. Sempere, S. Freemantle, I. Pitha-Rowe, E. Moss, E. Dmitrovsky and V. Ambros, *Genome Biol.*, 2004, **5**, R13.

107. L. Smirnova, A. Grafe, A. Seiler, S. Schumacher, R. Nitsch and F. G. Wulczyn, *Eur. J. Neurosci.*, 2005, **21**, 1469.

108. J. Kim, A. Krichevsky, Y. Grad, G. D. Hayes, K. S. Kosik, G. M. Church and G. Ruvkun, *Proc. Natl. Acad. Sci. U. S. A.*, 2004, **101**, 360.

109. G. M. Schratt, F. Tuebing, E. A. Nigh, C. G. Kane, M. E. Sabatini, M. Kiebler and M. E. Greenberg, *Nature*, 2006, **439**, 283.

110. B. P. Lewis, C. B. Burge and D. P. Bartel, *Cell*, 2005, **120**, 15.

111. J. Brennecke, A. Stark, R. B. Russell and S. M. Cohen, *PLoS Biol.*, 2005, **3**, e85.

112. A. Krek, D. Grun, M. N. Poy, R. Wolf, L. Rosenberg, E. J. Epstein, P. MacMenamin, P. da, I. K. C. Gunsalus, M. Stoffel and N. Rajewsky, *Nat. Genet.*, 2005, **37**, 495.

113. K. K. Farh, A. Grimson, C. Jan, B. P. Lewis, W. K. Johnston, L. P. Lim, C. B. Burge and D. P. Bartel, *Science*, 2005, **310**, 1817.

114. L. P. Lim, N. C. Lau, P. Garrett-Engele, A. Grimson, J. M. Schelter, J. Castle, D. P. Bartel, P. S. Linsley and J. M. Johnson, *Nature*, 2005, **433**, 769.

115. A. Stark, J. Brennecke, N. Bushati, R. B. Russell and S. M. Cohen, *Cell*, 2005, **123**, 1133.

116. S. Paushkin, A. K. Gubitz, S. Massenet and G. Dreyfuss, *Curr. Opin. Cell Biol.*, 2002, **14**, 305.

117. J. F. Abelson, K. Y. Kwan, B. J. O'Roak, D. Y. Baek, A. A. Stillman, T. M. Morgan, C. A. Mathews, D. L. Pauls, M. R. Rasin, M. Gunel, N. R. Davis, A. G. Ercan-Sencicek, D. H. Guez, J. A. Spertus, J. F. Leckman, L. S. Dure, R. Kurlan, H. S. Singer, D. L. Gilbert, A. Farhi, A. Louvi, R. P. Lifton, N. Sestan and M. W. State, *Science*, 2005, **310**, 317.

118. C. L. Jopling, M. Yi, A. M. Lancaster, S. M. Lemon and P. Sarnow, *Science*, 2005, **309**, 1577.

119. P. Jin, R. S. Alisch and S. T. Warren, *Nat. Cell Biol.*, 2005, **6**, 1048.

120. P. Nelson, M. Kiriakidou, A. Sharma, E. Maniataki and Z. Mourelatos, *Trends Biochem. Sci.*, 2003, **28**, 534.

121. M. Landthaler, A. Yalcin and T. Tuschl, *Curr. Biol.*, 2004, **14**, 2162.

122. R. Triboulet, B. Mari, Y. L. Lin, C. Chable-Bessia, Y. Bennasser, K. Lebrigand, B. Cardinaud, T. Maurin, P. Barbry, V. Baillat, J. Reynes, P. Corbeau, K. T. Jeang and M. Benkirane, *Science*, 2007, **315**, 1579.

123. B. Yang, H. Lin, J. Xiao, Y. Lu, X. Luo, B. Li, Y. Zhang, C. Xu, Y. Bai, H. Wang, G. Chen and Z. Wang, *Nat. Med.*, 2007, **13**, 486.
124. G. A. Calin, C. D. Dumitru, M. Shimizu, R. Bichi, S. Zupo, E. Noch, H. Aldler, S. Rattan, M. Keating, K. Rai, L. Rassenti, T. Kipps, M. Negrini, F. Bullrich and C. M. Croce, *Proc. Natl. Acad. Sci. U. S. A.*, 2002, **99**, 15524.
125. G. A. Calin, M. Ferracin, A. Cimmino, G. Di Leva, M. Shimizu, S. E. Wojcik, M. V. Iorio, R. Visone, N. I. Sever, M. Fabbri, R. Iuliano, T. Palumbo, F. Pichiorri, C. Roldo, R. Garzon, C. Sevignani, L. Rassenti, H. Alder, S. Volinia, C. G. Liu, T. J. Kipps, M. Negrini and C. M. Croce, *N. Engl. J. Med.*, 2005, **353**, 1793.
126. G. A. Calin, C. Sevignani, C. D. Dumitru, T. Hyslop, E. Noch, S. Yendamuri, M. Shimizu, S. Rattan, F. Bullrich, M. Negrini and C. M. Croce, *Proc. Natl. Acad. Sci. U. S. A.*, 2004, **101**, 2999.
127. J. Lu, G. Getz, E. A. Miska, E. Alvarez-Saavedra, J. Lamb, D. Peck, A. Sweet-Cordero, B. L. Ebert, R. H. Mak, A. A. Ferrando, J. R. Downing, T. Jacks, H. R. Horvitz and T. R. Golub, *Nature*, 2005, **435**, 834.
128. M. V. Iorio, M. Ferracin, C. G. Liu, A. Veronese, R. Spizzo, S. Sabbioni, E. Magri, M. Pedriali, M. Fabbri, M. Campiglio, S. Menard, J. P. Palazzo, A. Rosenberg, P. Musiani, S. Volinia, I. Nenci, G. A. Calin, P. Querzoli, M. Negrini and C. M. Croce, *Cancer Res.*, 2005, **65**, 7065.
129. S. M. Johnson, H. Grosshans, J. Shingara, M. Byrom, R. Jarvis, A. Cheng, E. Labourier, K. L. Reinert, D. Brown and F. J. Slack, *Cell*, 2005, **120**, 635.
130. M. Z. Michael, S. M. O'Connor, N. G. Holst Pellekaan, G. P. Young and R. J. James, *Mol. Cancer Res.*, 2003, **1**, 882.
131. P. S. Eis, W. Tam, L. Sun, A. Chadburn, Z. Li, M. F. Gomez, E. Lund and J. E. Dahlberg, *Proc. Natl. Acad. Sci. U. S. A.*, 2005, **102**, 3627.
132. L. He, J. M. Thomson, M. T. Hemann, E. Hernando-Monge, D. Mu, S. Goodson, S. Powers, C. Cordon-Cardo, S. W. Lowe, G. J. Hannon and S. M. Hammond, *Nature*, 2005, **435**, 828.
133. A. Valoczi, C. Hornyik, N. Varga, J. Burgyan, S. Kauppinen and Z. Havelda, *Nucleic Acids Res.*, 2004, **32**, e175.
134. W. P. Kloosterman, E. Wienholds, E. de Bruijn, S. Kauppinen and R. H. Plasterk, *Nat. Methods*, 2006, **3**, 27.
135. G. Hutvagner, M. J. Simard, C. C. Mello and P. D. Zamore, *PLoS Biol.*, 2004, **2**, e98.
136. D. Leaman, P. Y. Chen, J. Fak, A. Yalcin, M. Pearce, U. Unnerstall, D. S. Marks, C. Sander, T. Tuschl and U. Gaul, *Cell*, 2005, **121**, 1097.
137. J. A. Chan, A. M. Krichevsky and K. S. Kosik, *Cancer Res.*, 2005, **65**, 6029.
138. C. H. Lecellier, P. Dunoyer, K. Arar, J. Lehmann-Che, S. Eyquem, C. Himber, A. Saib and O. Voinnet, *Science*, 2005, **308**, 557.
139. U. A. Ørom, S. Kauppinen and A. H. Lund, *Gene*, 2005, **372**, 137.
140. B. John, C. Sander and D. S. Marks, *Methods Mol. Biol.*, 2006, **342**, 101.

141. P. Sethupathy, B. Corda and A. G. Hatzigeorgiou, *RNA*, 2006, **12**, 192.
142. J. Elmén, M. Lindow, A. Silahtaroglu, M. Bak, M. Christensen, A. Lind-Thomsen, M. Hedtjärn, J. B. Hansen, H. F. Hansen, E. M. Straarup, K. McCullagh, P. Kearney and S. Kauppinen, *Nucleic Acids Res.*, 2007, Dec, 23, [Epub ahead of print].
143. J. Krutzfeldt, N. Rajewsky, R. Braich, K. G. Rajeev, T. Tuschl, M. Manoharan and M. Stoffel, *Nature*, 2005, **438**, 685.
144. C. Esau, S. Davis, S. F. Murray, X. X. Yu, S. K. Pandey, M. Pear, L. Watts, S. L. Booten, M. Graham, R. McKay, A. Subramaniam, S. Propp, B. A. Lollo, S. Freier, C. F. Bennett, S. Bhanot and B. P. Monia, *Cell Metab.*, 2006, **3**, 87.

CHAPTER 6

Immune Stimulatory Oligonucleotides

EUGEN UHLMANN

Coley Pharmaceutical GmbH, Merowingerplatz 1a, D-40225 Düsseldorf, Germany

6.1 Introduction

Vertebrates have developed an ingenious defence response system to support the survival of the host organism. The underlying evolutionary conserved mechanisms are known as innate and adaptive immunity. The adaptive immune system is the more recently evolved and highly sophisticated system. It is based on lymphoid cell-surface receptors that recognize more-or-less an infinite number of antigens and also provide a memory system in case of repeated challenges. The evolutionary older innate immune response is based on a relatively small set of so-called pattern recognition receptors (PRRs) which recognize certain molecular structures that are present in pathogens [pathogen-associated molecular patterns (PAMPs)].[1] Specialized immune cells are activated upon recognition of PAMPs and trigger the generation of optimal immune responses. In recent years, significant progress in the understanding of innate immunity has been made through the identification of several families of pathogen sensors (Figure 6.1), including toll-like receptors (TLRs), nucleotide-binding oligomerization domain (NOD)-like receptors (NLRs) and retinoic acid inducible gene I (RIG-I)-like receptors (RLRs).[2–6] Members of the TLR family recognize bacteria, viruses and fungi, while RLRs are specialized on antiviral defence. The TLRs are a large family of PRRs consisting of at least 12 different TLR subtypes, TLR1 to TLR12. Recently, TLRs were shown to be essential for the recognition of double-stranded RNA (dsRNA; TLR3),

RSC Biomolecular Sciences
Therapeutic Oligonucleotides
Edited by Jens Kurreck
© Royal Society of Chemistry 2008

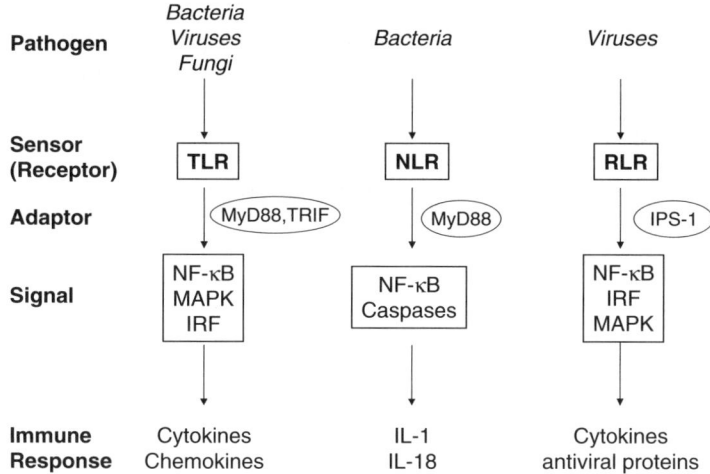

Figure 6.1 The three families of pathogen sensors that mediate immune responses against various pathogens, such as bacteria, viruses, fungi, *etc.*: toll-like receptors (TLRs), NOD-like receptors (NLRs) and RIG-like receptors (RLRs). (IL, interleukin; IPS-1, interferon-beta promoter stimulator 1; IRF, interferon regulatory factor; MAPK, mitogen-activated protein kinases; MyD88, myeloid differentiation primary response protein 88; NF-κB, nuclear factor kappa beta.)

lipopolysaccharide (LPS; TLR4), bacterial flagellin (TLR5), single-stranded RNA (ssRNA) and small synthetic antiviral compounds (TLR7 and -8) and bacterial DNA or synthetic cytidine–phosphate–guanosine (CpG) oligodeoxynucleotides (ODNs; TLR9).[7–11] Stimulation of these receptors results in immune response following specific signal transduction pathways. One of the best investigated immune stimulators in this field are CpG ODNs,[12,13] whose high therapeutic potential as TLR9 agonists has been demonstrated by many preclinical studies as well as by early human clinical trials in the areas of cancer, infectious diseases and allergy–asthma.[14–16] This review discusses the different oligonucleotide-based strategies for immune stimulation with major emphasis on CpG ODN TLR9 agonists.

6.2 Mechanism of Action

The best characterized PRRs are likely the TLRs, which are either expressed on the cell surface (TLR1, 2, 4, 5, 6 and 11) and recognize components of extracellular pathogens, such as bacterial lipopolysaccharides (LPS), or are expressed intracellularly within immune cells (TLR3, 7, 8 and 9) and are specific for nucleic acids (Figure 6.2).[5] Thus, TLR9 detects unmethylated CpG dinucleotides,[12] which appear frequently in bacterial and viral DNA, but are suppressed and largely methylated in vertebrate genes. TLR9 is located in the endosomal compartment together with TLR7 and/or TLR8 and TLR3,

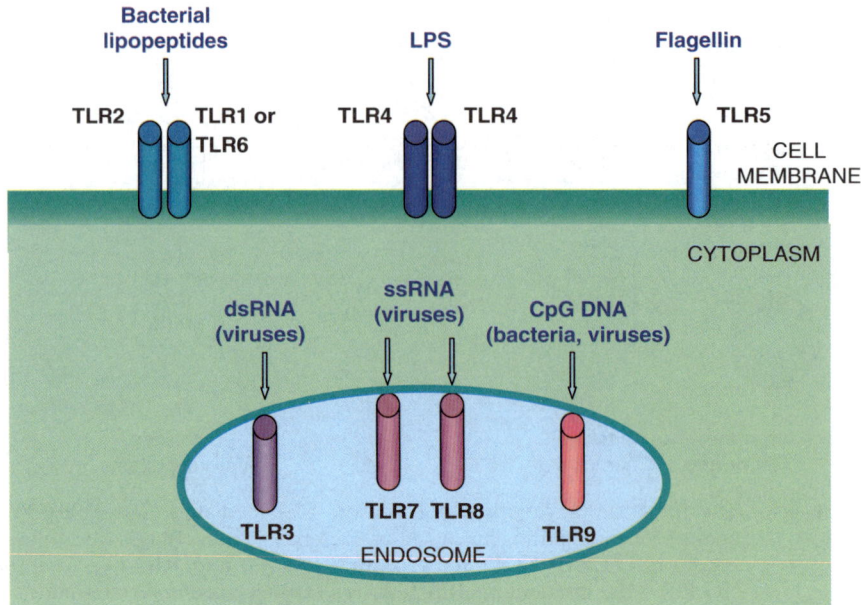

Figure 6.2 Schematic representation of the toll-like receptor (TLR) family. TLR1, TLR2 and TLR4 to TLR6 are located at the cell surface, while TLR3, TLR7 to TLR9 are located in the endosomal compartment. The ligands that activate the various TLRs are also indicated. (CpG, cytidine–phosphate–guanosine; dsRNA, double-stranded RNA; LPS, lipopoly-saccharide; ssRNA, single-stranded RNA.)

respectively, recognizing ssRNA and dsRNA, respectively. TLR7 and TLR9 are primarily expressed in plasmacytoid dendritic cells (pDCs) and B-cells (TLR9 only), while TLR3 and TLR8 are primarily found in myeloid dendritc cells (mDCs).[17] This assignment is true for human cells, but may vary between species, which can potentially complicate preclinical product development.

The endosomal location of the TLRs that recognize nucleic acid ligands is of special note, since this is where the immune stimulatory oligodeoxynucleotides (the ODNs) should be delivered to. This is in contrast to oligoribonucleotide (ORN) and ODN therapeutics, which have a mechanism based on the inhibition of gene expression, such as short interfering RNA (siRNA) and antisense ODN. For the efficient inhibition of gene expression antisense ODN and siRNA have to be de-livered to the cytosol by means of special delivery vehicles (*e.g.* cationic lipids). Antisense ODN may also be delivered to the nucleus to target pre-messenger RNA (mRNA), *e.g.* to redirect mRNA splicing. Since ODNs, when taken up by endocytosis or receptor-mediated endocytosis, accumulate in the endosomal–lysosomal compartment, CpG ODNs activate endosomal TLR9 without the need of cationic lipids. After the CpG ODNs have been taken up by an endocytotic pathway, they can bind to TLR9. Immune stimulatory RNA is usually delivered by means of cationic lipids or other particles, which prevents rapid degradation by ribonucleases and additionally improves cellular uptake. The binding of the

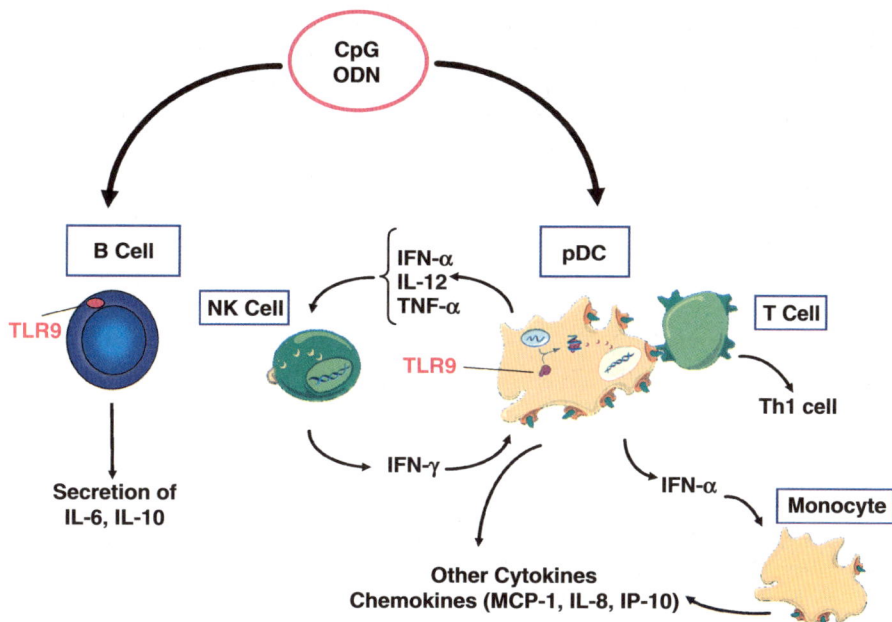

Figure 6.3 B-cells and plasmacytoid dendritic cells (pDC) express TLR9 and are directly activated by CpG ODNs resulting in the secretion of cytokines and chemokines. Other immune cells, such as T-cells, NK cells and monocytes, are indirectly activated by cytidine–phosphate–guanosine oligodeoxynucleotide (CpG ODN). (IFN, interferon; IL, interleukin; MCP, monocyte chemotactic protein-1; NK cell, natural killer cell; pDC, plasmacytoid dendritic cells; Th1, helper T cell 1; TNF, tumor necrosis factor.)

agonists to TLR7, 8 or 9 triggers the recruitment of an adaptor protein, called myeloid differentiation primary response protein 88 (MyD88),[18] resulting in the rapid induction of cytokine and chemokine secretion through various signalling pathways (Figure 6.3). MyD88 is essential for immune activation through TLR7, 8 and 9, but not for TLR3.[19] MyD88 binding seems to be followed by association of interleukin (IL) receptor associated kinase-1 (IRAK1), interferon (IFN) regulatory factor-7 (IRF7) and TNF receptor-associated factor-6 (TRAF6) resulting in activation of downstream kinases, including mitogen-activated protein kinases (MAPKs), extracellular receptor kinase (ERK), p38 and jun kinase (JNK). This finally leads to the activation of transcription factors, such as activator protein-1 (AP-1) and nuclear factor kappa beta (NF-κB), which turn on cytokine and chemokine production.

Only recently, a new sensor mechanism to detect viral RNA was revealed which, in contrast to TLR3, 7, 8 and 9, is not an endosomal but a cytosolic surveillance system. Interestingly, the structural feature for recognition of the viral RNA is a 5′-triphosphate moiety, which is either missing or is totally masked (*e.g.* by a 7-methyl guanosine cap structure) in vertebrate RNA. Viral RNA-5′-triphosphates are recognized by cytosolic RIG-I (also called DDX58).[20]

Subsequent to TLRs, RIG-I and melanoma-differentiation-associated gene 5 (MDA5) were discovered, which both share a RNA-helicase domain and trigger efficient induction of type I IFNs *via* NK-κB and IRF3 and/or IRF7.[3] RIG-I appears to be constitutively expressed in the cytoplasm of many cells and is inducible by viruses (*e.g.* influenza virus, vesicular stomatitis virus, Newcastle disease virus and Japanese encephalitis virus) and by cytokines (*e.g.* IL-1ß).[21] RIG-I has a very broad and significant role in mediating IFN and cytokine production in response to infection by viruses. While TLR7 and/or 8 activation is restricted to pDCs and monocytes, RIG-I is ubiquitously expressed and probably mediates the local IFN production more than TLR7, which controls systemic IFN levels. Agents which activate both RIG-I and TLR7 could potentially lead to very potent anticancer drugs. The second intracellular sensor of RNA is MDA5, which recognizes polyinosylic–polycytidylic acid [poly(I:C)][22] and has been shown to be essential for picornavirus recognition *in vivo*. Another cytoplasmic dsRNA-binding protein is dsRNA-dependent protein kinase receptor (PKR).[23] In normal cells, PKR is present at low level, but is upregulated by IFN. On binding to dsRNA, PKR appears to function by phosphorylating the alpha subunit of the translation initiating factor eIF2, resulting in inhibition of viral protein synthesis.

6.3   TLR9 Activation by CpG Oligonucleotides

6.3.1   Classes of CpG Oligonucleotides

Although unmethylated CpG dinucleotide is central to recognition by TLR9, the actual recognition motif is thought to be the hexanucleotide GTCGTT for human cells and GACGTT for mouse cells, respectively.[12,24] While one hexanucleotide motif is sufficient for recognition by TLR9, two or more motifs with appropriate spacing by intervening nucleotides, preferably thymidylates, result in much more potent immune activation. One of the early and very potent TLR9 agonists is PF-3512676 (formerly CpG 7909; 5′-TCGTCGTTTTGTCGTTTTGTCGTT; all-phosphorothioate), which has three hexanucleotide motifs and a total of four CpG dinucleotide motifs (Figure 6.4).[25] This 24-mer is linked by 23 phosphorothioate linkages, which renders it reasonably stable towards nucleases present in serum and in tissue. Secondary structure prediction and experimental investigation suggest no stable secondary structure for this ODN under physiological conditions (*e.g.* 37 °C, 130 mMol saline buffer pH 7). This class of CpG ODNs is designated as the B- (or K-) class CpG ODN (Figure 6.5) and is characterized by the efficient induction of B-cell and NK-cell activation, but moderate type I IFN induction.[26] The length of the ODN also appears to play a role in that optimal stimulation is obtained with CpG ODNs that have 16 to 24 nucleotides in length in the absence of delivery agents. A different immune stimulatory profile has been observed for the A- (or D-) class CpG ODNs, which exhibit very high type I IFN induction, but moderate B-cell stimulatory activity. A-class ODNs[26]

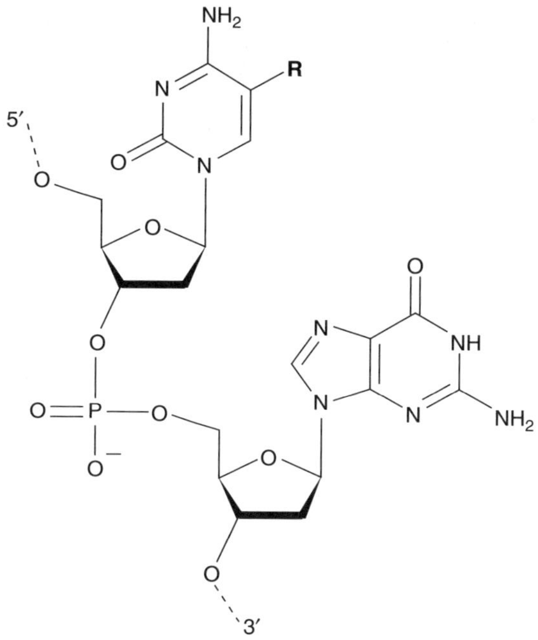

Figure 6.4 Chemical structure of the CpG dinucleotide in the unmethylated form (R is H), which activates TLR9, and in its methylated form (R is CH$_3$), which is poorly recognized by TLR9. This provides the basis for discriminating 'self' from 'non-self' DNA.

are phosphodiester–phosphorothioate mixed backbone oligomers having a central palindrome as a natural phosphodiester and partially phosphorothioate-modified G-tails at the 5'- and 3'-ends. The G-tails consist of four or more consecutive deoxyguanylate residues and support the formation of G-tetrads through Hoogsteen base-pairing. As known from antisense ODN, G-tetrads can lead to strong aggregation of the oligomer, which in turn appears to enhance cellular uptake. A third type of CpG ODN combines the advantages of the two described classes and strongly activates B-cells and, additionally, induces high levels of type I IFNs. These ODNs are termed C-class CpG ODNs[26] and were found to form well-defined secondary structures based on Watson–Crick base-pairing. C-class ODNs are usually phosphorothioate-modified and harbor a palindrome or partial palindrome, leading to duplex or hairpin structure formation, respectively.

It is stressed that all three classes of CpG ODN are recognized by TLR9 and appear to work *via* MyD88 as the adaptor protein, which poses the question as to which factors determine the observed distinct immune stimulatory profiles. At least two hypothesis explain the different immune stimulatory profiles of the three classes. One model suggests that the multimerization of A-class ODN or dimerization of C-class ODN may lead to receptor cross-linking,[27] which results in improved induction of IFN secretion. This observation would also be

A-Class (CPG 2216)

G*G*G-G-<u>G-A-C-G-A-T-C-G-T-C</u>-G*G*G*G*G
palindrome (underlined) including CpG
multimerization *via* G-tetrads
Strong induction of IFN secretion, but moderate induction of B-cell proliferation

B-Class (PF-3512676, formerly CPG 7909)

T*C*G*T*C*G*T*T*T*T*G*T*C*G*T*T*T*T*G*T*C*G*T*T
phosphorothioate, linear
Moderate induction of IFN secretion, but strong induction of B-cell proliferation

C-Class (CPG 2395)

T*C*G*T*C*G*T*T*T*T*C*G*G*C*G*C*G*C*G*C*C*G
    G*C*C*G*C*G*C*G*C*G*G*C*T*T*T*T*G*C*T*G*C*T

phosphorothioate, dimer formation due to 3′ palindrome
Strong induction of IFN secretion and potent induction of B-cell proliferation

Figure 6.5   The three distinct classes of CpG ODN. A-class CpG ODNs (also referred to as D-type) have poly-dG tails of four or more consecutive guanosine residues at the 3′- and 5′-termini, leading to aggregation *via* G-tetrads. Typical B-class CpG ODNs (also referred to as K-type) do not form secondary structures at physiological conditions. C-class CpG ODNs have TCG motifs at the 5′-end and a palindrome at the 3′-end, resulting in dimer formation under physiological conditions (where ∗ is phosphorothioate and - is phosphodiester).

consistent with the enhanced IFN induction, when CpG B-class ODNs are condensed with cationic lipids to higher ordered structures. However, this observation could also be explained by the second model, which suggests that different intracellular compartmentalization, in particular endosomal localization, is responsible for the distinct immune stimulatory profiles.[28,29] Using marker proteins, it has been shown that activation of TLR9 by the multimeric A-class CpG ODNs occurs in transferrin receptor-positive endosomes ('early endosomes') and leads exclusively to type I IFN production, while monomeric B-class ODNs localize to lysosome-associated membrane protein (LAMP)-1 positive endosomes ('late endosomes'), resulting in poor induction of IFN secretion. Interestingly, C-class CpG ODNs, which usually form dimers, are distributed to both early and late endosomes and can therefore combine the advantages of B- and A-class CpG ODNs. The distinct immune responses of the three different classes of CpG ODN are shown in Figures 6.6 and 6.7.

Further, the location of the CpG motifs within the sequence seems to play an important role for the potency to activate TLR9. The most potent TLR9 agonists usually have one or two CpG motifs at the 5′-ends and additional CpG dinucleotides towards the middle or the 3′-end. Two observations support a model in which TLR9 recognizes the CpG ODN preferentially from the 5′-end. Firstly, the more 5′ the CpG dinucleotide is positioned, the higher the

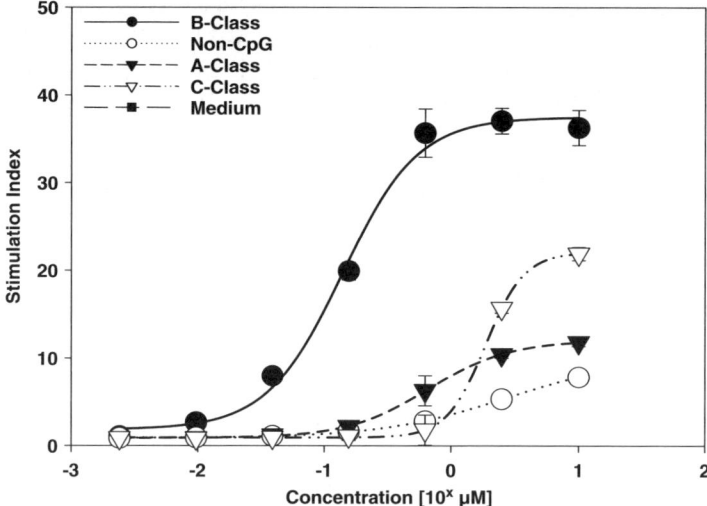

Figure 6.6 Stimulation of TLR9 by CpG ODN. HEK293 cells stably transfected with a vector expressing human TLR9 and an NF-κB–luciferase construct were incubated for 16 hours with indicated ODN. NF-κB stimulation was measured through luciferase activity. Stimulation indices (± standard deviation) were calculated in reference to luciferase activity of medium alone. Stimulation of TLR9 by B-, C- and A-class CpG ODN relative to a non-CpG ODN is shown.

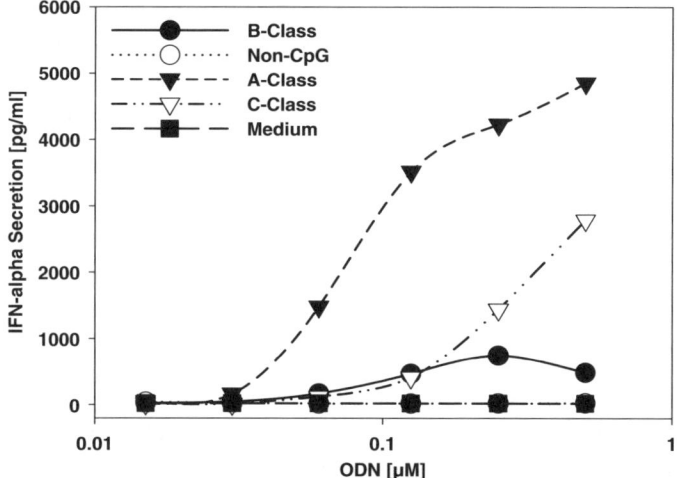

Figure 6.7 IFN-α induction by CpG ODN. Human peripheral blood mononuclear cells (PBMCs) are stimulated with the various CpG ODN for 48 hours. Culture supernatants were harvested and IFN-α was measured by enzyme-linked immunosorbent assay (ELISA).

stimulation by TLR9.[6] Secondly, 3′–3′-linked CpG ODNs with free 5′-ends show much better TLR9 activation than 5′–5′-linked CpG ODNs.[30]

Finally, certain phosphorothioate ODNs can also exhibit immune suppressive effects.[31,32] These so-called S-class ODNs do not contain CpG motifs and some of them are closely related in sequence to B-class ODNs in which G of the CpG motif is replaced by C. The consensus motif suppressing hTLR9-mediated immune effects has been identified as 5′-NCCNGGGN-3′.[32]

6.3.2 Chemical Modifications of CpG Oligonucleotides

Chemical modification of CpG ODNs can be used to enhance stability against nucleases, to improve cellular uptake and compartmentalization, and to modulate binding to TLR9. While there is a plethora of chemical modifications known from antisense ODN research and also on the impact of such modifications on nuclease stability and cell uptake, modulation of binding to and activation of TLR9 has to be elaborated specifically for CpG ODNs. Alterations of the sugar–phosphate backbone have significant influence on the nuclease stability and on receptor recognition, while base modifications are expected to have a major impact on recognition. Modifications have been introduced at the CpG dinucleotide, the hexanucleotide GTCGTT recognition motif or throughout the entire sequence. Modifications to improve cellular uptake or to target ODNs to certain cell types are preferentially introduced at the 3′- or 5′-termini of the ODNs (*e.g.* lipophilic end modifications). However, some of the reported structure–activity relationship (SAR) data in the literature are conflicting, which may be explained by the fact that different sequences were used to study the influence of chemical modifications on immune stimulation. Firstly, species-specific differences offer the possibility to use either the human (5′-GTCGTT) or mouse (5′-GACGTT) consensus hexanucleotide motif. Many CpG ODNs contain multiple CpG motifs within a sequence and the effect of the chemical modification can be position dependent. Sometimes, ODNs that contain the human consensus motif have been assayed in murine cells and *vice versa*.

To render the data of these studies more comparable, we have established a model assay system for SAR studies.[33] We chose a 20-nucleotide sequence (5′-T*G*T*C–G*T*T*T*T*T*T*T*T*T*T*T*T*T*T*T*T-3′, in which ∗ is phosphorothioate and – is phosphodiester) which contains just a single human hexanucleotide motif (GTCGTT) embedded in a homothymidylate for optimal recognition by hTLR9.[33] Since the natural ligand for TLR9 is a phosphodiester, the linkage at the CpG dinucleotide was maintained as phosphodiester, while all other linkages are modified as phosphorothioate, rendering the oligomer sufficiently stable in serum. The phosphodiester linkage at the CpG also avoids the formation of diastereoisomers at the CpG motif, which have previously been shown to be recognized differently by TLR9, whereby the R_p diastereoisomer of phosphorothioates showed higher immune stimulatory activity.[34] To shed some light on the species-specificity, the CpG ODNs have been investigated for their potency to stimulate human or mouse TLR9 (mTLR9) using HEK293 cells that express a luciferase gene controlled by a NK-κB promoter construct that constitutively expresses the genes for hTLR9 or mTLR9, respectively.

6.3.2.1 Modification of the Internucleotide Phosphodiester Linkage

In bacterial DNA, the natural phosphodiester linkage is well-recognized by TLR9 as long as it contains CpG motifs in the preferred sequence context that results in efficient immune stimulation. Since bacterial DNA is usually double-stranded and the DNA fragments are relatively large, it appears to be sufficiently stable to survive until it meets the TLR9 receptor *in vivo*. Although bacterial DNA could potentially be isolated and used for therapy, synthesis of ssDNA in the range of 12 to 30 nucleotides appears to be more attractive to overcome the regulatory requirements of immune stimulatory drugs. It is well-documented that unmodified ssDNA (phosphodiester ODN) is rapidly degraded in serum and in tissue. Although phosphodiester CpG ODNs have been shown to be immune stimulatory, the metabolically more stable phosphorothioate analogs usually give stronger effects, in particular for B- and C-class CpG ODNs.[12,35] However, if the ODN is stabilized against degradation by the formation of a secondary or tertiary structure which impairs with the attack by nucleases, then phosphodiester–phosphorothioate co-polymers may become more active than fully phosphorothioated analogs, as has been observed for A-class CpG ODNs.[36] The introduction of phosphorothioate linkages by standard phosphoramidite synthesis results in 2^n diastereomeric isomers, if n is the number of modified linkages. It has been found that after a short incubation time ($\sim$ one hour) the R_p diastereomers induce higher immune stimulation than the S_p diastereoisomers.[34] Of course, the higher activity might be either through better binding and activation of the receptor or, alternatively, through enhanced metabolic stability. Since the S_p diastereoisomer has been reported to be more resistant to endonuclease degradation,[37] the R_p diastereoisomer appears to better mimic the natural phosphodiester linkage, leading to better recognition and activation of TLR9. This is in agreement with the observation that after a much longer incubation time (two days), the S_p diastereoisomer becomes the more active compound.[34]

Furthermore, it was reported that the immune stimulatory potency is decreased if the naturally occurring 3′–5′ linkage at CpG is replaced by a 2′–5′ linkage.[38] Similarly, when the 2′–5′ linkage was introduced at 3′ to the CpG motif, immune stimulation was reduced, while modification of the phosphodiester linkage at 5′ to the CpG motif resulted in slightly increased cytokine production. It is difficult to determine if the observed effects are really position dependent or more sequence dependent. Replacement of the anionic phosphodiester linkage at CpG by the neutral methylphosphonate linkage has been reported to dramatically decrease immune stimulatory activity.[39] However,

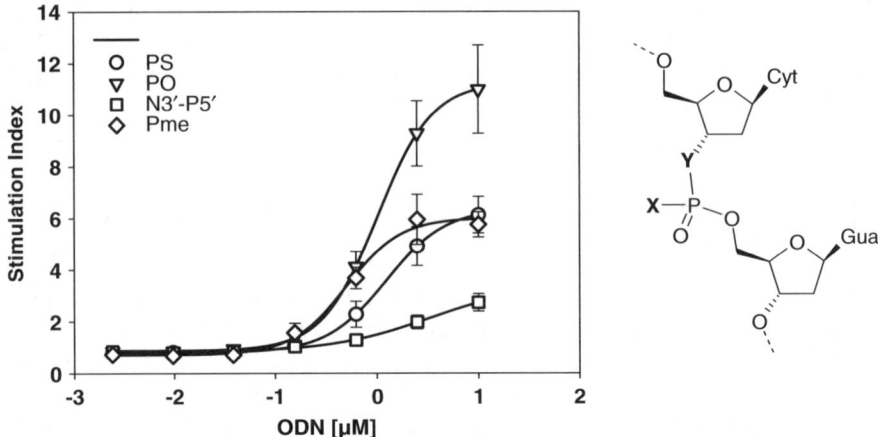

Figure 6.8 Stimulation of TLR9 by CpG ODNs with various phosphate modifications at the CpG dinucleotide. HEK293 cells stably transfected with a vector expressing human TLR9 and an NF-κB–luciferase construct were incubated for 16 hours with the indicated ODN. NF-κB stimulation was measured through luciferase activity. Stimulation indices (± standard deviation) were calculated in reference to luciferase activity of the medium alone. Stimulation of TLR9 by a B-class CpG ODN modified as phosphodiester (PO, $X = Y = O$), phosphorothioate (PS, $X = S$, $Y = O$), N3′-P5′-phosphoramidate (N3′-P5′, $X = O$, $Y = NH$) and methylphosphonate (Pme, $X = CH_3$, $Y = O$).

when we replaced a single phosphorothioate linkage in an all-phosphorothioate CpG ODN with just a single methylphosphonate at CpG, the activity of the methylphosphonate derivative was equivalent to the analogous compound having a phosphorothioate at CpG (Figure 6.8).[40] Compared to the analogous CpG ODN having a single phosphodiester linkage at CpG, both phosphorothioates and methylphosphonates as diastereomeric mixtures showed significantly reduced immune stimulator efficacy in our human TLR9 (hTLR9) assay.

6.3.2.2 Modification of the Heterocyclic Nucleobases

As mentioned above, activation of TLR9 by a CpG ODN is sequence-specific with unmethylated 5′-CpG being the minimal recognition motif. Simple inversion of the motif to 5′-GpC strongly reduces immune stimulatory activity, although there is still some remaining activity measured in a cellular hTLR9 assay. Similarly, substitution of cytosine by 5-methylcytosine (Figure 6.9) significantly reduces activity in a TLR9 assay, but does not completely abolish it.[41] Since substitution at the 5-position in cytosine is the basis for recognition between pathogen and host, decreased immune stimulation of 5-substituted cytosine at CpG is not unexpected. Thus, most substitutions at C(5) of cytosine have been reported to decrease activity in a mouse lymphocyte assay,

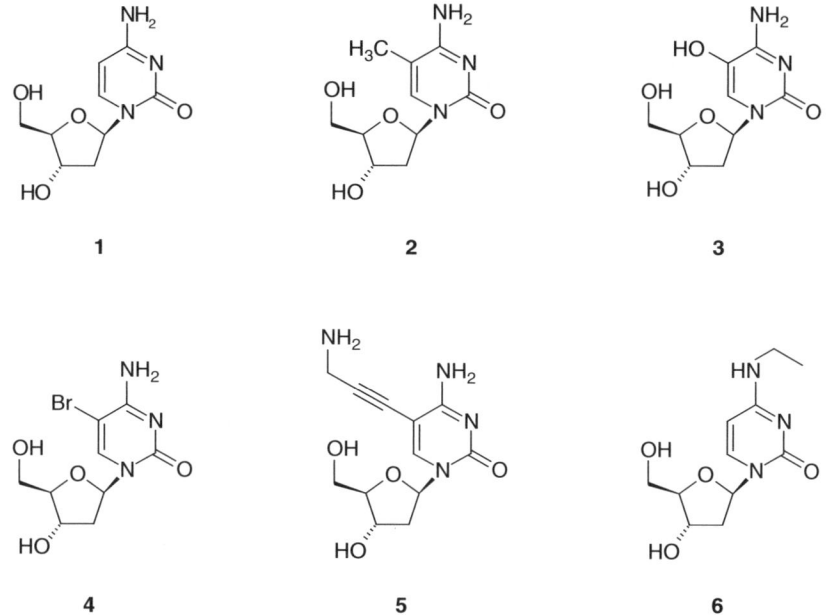

Figure 6.9 Chemical modifications of the cytosine base at the CpG motif. Deoxy-cytidine (**1**), in which the cytosine base is replaced by 5-methylcytosine (**2**), 5-hydroxycytosine (**3**), 5-bromocytosine (**4**), 5-aminopropargylcytosine (**5**) or N(4)-ethylcytosine (**6**).

with the exception of 5-hydroxycytosine, which has been reported to be a perfect replacement for cytosine when evaluated in a mouse lymphocyte assay.[42] In contrast to this report,[42] we found a strong reduction (85%) when cytosine was changed to 5-hydroxycytosine at CpG using a hTLR9 assay system.[33]

This negative impact of the 5-hydroxy group is even stronger than substitution of the hydrogen at C(5) of cytosine by a methyl group, which resulted in the expected drop in hTLR9 stimulation by about 70% in maximal stimulation. It is interesting that substitution of C(5) of cytosine by bromine results in a 14% and 74%, respectively, drop in maximal stimulation in the mTLR9 and hTLR9 assays, respectively.[33] These unexpected findings suggest that the spatial requirements at the murine receptor differ substantially from those of the human receptor. Furthermore, substitution of cytosine with 5-aminopropargyl residue resulted in virtually complete loss in activity both for the hTLR9 and mTLR9. While replacement of C by N(4)-ethyl-C still resulted in moderate activity (24% maximal stimulation) for hTLR9, the same modification completely abolished recognition by mTLR9. In summary, the results with the modifications at cytosine suggest that the primary exocyclic amino group and the spatial requirements at C(5) of cytosine are very important for the observed immune stimulatory effects of CpG ODNs. The requirement for correct space-filling at the cytosine position is also demonstrated by the

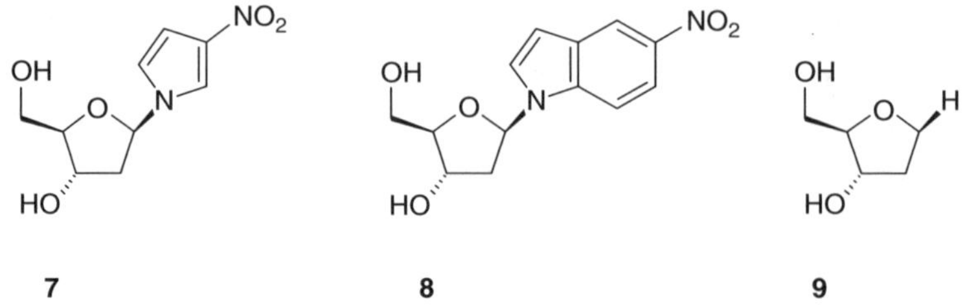

Figure 6.10 Substitution of dC or dG by universal base nucleoside analogs that contain 3-nitropyrrole (**7**) or 5-nitroindole (**8**), or by an abasic residue (**9**).

observation that deletion of cytosine in the CpG motif (abasic residue preceding G) or substitution by a universal base, such as 3-nitropyrrole or 5-nitroindole, results in complete loss of activity.[33] In contrast, deletion of the guanine base in the CpG motif results in a drop by approximately 70% in the maximal hTLR9 activation of the parent molecule. The observed residual activity of abasic CpD (G replaced by abasic spacer) and replacements by universal bases nitroindole and nitropyrrole (Figure 6.10) in the hTLR9 assay again contrasts with the results in the murine TLR9 assay, in which all three derivatives are completely inactive. These results are in contrast to a previous report, which suggested that a nucleobase is absolutely required at both C and G positions in CpG ODNs.[43]

Furthermore, we have investigated another series of CpG derivatives in which the hydrogen donor and acceptor function at guanine were either deleted or exchanged by other functional groups.[33] It had been reported previously by others that 7-deazaguanine-modified CpG ODNs exhibit similar activity to unmodified CpGs, while replacing guanine by hypoxanthine results in an approximately 65% reduction in proliferation in a murine lymphocyte proliferation assay. Other investigated modifications of guanine, including purine, 2,6-diaminopurine and 2-aminopurine (Figure 6.11) have been reported to induce only insignificant immune stimulation (8–19%) in the mouse cell assay.[42,44] These results are in contrast to our findings using the 20-mer model sequence and the hTLR9 assay system. In our system, replacement of guanine by 7-deazaguanine resulted in an 81% reduction in maximal hTLR9 activation. Interestingly, TLR9 activation was almost totally retained when guanine was exchanged by hypoxanthine (deoxyinosine as nucleoside), which is in agreement with our previous investigation using a different sequence and fully phosphorothioate-modified ODNs. In addition, hTLR stimulation was fully retained when O(6) of guanine was replaced by sulfur (6-thioguanine). However, replacement of guanine by 2-aminopurine, purine, 2,6-diaminopurine or 8-oxoguanine resulted in about 40–60% of the parent ODN efficacy in hTLR9 activation. In conclusion, these results suggest that recognition of guanine base in the CpG motif is strongly determined by the N(7) and exocyclic O(6) functions. Since N(7) and O(6) are involved in Hoogsteen base-pairing, it is very likely that guanine is recognized from the 'Hoogsteen base pairing

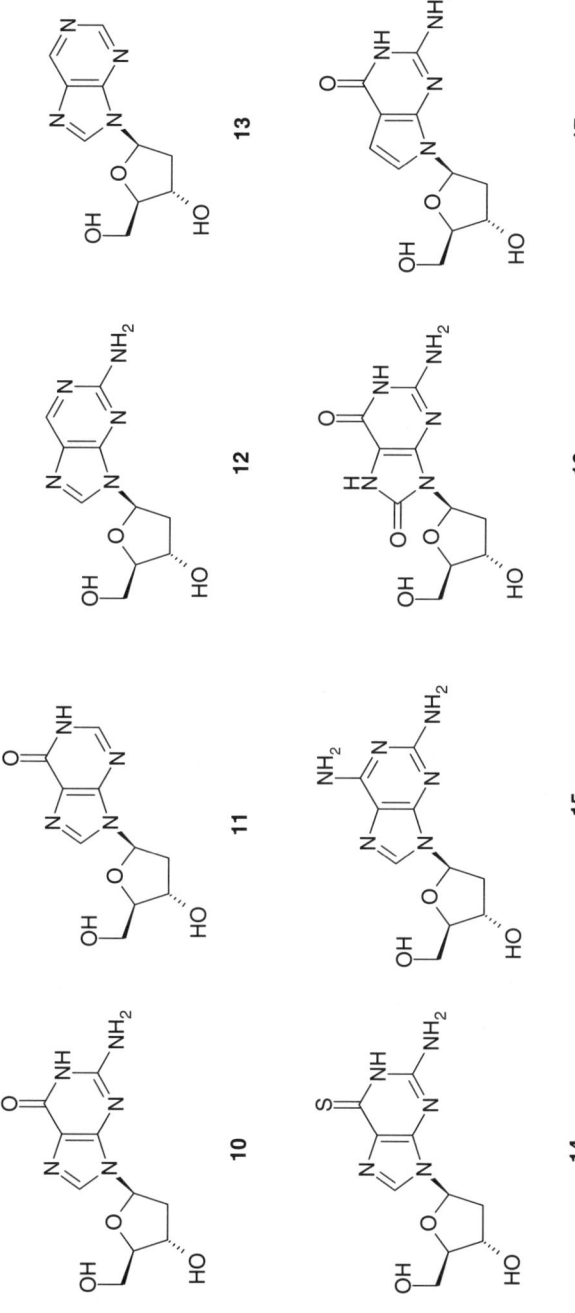

Figure 6.11 Chemical modifications of the guanine base at the CpG motif. Deoxyguanosine (**10**), in which guanine base is replaced by hypoxanthine (**11**), 2-aminopurine (**12**), purine (**13**), 6-thioguanine (**14**), 2,6-diaminopurine (**15**), 8-oxo-7,8-dihydroguanine (**16**) or 7-deazaguanine (**17**).

site' by hTLR9. As 8-oxo-7,8-dihydroguanine is preferentially in the syn-conformation, the decreased activity of 8-oxo-7,8-dihydroguanine could be caused by a shift of the equilibrium from the normal anti- to the syn-orientation of the guanine base, suggesting that the guanine base is preferably in the anti-conformation for recognition by TLR9. Also, only one modification at G, namely replacement of G by inosine, resulted in modest activation (28%) of murine TLR9, while all other modifications lead to a drop by more than 80% in mTLR9 activation.[33]

In summary, our SAR data suggest that all investigated base modifications at C(5) and N(4) of cytosine are poorly compatible with recognition by hTLR9. The following ranking order in maximal stimulation of hTLR9 was observed for modifications at the CpG motif: cytosine > 5-methylcytosine ~ 5-bromocytosine ~ N(4)-ethylcytosine > 5-hydroxycytosine > 5-amino-propargylcytosine. In the case of guanine modifications, N(7) appears to be most important for recognition by hTLR9, while 2-amino, 6-amino and 6-oxo seem to be less critical for optimal activation. Full compatibility for hTLR9 recognition has been found on substitution of guanine by hypoxanthine and 6-thioguanine. The ranking in hTLR9 stimulation is guanine ~ 6-thioguanine ~ hypoxanthine > 2-aminopurine > purine ~ 2,6-diaminopurine ~ 8-oxo-guanine > 7-deazaguanine. The importance of using human cell-based assay systems for the optimization of potential drug candidates in CpG ODN drug development is supported by the different results obtained from the human versus murine TLR9 in *in vitro* assay systems. It must be taken into account that the use of *in vivo* mouse models for ranking CpG ODNs in the selection of drug candidates for human use might also be of limited value. For drug-screening purposes it is therefore also mandatory to determine the potency of the CpG ODN to induce cytokine and chemokine secretion in human peripheral blood mononuclear cells (PBMCs) or isolated pDCs. Figure 6.12 summarizes the impact of the different chemical modifications at the CpG dinucleotide on TLR9 activation.

6.3.2.3 Modifications of the Sugar Moiety

Modifications of the sugar frequently involve 2′-modifications, which may result in a change of the sugar conformation. However, recognition of CpG ODNs by TLR9 clearly favors a DNA-like conformation (*S*-type or 2′-endo) at the recognition motif. Therefore, it is not surprising that 2′-*O*-methyl-modified CpG ODNs show significantly decreased immune stimulatory activity *in vitro* and *in vivo*.[45] However, if 2′-*O*-alkyl modifications were introduced outside the immune stimulatory motif, immune stimulatory activity was reported to be either increased or decreased, depending on the sequence, the position, the number of modifications and the assay system used. Using our 20-mer–/hTLR9 model described above, we found about a 50% reduction when C at CpG was modified as 2′-*O*-methyl, while modification of C as a locked nucleic acid (LNA) nearly completely abolished activity.[46] Modifications at G were generally better tolerated as compared to those at C in the CpG dinucleotide.

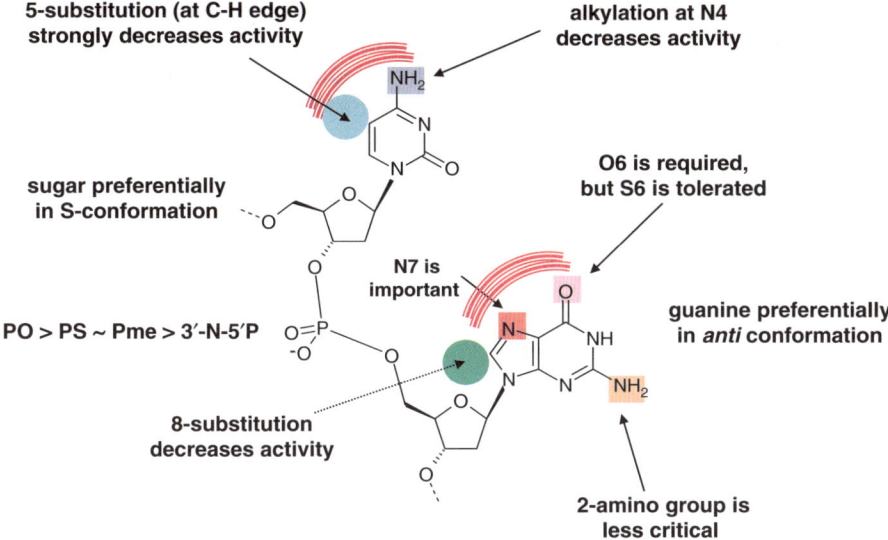

Figure 6.12 Summary of the results from structure–activity relationship studies with various chemical modifications at the CpG motif. Impact of different modifications at cytosine, guanine, the sugar moiety and the inter-nucleotide linkage on recognition by human TLR9.

Similarly, modification of C as arabino-C or 2′-fluoro-C resulted in a strong decrease in TLR9 activation, although the arabino nucleoside has more of a DNA-like sugar conformation compared to the other investigated sugar modifications, with a more RNA-like sugar conformation.

6.4 Therapeutic Applications of CpG Oligonucleotides

The first generation of CpG ODN drug candidates are fully phosphorothioate modified ODNs of about 22–24 nucleotides in length. They can be prepared by standard phosphoramidite synthesis at multi-kilogram scale and appropriate analytical methods have been established to meet the regulatory requirements. Since antisense ODNs of the first generation belong to the same class of chemical substances, the ADME (adsorption, distribution, metabolism, elimination) properties have been well-investigated and the phosphorothioate class effects are pretty much understood, at least for the B-class CpG ODNs which do not form stable secondary structures in solution. The metabolism of C-class CpG ODNs differs from that of B-class ODNs in that the usually observed degradation by 3′-exonucleases is slowed down because of duplex formation of the 3′-palindromic sequence.[47,48] The CpG ODNs are readily soluble in aqueous buffers and can be administered by most established routes (*e.g.* intravenously, subcutaneously, intraperitoneally, intrapulmonary, intranasally, *etc.*). Subcutaneous application of CpG ODNs turned out to be particularly advantageous, as the ODN is detected at relatively high levels in the lymph nodes, resulting in excellent cytokine induction.[49]

As a result of the activation of both innate and adaptive immunity, CpG ODNs can be applied in the therapy of cancer, infectious diseases, and asthma and/or allergy, but also as highly effective adjuvants in vaccination.[50] The therapeutic approaches in cancer and antiviral therapy involve both mono-therapy and combination therapies. TLR9 agonists are attractive drug candidates capable of triggering T helper cell 1 (Th1)-type immune responses (*e.g.* secretion of Th1-promoting chemokines and cytokines).[14,51]

Since CpG ODNs have shown antitumor activity in many mouse models, they have been introduced into clinical trials to treat human cancer patients. After isolation of dendritic cells from patients treated with CpG ODNs, a Th1-like cytokine response could be detected, providing evidence for the immune stimulatory potency of the new drug candidates.[52] B-class CpG ODNs have been studied in various cancer indications, including non-Hodgkin's lympho-ma, cutaneous T cell lymphoma, basal cell carcinoma, breast cancer, renal cell cancer, melanoma and non-small-cell lung cancer (NSCLC).[15,53] Furthermore, a B-class CpG ODN has shown promising results in a Phase II randomized controlled human clinical trial in patients with NSCLC. The response rate was improved from 19% in patients receiving standard chemotherapy to 37% in patients receiving a combination therapy of taxane–platinum plus CpG ODN.[54] The one-year survival was improved from 33% to 50% when chemotherapy was combined with the CpG ODN. However, subsequent Phase III trials using a combination therapy of CpG ODN and chemotherapy to treat NSCLC have been discontinued recently after a scheduled interim analysis by an independent Data Safety Monitoring Committee concluded that the risk–benefit profile did not justify continuation of these trials. Despite this, CpG TLR9 agonists are still considered to have potential utility in anticancer ther-apy, and clinical trials in other settings are ongoing. Thus, a B-Class CpG ODN is currently under investigation in combination with Tarceva® in a Phase II refractory NSCLC clinical trial as well as in combination with anti-CTLA-4 in a Phase I advanced melanoma clinical trial. Previous clinical trials with TLR9 agonists have shown objective responses in six different types of cancer, both solid and liquid tumors, either as monotherapy or in combination with chemotherapy.

CpG ODNs are also candidates for the therapy of infectious diseases. It has been demonstrated in preclinical experiments that CpG ODNs can protect mice against lethal challenges with a variety of pathogens, such as *Listeria mono-cytogenes*, malaria, anthrax and Ebola virus. BALB/c mice, which are highly susceptible to *L. monocytogenes*, were challenged with approximately $10\,LD_{50}$ of *Listeria*.[55] Interestingly, the mice were protected if they were treated by a single dose of CpG ODN prior to the challenge. Recently, a human clinical trial was performed with C-class ODNs, which induce potent secretion of type I IFN. Thus, increased IFN-α serum levels, which correlated with CpG ODN serum levels, could be detected in a Phase I safety study with C-class ODN CPG 10101. Additionally, activation of immune cells, such as B-cells, natural killer (NK) and natural killer-T (NKT) cells, has been found after treatment of hepatitis C virus (HCV)-infected patients.[56]

B-class CpG ODNs have also been studied for their adjuvant effects when combined with commercial vaccines, such as the hepatitis B virus (HBV) vaccine Engerix-B.[25] Vaccination using a combination of Engerix-B vaccine and CpG ODN resulted in a significant improvement as compared to the HBV vaccine alone, since the HBsAG-specific Ab responses appeared sooner and levels were significantly higher in volunteers receiving CpG ODN as adjuvant. Vaximmune™ is currently under investigation as a cancer vaccine adjuvant and is planned to enter a Phase III clinical trial in resectable stage I, II and III lung cancer.

It is known that the Th2-type cytokines IL-4, IL-5, IL-6, IL-10 and IL-13 play a central role in the pathogenesis of allergic diseases, including asthma. While the Th2-type cytokines are secreted by activated $CD4^+$ T-cells, the Th1-type cytokines IL-2 and IFN-γ are produced by Th1 cells (Figure 6.3). Th1 and Th2 cells interact in a counter-regulatory fashion, since the Th2-type cytokines Il-4 and IL-10 promote Th2 development and inhibit Th1 cell and cytokine production. In contrast, IFN-γ inhibits the proliferation of Th2 cells and induces a Th1-type environment.[57] Therefore, if asthma is caused by a Th2 immune response to inhaled environmental antigens and childhood infection is able to protect against this, then induction of a Th1-like response should prevent the Th2-type immune stimulatory effects causing allergic asthma. As noted above, CpG ODNs are very effective agents to induce a Th1-like immune response and thus might enable the development of therapeutics which interfere with the cause of the allergic disease rather than curing the symptoms only. It has been reported that co-administration of CpG ODNs with an antigen efficiently prevented airway eosinophilia, Th2 cytokine induction, IgE production and bronchial hyper-reactivity in a murine model of asthma.[58] In addition, CpG ODNs were able to prevent allergen-induced airway inflammation in a previously sensitized mouse, suggesting that exposure to CpG ODN may protect against asthma. It was also shown that a single dose of CpG ODN inhibited airway eosinophilia at least as efficiently as daily injections of corticosteroids, the standard treatment for allergic airway disease.[59] A CpG ODN (AVE0675, developed by Coley for Sanofi-Aventis) delivered by inhalation is currently in a Phase I clinical trial in asthma–allergy.

In conclusion, the available results from preclinical and clinical studies suggest that TLR9-targeted therapy can stimulate both innate and adaptive immunity *in vivo*. It also appears that CpG ODNs are generally safe, although moderate mechanism-based as well as backbone-related side effects have been reported. Vaccination trials suggest that CpG can effectively boost the immunogenicity of vaccines. Since most studies so far have been performed with first-generation CpG ODNs, it will be exciting to see how second-generation immune stimulatory drug candidates with enhanced potency and efficacy can further improve the utility of this class of therapeutic agents.

Acknowledgements

I would like to thank my colleagues Dr Marion Jurk and Dr Joerg Vollmer for kindly providing the biological activity graphs. Further, careful reading of the manuscript by Dr Christian Schetter and David Schubert is highly appreciated.

References

1. S. Akira, K. Takeda and T. Kaisho, *Nat. Immunol.*, 2001, **2**, 675.
2. A. G. Bowie, *Clin. Exp. Immunol.*, 2007, **147**, 217.
3. A. G. Bowie and K. A. Fitzgerald, *Trends Immunol.*, 2007, **28**, 147.
4. E. M. Creagh and L. A. O'Neill, *Trends Immunol.*, 2006, **27**, 352.
5. G. Trinchieri and A. Sher, *Nat. Rev. Immunol.*, 2007, **7**, 179.
6. G. Hartmann, R. D. Weeratna, Z. K. Ballas, P. Payette, S. Blackwell, I. Suparto, W. L. Rasmussen, M. Waldschmidt, D. Sajuthi, R. H. Purcell, H. L. Davis and A. M. Krieg, *J. Immunol.*, 2000, **164**, 1617.
7. L. Alexopoulou, A. C. Holt, R. Medzhitov and R. A. Flavell, *Nature*, 2001, **413**, 732.
8. H. Hemmi, O. Takeuchi, T. Kawai, T. Kaisho, S. Sato, H. Sanjo, M. Matsumoto, K. Hoshino, H. Wagner, K. Takeda and S. Akira, *Nature*, 2000, **408**, 740.
9. H. Hemmi, T. Kaisho, O. Takeuchi, S. Sato, H. Sanjo, K. Hoshino, T. Horiuchi, H. Tomizawa, K. Takeda and S. Akira, *Nat. Immunol.*, 2002, **3**, 196.
10. M. Jurk, F. Heil, J. Vollmer, C. Schetter, A. M. Krieg, H. Wagner, G. Lipford and S. Bauer, *Nat. Immunol.*, 2002, **3**, 499.
11. A. Poltorak, X. He, I. Smirnova, M. Y. Liu, C. V. Huffel, X. Du, D. Birdwell, E. Alejos, M. Silva, C. Galanos, M. Freudenberg, P. Ricciardi-Castagnoli, B. Layton and B. Beutler, *Science*, 1998, **282**, 2085.
12. A. M. Krieg, A. K. Yi, S. Matson, T. J. Waldschmidt, G. A. Bishop, R. Teasdale, G. A. Koretzky and D. M. Klinman, *Nature*, 1995, **374**, 546.
13. A. M. Krieg, *Annu. Rev. Immunol.*, 2002, **20**, 709.
14. E. Uhlmann and J. Vollmer, *Curr. Opin. Drug Discov. Dev.*, 2003, **6**, 204.
15. J. Vollmer, *Expert Opin. Biol. Ther.*, 2005, **5**, 673.
16. C. Schetter and J. Vollmer, *Curr. Opin. Drug Discov. Dev.*, 2004, **7**, 204.
17. W. Cao and Y. J. Liu, *Curr. Opin. Immunol.*, 2007, **19**, 24.
18. M. Schnare, A. C. Holt, K. Takeda, S. Akira and R. Medzhitov, *Curr. Biol.*, 2000, **10**, 1139.
19. K. Honda, H. Yanai, T. Mizutani, H. Negishi, N. Shimada, N. Suzuki, Y. Ohba, A. Takaoka, W. C. Yeh and T. Taniguchi, *Proc. Natl. Acad. Sci. U. S. A.*, 2004, **101**, 15416.
20. V. Hornung, J. Ellegast, S. Kim, K. Brzozka, A. Jung, H. Kato, H. Poeck, S. Akira, K. K. Conzelmann, M. Schlee, S. Endres and G. Hartmann, *Science*, 2006, **314**, 994.

21. T. Saito and M. Gale Jr, *Curr. Opin. Immunol.*, 2007, **19**, 17.
22. H. Kato, O. Takeuchi, S. Sato, M. Yoneyama, M. Yamamoto, K. Matsui, S. Uematsu, A. Jung, T. Kawai, K. J. Ishii, O. Yamaguchi, K. Otsu, T. Tsujimura, C. S. Koh, C. Reis e Sousa, Y. Matsuura, T. Fujita and S. Akira, *Nature*, 2006, **441**, 101.
23. B. R. Williams, *Oncogene*, 1999, **18**, 6112.
24. G. Hartmann and A. M. Krieg, *J. Immunol.*, 2000, **164**, 944.
25. C. L. Cooper, H. L. Davis, M. L. Morris, S. M. Efler, M. A. Adhami, A. M. Krieg, D. W. Cameron and J. Heathcote, *J. Clin. Immunol.*, 2004, **24**, 693.
26. J. Vollmer, R. Weeratna, P. Payette, M. Jurk, C. Schetter, M. Laucht, T. Wader, S. Tluk, M. Liu, H. L. Davis and A. M. Krieg, *Eur. J. Immunol.*, 2004, **34**, 251.
27. S. L. Fanning, T. C. George, D. Feng, S. B. Feldman, N. J. Megjugorac, A. G. Izaguirre and P. Fitzgerald-Bocarsly, *J. Immunol.*, 2006, **177**, 5829.
28. C. Guiducci, G. Ott, J. H. Chan, E. Damon, C. Calacsan, T. Matray, K. D. Lee, R. L. Coffman and F. J. Barrat, *J. Exp. Med.*, 2006, **203**, 1999.
29. C. C. Wu, J. Lee, E. Raz, M. Corr and D. A. Carson, *J. Biol. Chem.*, 2004, **279**, 33071.
30. D. Yu, Q. Zhao, E. R. Kandimalla and S. Agrawal, *Bioorg. Med. Chem. Lett.*, 2000, **10**, 2585.
31. R. F. Ashman, J. A. Goeken, J. Drahos and P. Lenert, *Int. Immunol.*, 2005, **17**, 411.
32. M. Jurk, *Clin. Invest. Med.*, 2005, **27**, 2333.
33. M. Jurk, A. Kritzler, H. Debelak, J. Vollmer, A. M. Krieg and E. Uhlmann, *ChemMedChem.*, 2006, **1**, 1007.
34. A. M. Krieg, P. Guga and W. Stec, *Oligonucleotides*, 2003, **13**, 491.
35. D. P. Sester, S. Naik, S. J. Beasley, D. A. Hume and K. J. Stacey, *J. Immunol.*, 2000, **165**, 4165.
36. Z. K. Ballas, W. L. Rasmussen and A. M. Krieg, *J. Immunol.*, 1996, **157**, 1840.
37. M. Koziolkiewicz, A. Owczarek and E. Gendaszewska, *Antisense Nucleic Acid Drug Dev.*, 1999, **9**, 171.
38. D. Yu, E. R. Kandimalla, Q. Zhao, Y. Cong and S. Agrawal, *Nucleic Acids Res.*, 2002, **30**, 1613.
39. Q. Zhao, J. Temsamani, P. L. Iadarola, Z. Jiang and S. Agrawal, *Biochem. Pharmacol.*, 1996, **51**, 173.
40. D. Yu, E. R. Kandimalla, Q. Zhao, Y. Cong and S. Agrawal, *Bioorg. Med. Chem.*, 2001, **9**, 2803.
41. A. M. Krieg, *Biochim. Biophys. Acta*, 1999, **1489**, 107.
42. E. R. Kandimalla, D. Yu, Q. Zhao and S. Agrawal, *Bioorg. Med. Chem.*, 2001, **9**, 807.
43. D. Yu, E. R. Kandimalla, Q. Zhao, L. Bhagat, Y. Cong and S. Agrawal, *Bioorg. Med. Chem.*, 2003, **11**, 459.
44. E. R. Kandimalla, L. Bhagat, Y. Li, D. Yu, D. Wang, Y. P. Cong, S. S. Song, J. X. Tang, T. Sullivan and S. Agrawal, *Proc. Natl. Acad. Sci. U. S. A.*, 2005, **102**, 6925.

45. S. Henry, K. Stecker, D. Brooks, D. Monteith, B. Conklin and C. F. Bennett, *J. Pharmacol. Exp. Ther.*, 2000, **292**, 468.
46. J. Vollmer, J. S. Jepsen, E. Uhlmann, C. Schetter, M. Jurk, T. Wader, M. Wullner and A. M. Krieg, *Oligonucleotides*, 2004, **14**, 23.
47. B. O. Noll, M. J. McCluskie, T. Sniatala, A. Lohner, S. Yuill, A. M. Krieg, C. Schetter, H. L. Davis and E. Uhlmann, *Biochem. Pharmacol.*, 2005, **69**, 981.
48. B. O. Noll, H. Debelak and E. Uhlmann, *J. Chromatogr. B Analyt. Technol. Biomed. Life Sci.*, 2007, **847**, 153.
49. A. M. Krieg, *Nat. Rev. Drug Discov.*, 2006, **5**, 471.
50. J. Vollmer, *Int. Rev. Immunol.*, 2006, **25**, 155.
51. D. M. Klinman, *Expert Opin. Biol. Ther.*, 2004, **4**, 937.
52. A. M. Krieg, S. M. Efler, M. Wittpoth, M. J. Al Adhami and H. L. Davis, *J. Immunother.*, 2004, **27**, 460.
53. S. Paul, *Curr. Opin. Mol. Ther.*, 2003, **5**, 553.
54. C. Manegold, G. Leichman, D. Gravenor, D. Woytowitz, J. Mezger, C. Haarmann, M. Al-Adhami, T. Schmalbach and J. Whisnant, *Eur. J. Cancer*, 2005, **3** (Suppl.), 326.
55. A. M. Krieg, L. Love-Homan, A. K. Yi and J. T. Harty, *J. Immunol.*, 1998, **161**, 2428.
56. J. G. McHutchison, B. R. Bacon, E. Gordon, E. Lawitz, M. Shiffman, N. H. Afdhal, A. Jacobson, A. Muir, S. M. Efler, A. Vicari, K. Myette, P. Ahluwalia, G. Murzenok, G. Lipford and T. Schmalbach, *Hepatology*, 2005, **42** (Suppl 1), 539A.
57. J. N. Kline, *Curr. Top. Microbiol. Immunol.*, 2000, **247**, 211.
58. J. N. Kline, T. J. Waldschmidt, T. R. Businga, J. E. Lemish, J. V. Weinstock, P. S. Thorne and A. M. Krieg, *J. Immunol.*, 1998, **160**, 2555.
59. D. Broide, J. Schwarze, H. Tighe, T. Gifford, M. D. Nguyen, S. Malek, J. Van Uden, E. Martin-Orozco, E. W. Gelfand and E. Raz, *J. Immunol.*, 1998, **161**, 7054.

CHAPTER 7
Decoy Oligodeoxynucleotides to Treat Inflammatory Diseases

MARKUS HECKER,*,[1] SWEN WAGNER,[2]
STEFAN W. HENNING[2] AND ANDREAS H. WAGNER[1]

[1] Institute of Physiology and Pathophysiology, Division of Cardiovascular Physiology, University of Heidelberg, Germany; [2] AVONTEC GmbH, Martinsried, Germany

7.1 Introduction

Apart from epigenetic causes an increased genetic risk for diseases typically stems from mutations or polymorphisms in one or several genes (monogenetic and/or polygenetic diseases) that result either in erroneous transcription of these genes into the corresponding messenger RNA (mRNA) or translation of mRNA into an erroneous peptide or protein. Despite some 30 years of intense research, it is still a major challenge, both from a technical and from a regulatory point of view, to replace a defective gene by way of gene transfer. Interfering with genetically caused changes in gene expression, on the other hand, appears to be less challenging. Especially in terms of a disease-related profuse expression of a given gene, one might employ inhibitory molecules directed against the protein to block or attenuate its activity, such as neutralizing antibodies (the so-called biologicals) or small molecules which mark the latest stage in drug development. Alternatively, nucleic acid-based drugs, such as antisense oligodeoxynucleotides (ODNs) or small interfering RNA (siRNA), might be used, which primarily block translation of the target gene mRNA into protein.

RSC Biomolecular Sciences
Therapeutic Oligonucleotides
Edited by Jens Kurreck
© Royal Society of Chemistry 2008

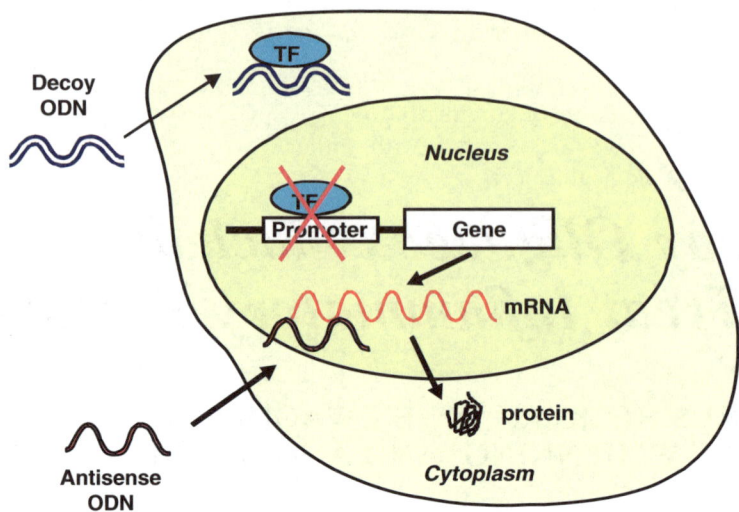

Figure 7.1   The principal mechanism of action of a decoy oligodeoxynucleotide (ODN) as compared to that of an antisense ODN.

Yet another possibility to interfere with the aberrant expression of a given gene is to block the activity of the transcription factor(s) that drive(s) its transcription. To this end, decoy ODNs have been developed. These are short doubled-stranded DNA (dsDNA) molecules which mimic the consensus binding site of their target transcription factor in the promoter of such genes whose expression is governed by the said transcription factor. As a result, decoy ODNs inhibit the target transcription factor from binding to DNA and thus modulate the expression of all genes which are controlled by it; however, they do not normally block expression of the transcription factor itself (Figure 7.1).

7.2   Decoy Oligonucleotide Synthesis

Typically, the double-stranded decoy ODN is synthesized from the complementary single-stranded oligodeoxynucleotides by way of hybridization. The chimeric duplex can be dissolved in aqueous solution or lyophilized as the corresponding sodium salt. All residues are deoxyribonucleosides, and all internucleotide linkages are phosphate diesters, except a varying number of modified phosphorothioate linkages. Non-bridging oxygen atoms in phosphate linkages are typically replaced by sulfur to form phosphorothioate-modified oligodeoxynucleotides which are highly resistant to degradation by nucleases.[1] Initially only synthesized at laboratory scale in quantities up to 10 mg or so, good manufacturing practices (GMP)-certified decoy ODNs are now commercially available in the low-to-mid kilogram range (see www.avecia.com).

7.3 Decoy Oligonucleotide Administration

Some of the most important aspects for the applicability of decoy ODNs as drugs relate to the route of administration, *i.e.* topical or systemic drug delivery, the pharmacokinetics associated with either route of administration and potential toxicological aspects.

7.3.1 Delivery

Probably one of the greatest advantages of decoy ODNs is that they do not need to be transfected into their target cells or require any other auxiliary means to facilitate their uptake into cells to elicit a biological function. This is especially important in the context of a topical or local administration. However, typical lipid-based transfection reagents that are a must for effective delivery of antisense ODNs or siRNAs into cells, even the most advanced reagents available to date, do not facilitate the uptake of decoy ODNs into their target cells. On the contrary, transfection efficiency is much less and most of the DNA appears to be lost in endosomes or lysosomes so that the biological activity is low or even lost.[2] The same might be true for liposomal drug delivery, but very much seems to depend on the liposomal formulation and the density at which the dsDNA is packed into the liposomes, the route of administration and the target cell population (see www.novosom.com).[3] Other approaches tested so far to improve cellular delivery of decoy ODNs include physical pressure,[4] linkage to a transport peptide,[5] microspheres[6] or nanoparticles.[7] Liposomal or artificial viral envelope-based decoy ODN delivery or organ-specific ultrasound destruction of microbubble-encapsulated nucleic acids,[8,9] however, is a must for systemic administration, for example by way of intravenous infusion, and probably also through other compartments of the body, such as with a suppository.

The reason why dsDNA molecules, in contrast to dsRNA molecules and (in the majority of cases) also to single-stranded DNA (ssDNA) molecules are taken up by their target cells with relative ease is best explained by the fact that they are transported into the cells (Figure 7.2). This transport mechanism is clearly distinct from the endocytotic pathway that has been described for antisense ODNs[3] and involves both a saturable carrier-mediated component and a receptor-mediated podocytosis with subsequent intracellular release from caveloae vesicles. While the first pathway seems to be operating primarily under physiological conditions, in particular at neutral pH, the second pathway gains importance under conditions of a decreased pH.[10] In addition, the length (*i.e.* uptake of decoy ODNs greater than 25-mer in size) tends to be rather poor, but also certain sequence characteristics of the dsDNA molecules seem to be important for the speed by which they are taken up by the target cell (see Section 7.4). Kinetic measurements with a ^{35}S-labelled 18-mer decoy ODN against the transcription factor signal transducer and activator of transcription (Stat-1) suggest that this uptake is relatively fast (minutes), with a maximum attained within less than one hour and declining towards a plateau thereafter,

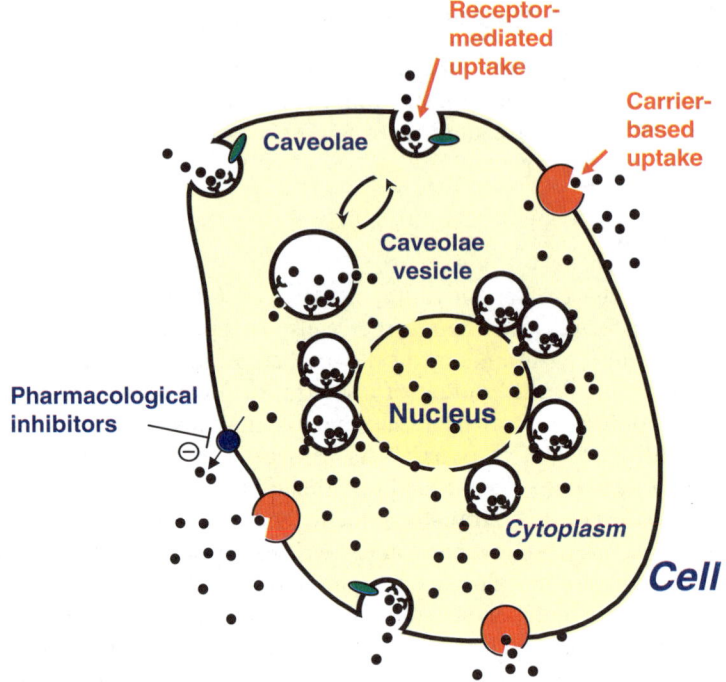

Figure 7.2 The principal uptake mechanisms by which naked, short double-stranded DNA molecules, *i.e.* decoy ODNs, are transported into eukaryotic cells.

which corresponds to approximately one-third of the maximum value (Figure 7.3). Moreover, these measurements also indicate that the concentration of the decoy ODN inside the cell – in this case human endothelial cells were employed – equals its concentration in the cell supernatant, suggesting that after this time a steady state has been obtained. Moreover, the optimal medium concentration of the decoy ODN required to achieve a maximum accumulation inside the target cell is estimated at about $10\,\mu\mathrm{mol}\,\mathrm{L}^{-1}$. Another important aspect of these studies was that the uptake of the decoy ODN, provided that it has a reasonable fit to the targeted transcription factor, determines the strength of its biological activity, which is typically assessed by the relative degree of expression of the main target genes of the transcription factor.

7.3.2 Pharmacokinetics

There is only rather limited information available on this topic from the literature and, until recently, most of this was conclusion by analogy derived from published work on antisense ODNs. However, as explained above, the uptake of dsDNA and ssDNA into cells does not always follow the same principles. Moreover, monitoring the biodistribution, metabolism and elimin-ation of short dsDNA molecules after topical application requires a rather

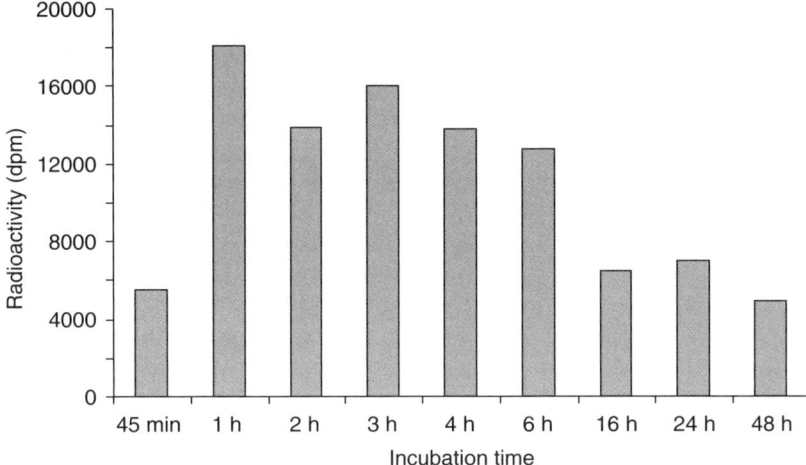

Figure 7.3 Kinetics of the uptake of a ^{35}S-labelled Stat-1 decoy ODN (18-mer) into human-cultured endothelial cells. Note that the calculated maximum concentration of the decoy ODN inside the endothelial cells is approximately three times higher than that in the medium ($10\,\mu\text{mol L}^{-1}$).

sensitive detection method. This sensitivity is neither provided by classic high-performance liquid chromatography or capillary gel electrophoresis followed by ultraviolet (UV) detection, since for these methods the limit of detection for decoy ODNs in aqueous solution is in the low $\mu\text{g mL}^{-1}$ range. When considering the need for a quantitative extraction method from serum, for example, the limit of quantification is even higher by one order of magnitude.

Hybridization assays are much more sensitive and typically do not require extensive sample preparation so that their limit of quantification for decoy ODNs in the various body fluids is in the low pg mL^{-1} range. They all utilize binding of a complementary probe to one of the single strands of the decoy ODN, and in one variation employ a capture probe to immobilize one ssDNA molecule of the duplex and a detection probe that can be labelled with, for example, biotin, thus allowing highly sensitive detection through the formation of a horseradish peroxidase–streptavidin conjugate (see www.criver.com).[11]

With this technique plasma levels down to the low ng mL^{-1} range have been monitored, for example in dogs after intravenous injection of the aforementioned Stat-1 decoy ODN. Compared to intravenous injection, inhalative application of the Stat-1 decoy ODN in the same dose range resulted in peak plasma levels that were two orders of magnitude lower. In either case, the nucleic acid was cleared from the circulation within eight hours. Moreover, topical application of an NF-κB decoy ODN ointment at doses of up to $20\,\text{mg kg}^{-1}$ per day for four weeks in pigs did not result in an accumulation of the decoy ODN in plasma at a lower limit of quantification of $1.3\,\text{ng ODN mL}^{-1}$ plasma.[12] Further absorption, distribution, metabolism, and excretion (ADME) studies with the ^{35}S-labelled Stat-1 decoy ODN in rats suggest that tissue distribution is fairly rapid, with the majority of the nucleic acid being excreted within the first 24 hours via the urine and the faeces.

7.3.3 Toxicology

Again, there is not much information available from the literature on this topic. From the limited knowledge on the toxicity of chemically modified (phosphorothioate) decoy ODNs, one can deduce that such molecules exert a similar toxicology profile to chemically modified antisense ODNs. Their toxicology is mainly governed by non-specific toxicity emerging from the chemical modification. For instance, single intravenous application of the above-mentioned Stat-1 decoy ODN in rats and dogs did not produce any toxicological findings up to double-digit $mg\,kg^{-1}$ doses. Upon repeated administration via the airways, some reversible treatment-related adverse findings were observed at the high doses in the rat, but not in the dog. These were inflammatory cell infiltrates in the lung and spleen, a well-known response of rodents to chemically modified ODNs. These treatment-related adverse findings were substantially reversed five weeks after cessation of treatment. Repeated administration of the same decoy ODN as an ointment for four weeks, however, did not produce any drug-related adverse findings or local dermal reactions in minipigs.

As generally known from other ODNs, the tested decoy ODNs also did not exert any genotoxic effects: According to an *in vitro* Ames test in bacteria and an *in vivo* micronucleus test in mouse bone-marrow cells, no toxic or mutagenic potential of the aforementioned Stat-1 decoy ODN has been observed. Also, various safety pharmacological assessments in the dog and rat did not reveal any adverse effects of the Stat-1 decoy ODN.

Repeated administration of an NF-κB decoy ODN ointment at doses of up to $20\,mg\,kg^{-1}$ day for four weeks in pigs also did not produce any drug-related adverse findings or local dermal reactions. The toxicology data generated for decoy ODNs as yet thus indicate that these molecules are generally safe and their toxicology profile is mainly determined by non-specific effects of the modifications of the phosphate backbone. It remains to be seen from chronic toxicology studies as to what extent long-term attenuation of the respective transcription factor might lead to specific toxicological effects.

7.4 Mechanism of Action

Initiated by numerous studies that demonstrated the importance of altered gene expression in disease initiation and/or progression, the modification of gene expression has emerged as an important therapeutic strategy and agents that inhibit the expression of disease-mediating genes have been developed for some 20 years or so, and continue to be developed. Regulation of gene expression is a complex biological process initiated by the interaction of DNA with transcription factors. Transcription factors are proteins that reside in the cell nucleus or cytoplasm and, upon activation, bind to specific DNA motifs, the consensus binding site, in the promoter region of the target gene. The activation process may involve phosphorylation, dimerization, proteolytic cleavage, ubiquitination and/or translocation of the transcription factor and is governed by signalling events emerging from stimulated membrane receptors. Accordingly, molecules that block one or more steps of the transcription machinery can potentially modify the expression of a specific gene or group of genes, yielding a functional response. It has been shown that short DNA molecules containing the consensus binding site can bind to transcription factors in a highly specific manner, rendering the transcription factor incapable of subsequent binding to the promoter region of the target genes and thus acting as 'decoys' for the targeted transcription factor.[13] This approach involves the intracellular delivery of such decoy ODNs and has been shown to be effective in modulating gene expression *in vitro* and *in vivo*, which has prompted the promotion of its therapeutic use (see Section 7.5). Decoy ODNs can be devised as therapeutic agents that either inhibit the expression of genes that are transactivated by the transcription factor in question, or upregulate genes which are transcriptionally suppressed by the binding of an inhibitory transcription factor.[14,15] In contrast to antisense ODNs (which hybridize with the transcribed and processed mRNA and hence block its translation) and double-stranded short hairpin RNA (shRNA), siRNA or microRNA (which act on the same level of gene expression by facilitating the destruction of the target mRNA), double-stranded decoy ODNs bearing the consensus binding sequence of a specific transcription factor inhibit the transcription of disease-mediating genes by direct interaction with the transcription factor. They thus act further 'upstream' in the cascade of events that lead to the expression of disease-mediating genes (see Figure 7.1).

7.4.1 Target Identification

Of paramount importance is the choice of the transcription factor that is to be targeted for the prevention and/or therapy of a given disease. To date less than 20 of the more than 2000 transcription factors encoded in the human genome[16] have been targeted in animal experimental models or in the clinic (see Section 7.5). They are mainly involved in the regulation of pro-inflammatory as well as growth and differentiation-promoting gene products. Typically, their role in the expression of a disease-mediating gene is first tested *in vitro* in an appropriate and, with respect to the chosen stimulus, well-controlled cell culture model. As an example, the concentration-dependent inhibition by a Stat-1 decoy ODN of CD40 expression in cultured human endothelial cells stimulated by interferon-γ (IFN-γ) and/or tumour necrosis factor-α (TNF-α), is shown in Figure 7.4. Subsequently, the efficacy of the decoy ODNs is verified in more complex models, such as isolated blood vessels or organs, and finally in the whole organism, which certainly represents the greatest challenge both in terms of complexity and drug delivery. Usually, the transcription factor target will be chosen because of its known role in the expression of gene(s) whose expression is relevant for the disease to be treated. However, it may be that this has to be

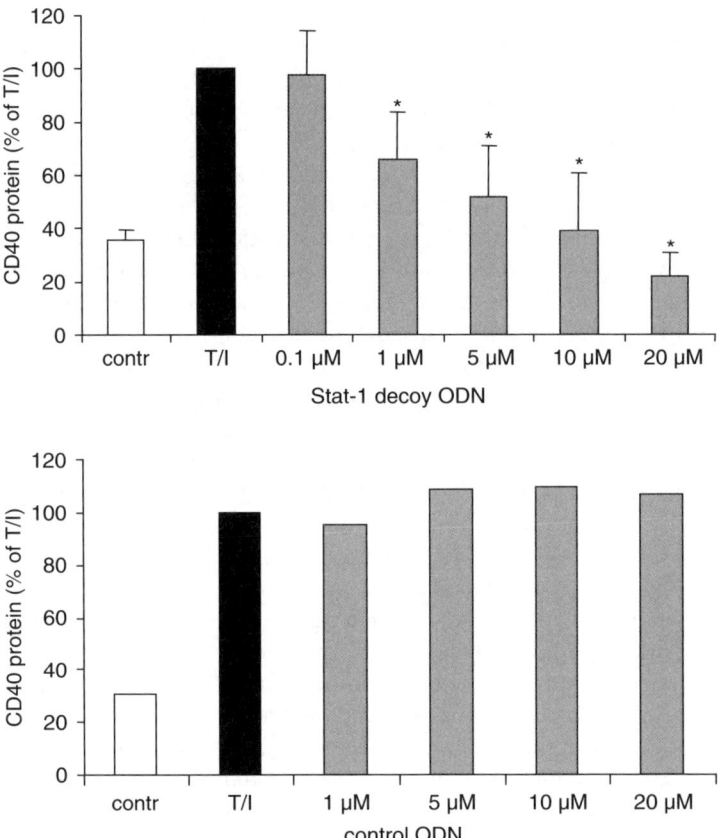

Figure 7.4 Concentration-dependent inhibition by a Stat-1 decoy ODN of CD40 protein expression in human cultured endothelial cells stimulated by interferon-γ (1000 U mL^{-1}) plus TNF-α (100 U mL^{-1}) (concentration as indicated, pre-incubation period was four hours). The corresponding mutated control ODN did not affect CD40 protein levels. (n = 6, $*P < 0.05$ *vs.* non-stimulated control).

verified first, for example by way of reporter gene analyses with appropriate deletion constructs of the target gene or antisense ODN and/or siRNA-mediated knockdown of the target transcription factor.

7.4.2 Target Specificity

Decoy ODNs can be designed in such a way that their sequence mimics the consensus binding site of their target transcription factor. Normally, this consensus sequence is derived as an integrated site from transcription binding sites in the promoter of several genes, and in the case of a known transcription factor can be retrieved from pertinent databases. A handy tool in this context is, for example, MatInspector,[17] a software that gives access to a large library of matrix descriptions for transcription-factor binding sites to locate matches in DNA sequences (see http://www.genomatix.de/products/MatInspector/). This software can also be utilized to design an appropriate mutated control ODN or to make sure that a scrambled control ODN of random sequence does not bind to and influence another transcription factor (off-target effect). In any case, the consensus sequence typically comprises six to ten bases and is highly specific for the targeted transcription factor. Exceptions to this rule are overlapping binding sites to which different transcription factors can bind [such as the cAMP response element (CRE), to which the CRE binding protein (CREB) and also members of the activator protein-1 (AP-1) family of transcription factors bind] or common bindings sites with which two different transcription factors interact in a competitive manner [such as, for instance, the STRE-binding transcription factors early growth response protein-1 (Egr-1) and Sp-1]. In this case, it may be reasonable to derive the sequence from the binding site in the promoter of the primary target gene rather than from the consensus binding site. Another problem is that members of relatively large families of transcription factors, such as AP-1 or the Stat proteins, may bind to the same or a closely related sequence. Thus, by using a decoy ODN design on the basis of the consensus sequence of these transcription factors, no discrimination between different members of these transcription factor families may be possible.

There are several experimental approaches to test for the specificity of a given decoy ODN sequence. One is electrophoretic mobility shift analyses (EMSA), in which nuclear cell extracts are exposed to a radioactively labelled (typically ^{32}P-labelled) dsDNA probe comprising a generic binding sequence for the transcription factor(s) to be tested followed by native polyacrylamide gel electrophoresis. Displacement of the labelled probe from the transcription factor in the extract by addition of the test decoy ODN would result in diminution of the labelled DNA–protein complex. Unequivocal identification of the targeted transcription factor can then be achieved by antibody-based supershift analyses. A second experimental possibility is to monitor changes in the expression of genes which are known to be governed by the target transcription factor and a closely related transcription factor in the presence of a given stimulus, and to test the selective effects of the decoy ODN on the transcription of certain genes only in this cellular model. A third possibility, which also allows an estimate of the binding affinity between transcription factor and decoy ODN, is a sandwich enzyme-linked immunosorbent assay (ELISA) system. In this assay, a nuclear extract from stimulated cells containing the activated transcription factor or the purified protein itself is allowed to bind to an immobilized dsDNA probe comprising the consensus sequence for the transcription factor(s) in question. The binding and the (possible) displacement of the transcription factor from this probe by the test decoy ODN is visualized by using a specific, preferably monoclonal, antibody against the transcription factor (Figure 7.5). The latter assay not only enables a precise definition of the decoy ODN's specificity, but can also be employed to improve its efficacy by selecting the sequence with the highest affinity. Moreover, it provides a means

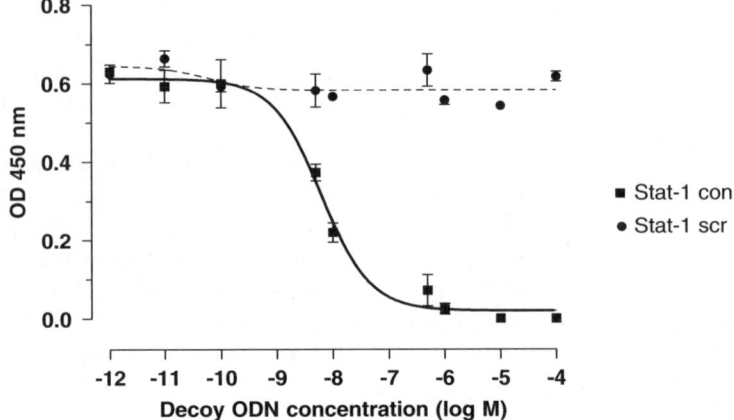

Figure 7.5 ELISA determination of the apparent affinity with which a Stat-1 decoy ODN binds to the target transcription factor (nuclear extract of COS7 cells stimulated by interferon-γ). The corresponding scrambled control ODN does not seem to bind to the transcription factor in the tested range of concentrations.

to determine to what extent chemical modifications of the ODN, such as the degree of phosphorothioate linkages in its backbone, affect its affinity and/or specificity to the transcription factor of interest.

7.4.3 Sequence Optimization for Cellular Uptake

The efficacy of a given decoy ODN not only depends on a best fit to the *cis*-element recognition sequence of the target transcription factor, but also on the ease and speed by which it is taken up by the target cells (see Section 7.3). This can be monitored best by using a fluorescent dye-labelled decoy ODN such as Texas Red or Alexa Fluor 594 followed by fluorescence microscopy or by means of a fluorescence reader. Such techniques are suitable not only for cultured cells *in vitro*, but also for tissues or organs *ex vivo* (Figure 7.6). Alternatively, radioactively labelled nucleic acids can be employed. In the case of phosphorothioate-modified decoy ODNs, this is best achieved by using ^{35}S as a label which is incorporated into at least one or several of the phosphorothioates. ^{32}P or ^{33}P-labelling of the nucleic acids, as is usually done with the EMSA ODN probes, appears less suitable because of rapid phosphodiesterase-mediated loss of the terminal label inside the cells.

At least four parameters influence uptake of the naked dsDNA into eukaryotic cells:

  (i) concentration,
  (ii) length,
 (iii) nucleotide sequence,
 (iv) nucleotide composition.

nuclear stain   Stat-1 decoy ODN
(DAPI)   (Texas Red)

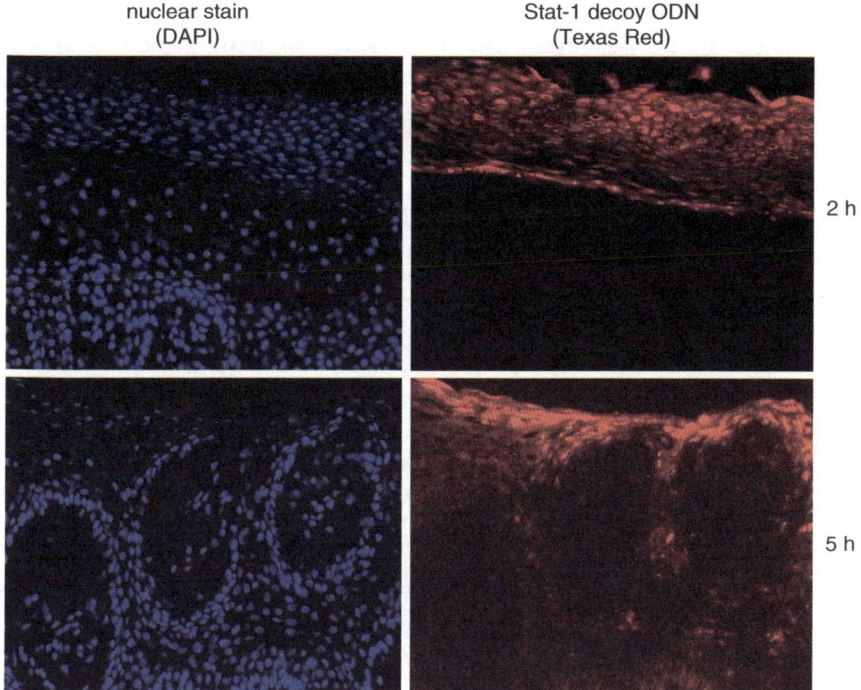

Figure 7.6 Fluorescence micrographs (magnification 200×) demonstrating the uptake of a Texas Red-labelled Stat-1 decoy ODN (18-mer) ointment into human skin samples (red) derived from a patient with psoriasis *ex vivo* at the indicated time points. Nuclear localization was assessed by DAPI staining (blue).

Maximum cellular uptake is achieved at an extracellular concentration of $10 \, \mu mol \, L^{-1}$ and a pre-incubation period of least one hour. In most cases, this will result in a satisfactory biological effect, *i.e.* maximum or near-maximum inhibition of target gene expression, which, however, rarely exceeds 60–80% inhibition relative to a mutated control ODN. Ideally, the length of the decoy ODN should not exceed 30 base pairs, uptake of a 25-mer is clearly more efficient and the lower limit, also in the context of specificity, appears to be a 10-mer. The majority of the decoy ODNs employed over recent years is in the 15 to 20-mer range. However, uptake kinetics of decoy ODNs in this range appear to differ with respect to their nucleotide sequence and possibly also their nucleotide composition, which may be interpreted as another indication for the existence of (a) specific transport system(s) for these short dsDNA molecules. The characteristics of these transport systems may be exploited, for example, by altering the extracellular milieu (altered ionic composition, pH), to speed up the uptake of decoy ODNs which, because of certain restrictions, *i.e.* lengthy consensus binding site, cannot be optimized according to the above-mentioned rules. These rules, however, only apply to the topical application of the decoy

ODNs – systemic administration in which the nucleic acids most likely are encapsulated in a liposomal envelope follows different principles that are mainly determined by the characteristics of the vehicle.

7.4.4 Stability

Decoy ODN duplexes dissolved in aqueous solution are relatively stable at neutral pH and ambient temperature. There are now sufficient real-time stability data available to support the use of decoy ODNs in clinical studies over extended periods. Decoy ODNs lyophilyzed as the corresponding sodium salt and stored at 5 °C are even more stable than those in aqueous solutions. Until recently, their stability upon uptake into eukaryotic cells could only be estimated according to the duration of their biological efficacy, *i.e.* effective binding of the target transcription factor according to EMSA. In some animal experimental models this seemed to be in excess of 48 hours[18] despite a rather moderate degree of phosphorothioate-modified linkages (only the 3'- and 5'-terminal three nucleotides on either strand).

With the aid of the ^{35}S-labelled Stat-1 decoy ODN it is now possible, by way of native polyacrylamide gel electrophoresis and autoradiography, to determine the stability of the nucleic acid in human umbilical vein endothelial cells. No degradation of the duplex was observed over a 24-hour period. It appears, therefore, that decoy ODNs as soon as they have been taken up by their target cells remain stable and retain their activity for quite some time. More work is necessary, though, to delineate whether chemical modifications of some other sort would achieve the same degree of stability or improve it even further. In rat serum at 37 °C, however, degradation of phosphorothioate-modified decoy ODNs is evident with a calculated half-life of 16 hours (Figure 7.7).

7.5   Therapeutic Applications

Thus far, the therapeutic potential of transcription-factor neutralization by decoy ODNs has been tested not only in a variety of animal experimental models, but also in humans (see Section 7.6). Most studies have assessed their therapeutic applicability in subacute and chronic inflammation, while there are a few reports on infectious diseases, including sepsis. In addition, there are a number of reports on the antiproliferative and antitumorigenic properties of decoy ODNs, mainly targeting the transcription factors AP-1, CRE, E2F, NF-κB, Sp-1 and Stat-3. However, the focus here is on inflammatory diseases.

7.5.1 Asthma

With a prevalence of 5–10% and an annual increment of some 10%, asthma is one of the most common diseases in developed countries. It is an inflammatory disease of the conducting airways characterized by bronchial oedema, bronchial hyper-responsiveness and reversible airway obstruction. In mild-to-moderate

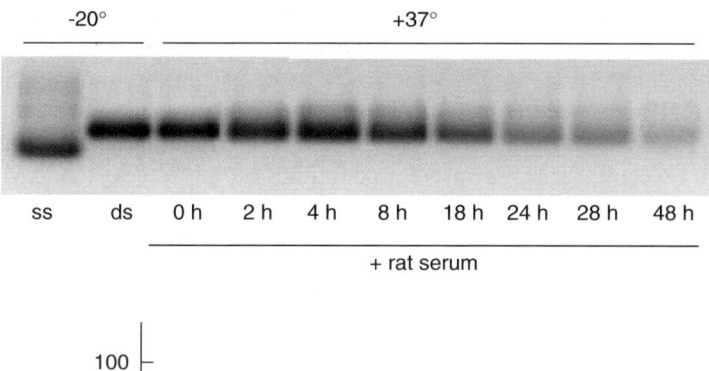

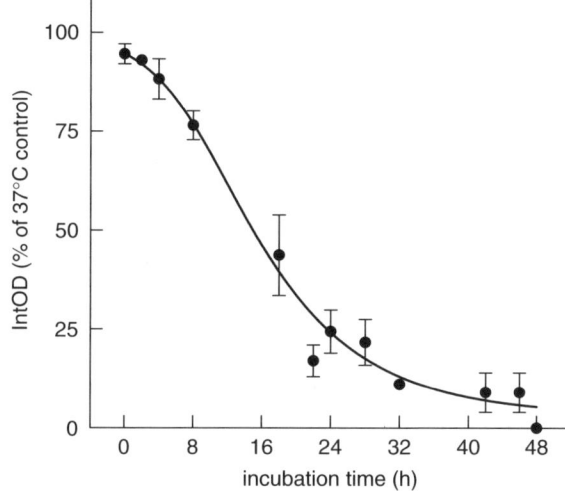

Figure 7.7 Time-dependent degradation of phosphorothioate-modified Stat-1 decoy ODN in fresh rat serum (derived from five different animals), as determined by agarose gel electrophoresis and UV light detection.

disease, standard therapies, such as short-acting beta-agonists and topical or systemic corticosteroids, are very effective. However, in severe asthma, affecting only 5–10% of the asthma population, but accounting for up to half of the total treatment costs, inflammation and clinical symptoms are under poor control; as a result morbidity and mortality are highest among these patients. Thus far, three transcription factors have been addressed as possible drug targets in asthma, *i.e.* AP-1,[19] NF-κB[20] and Stat-1.[21] Corresponding decoy ODNs have been tested in the same standard model of allergic airway inflammation, *i.e.* in ovalbumin-sensitized mice, where they were administered once or twice by intratracheal instillation prior to the ovalbumin aerosol challenge. It may not come as a surprise, therefore, that the three decoy ODNs exerted comparable inhibitory effects on leukocyte infiltration, airway hyperresponsiveness and expression of cytokines, such as interleukin-5 (IL-5), in this model, while the corresponding scrambled or mutated control ODNs were without effect. Despite this apparent similarity, there were some subtle differences between the three decoy ODNs. Thus, the NF-κB decoy ODN was somewhat less effective than the AP-1 decoy ODN with respect to suppression of bronchial hyper-reactivity – both decoy ODNs were administered intra-tracheally at a dose (15 nmol) which was twice as high as that of the Stat-1 decoy ODN (8 nmol), which was administered intranasally. Moreover, only the Stat-1 decoy ODN was shown to block expression of endothelial cell adhesion molecules, such as vascular cell adhesion molecule-1 (VCAM-1), and that of co-stimulatory molecules for T-helper (Th) cells, such as CD40.

Recently, the inhibitory effect of the Stat-1 decoy ODN was independently validated in a different model of airway inflammation and hyper-responsiveness in which mice are exposed to house-dust mite allergens (Professor Amy Simon, Tufts University Medical School, Boston, personal communication). These findings not only support a role for the Janus kinases (JAK)–Stat pathway in the pathogenesis of asthma,[22] but also point to the significance of the interaction between endothelial cells lining the bronchial blood vessels and infiltrating leukocytes, namely Th cells. The overlapping expression profile of pro-inflammatory genes controlled by AP-1, Stat-1 and NF-κB, at least in rodents, further suggests that these decoy ODNs, and in particular the Stat-1 decoy ODN, interfere with the innate immune response that eventually also boosts the specific type 2 Th cell-driven, IgE-mediated pulmonary immune response against a given allergen.

In this context, topical administration of decoy ODNs may provide a meaningful therapeutic alternative and/or addition to the inhaled or systemic corticosteroids that are administered to patients with intrinsic asthma. However, the powerful anti-inflammatory and anti-hyper-reactivity effect, especially of the Stat-1 decoy ODN, may be exploited for the treatment of other allergic airway diseases, such as allergic rhinitis or hay fever. Of the aforementioned decoy ODNs, only the Stat-1 decoy ODN is currently in clinical development in two extended Phase IIa trials for the treatment of patients with asthma (see Section 7.6).

7.5.2 Atherosclerosis and Related Vascular Remodelling Processes

Coronary heart disease (CHD) is the leading cause of death in industrialized nations. Although cellular events leading to the formation of coronary atherosclerotic lesions are not yet fully characterized, persistent dysfunction of the endothelium in affected arteries is an important aspect of this chronic inflammatory disease.[23] Apart from environmental factors that influence endothelial function, intrinsic impairment of the expression of endothelial gene products involved in the maintenance of vascular homeostasis may predispose to atherosclerosis and thus CHD. One such gene product is the endothelial isoform of nitric oxide (NO) synthase, the expression of which under physiological conditions is controlled by the level of shear stress, *i.e.* the viscous drag exerted on the luminal surface of the endothelial cells by the flowing blood. This ultimately determines the capacity of the endothelial cells to synthesize NO, which is a powerful anti-inflammatory and hence anti-atherosclerotic molecule that mainly acts on the endothelial cells themselves, but also on neighbouring cells, such as the smooth muscle cells of the vessel wall and infiltrating leukocytes.

In about 1998 a single nucleotide polymorphism (SNP) was identified in the promoter region of the endothelial NO synthase (eNOS) gene,[24] which predisposes homozygous Caucasian individuals to an increased risk of developing CHD.[14,25] Subsequent functional studies with clonally expanded human endothelial cells harbouring the various genotypes revealed that this T to C transition at position -786 of the eNOS gene not only affected the basal level of expression of the enzyme, but virtually abrogated the responsiveness of the gene to shear stress. Pre-treatment of homozygous C/C genotype cells with a decoy ODN comprising position -800 to -779 of the C-type, but not the T-type, NO synthase gene promoter fully reconstituted the responsiveness of the defective gene to shear stress.[14] It appears, therefore, that an inhibitory DNA binding protein, presumably a transcription factor, binds with high affinity to the C-type promoter of the eNOS gene, thereby blocking shear-stress dependent maintenance of an adequate level of expression of the enzyme and, as a consequence, attenuating the NO-synthesizing capacity of the endothelial cells (Figure 7.8). This, so far, is probably the only example that may also be extended to rheumatoid arthritis (see below) and possibly to different forms of vasculitides in which a decoy ODN is employed for deblocking rather than blocking the expression of a target gene. Moreover, with all the necessary precautions, it might be considered as a tool to repair defective genes functionally, thus moving the entire technology platform to what one would name

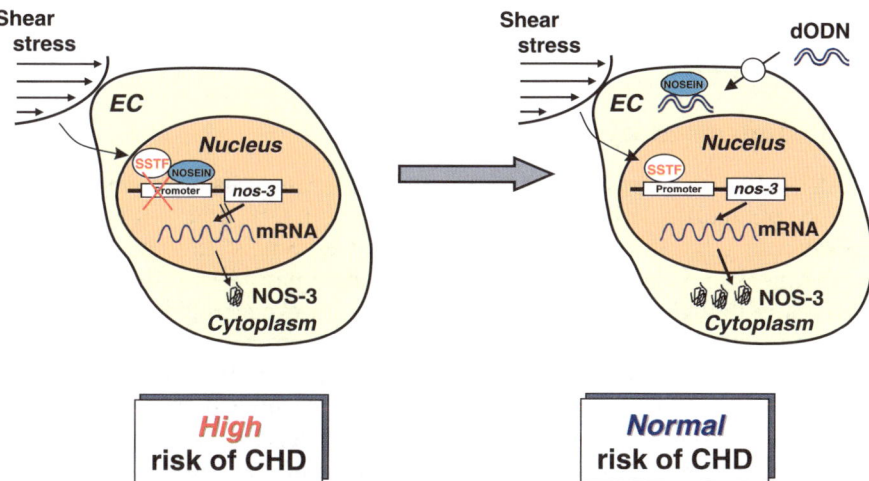

Figure 7.8 The putative mechanism of action of the NO synthase expression inhibiting protein (NOSEIN) and functional correction of the endothelial NO synthase T-786C SNP by using an anti-NOSEIN decoy ODN, thus lowering the risk of coronary heart disease (CHD). Note that SSTF stands for shear stress-sensitive transcription factor.

personalized medicine, *i.e.* the treatment of a patient according to individual genetic risk.

Other inflammation-driven remodelling processes of the vessel wall, apart from chronic transplant rejection as outlined below, include restenosis following angioplasty or venous bypass graft vasculopathy. Similar to transplant atherosclerosis these remodelling processes develop much faster than the classic form of atherogenesis, which is the source of myocardial infarction, stroke and peripheral arterial occlusive disease. In addition to an enhanced infiltration of leukocytes into the vessel wall, especially smooth muscle cell hypertrophy and/or hyperplasia as well as profound changes in the composition of the extracellular matrix are hallmarks of these remodelling processes. Therefore transcription factor targets have been chosen for attenuating restenosis after angioplasty that cannot be considered as prototypically pro-inflammatory, but rather as pro-mitogenic or pro-fibrotic, such as AP-1 and CCAAT enhancer binding protein (C/EBPβ; in humans NF-IL6). Moreover, these transcription factors are known to be activated by biomechanical stimuli, such as a supra-physiological increase in pressure or wall tension, which is inextricably connected with this method of intervention. Both the AP-1 decoy ODN in a pig coronary angioplasty model[18] and the C/EBP decoy ODN in a rabbit carotid artery injury model[26] exerted profound effects on, for example, neointima formation and smooth muscle cell proliferation when administered once peri-procedurally to the site of pressure traumatization (up to 16 atmospheres during inflation of the balloon) of the vessel wall. Vein graft disease can also be considered as a pressure trauma to which the venous bypass grafts are exposed to when positioned in the arterial part of the circulation. Here, decoy ODNs against AP-1,[27] E2F[28] and NF-κB[29] have been successfully tested in rabbit jugular vein to carotid artery interposition graft models with or without experimental dyslipidemia.

7.5.3   Inflammatory Skin Disease

T-Cell mediated skin diseases, mainly comprising psoriasis (prevalence >2%), allergic contact dermatitis (ACD, prevalence about 9%) and atopic dermatitis (prevalence >3% in developed countries and especially among Caucasians), are the most prevalent and severe chronic inflammatory skin diseases. While psoriasis is thought to be based primarily on an exaggerated Th1 or maybe Th17-type immune response, atopic dermatitis more likely represents an overt Th2-type immune response. ACD lies somewhere in the middle, revealing characteristics of either immune response, but seeming to be inclined more towards the Th1-type, especially during the induction or sensitization phase. ACD is an eczematous reaction to repeated exposure against an allergen and can be regarded as a delayed-type hypersensitivity response with a sensitization phase, generally asymptomatic, an effector and a resolution phase. Only superficial regions of the skin are affected with inflammation present in the outer dermis and epidermis, and the cellular infiltrate is mainly composed of mononuclear cells.

Generally, ACD has been regarded as a Th1-type immune response, and the crucial role of the Th1 cytokine IFN-γ, especially during the induction (sensitization) phase, remains undisputed. Thus, it does not come as a surprise that the transcription factor Stat-1, as the principal effector molecule in IFN-γ signalling, is considered to be a pivotal drug target not only in ACD but also in psoriasis. Topical corticosteroids, which have been widely used in ACD therapy, exert their anti-inflammatory effects mainly through inhibition of the transcription factors AP-1 and NF-κB. However, long-term treatment with these corticosteroids may produce significant adverse effects, such as skin atrophy at the site of application, and these have spurred the search for other treatment modalities, such as drugs based on nucleic acid.

In comparison with common topical steroids, the efficacy of a Stat-1 decoy ODN containing ointment on dinitrohalobenzene-induced ACD, both in guinea pigs and domestic pigs, has recently been examined. In guinea pigs, single application of the Stat-1 decoy ODN, but not that of an appropriate control ODN, in a rather simple ointment not only facilitated its penetration deep into the dermis (see Figure 7.6), but also resulted in a profound dose-dependent improvement of both the macroscopic and histopathological signs of inflammation. This was accompanied by a significant decrease in leukocyte (polymorphonuclear neutrophils, monocytes, T-cells) infiltration into the dermis (Figure 7.9) and reduced skin thickening. Moreover, the Stat-1 decoy ODN ointment strongly attenuated the expression of IL-1β, IL-8 and IL-12 as well as that of TNF-α and IFN-γ. When compared to the simultaneously tested topical steroids (clobetasol, hydrocortisone), the Stat-1 decoy ODN ointment appeared to be at least as effective. A related Stat-1 decoy ODN, when injected intra-articularly, also profoundly blocked the delayed-type hypersensitivity reaction to injection of the antigen seven days after the induction of arthritis.[30]

Another interesting target molecule for the treatment of inflammatory skin diseases is NF-κB. Thus, a 1–2% ointment of an NF-κB decoy ODN was successfully tested in a spontaneous atopic dermatitis model (NC/Nga mice) with[12] or without[31] additional dust-mite antigen challenge, as well as in a delayed-type hypersensitivity response model in mice.[32] In the chronic dust-mite antigen skin inflammation treatment model, the NF-κB decoy ODN ointment appeared to be equipotent to the corticosteroid betamethasone valerate, both in the prevention and treatment arm of the study, and down-regulated expression of a similar subset of pro-inflammatory gene products in the mouse skin.[12] However, administration of the NF-κB decoy ODN also resulted in a marked increase in the number of apoptotic cells, which may have comprised primarily inflamed skin cells. Nonetheless, administration of such decoy ODNs should probably be restricted to the topical route of application, since systemic suppression of NF-κB might be harmful, given that knockout mice for various NF-κB signalling components suffer from immune deficiency or lack lymphocyte activation.[33] However, when compared to topical steroids, repeated administration of the NF-κB decoy ODN ointment did not result in skin atrophy or any other local adverse effects.[12]

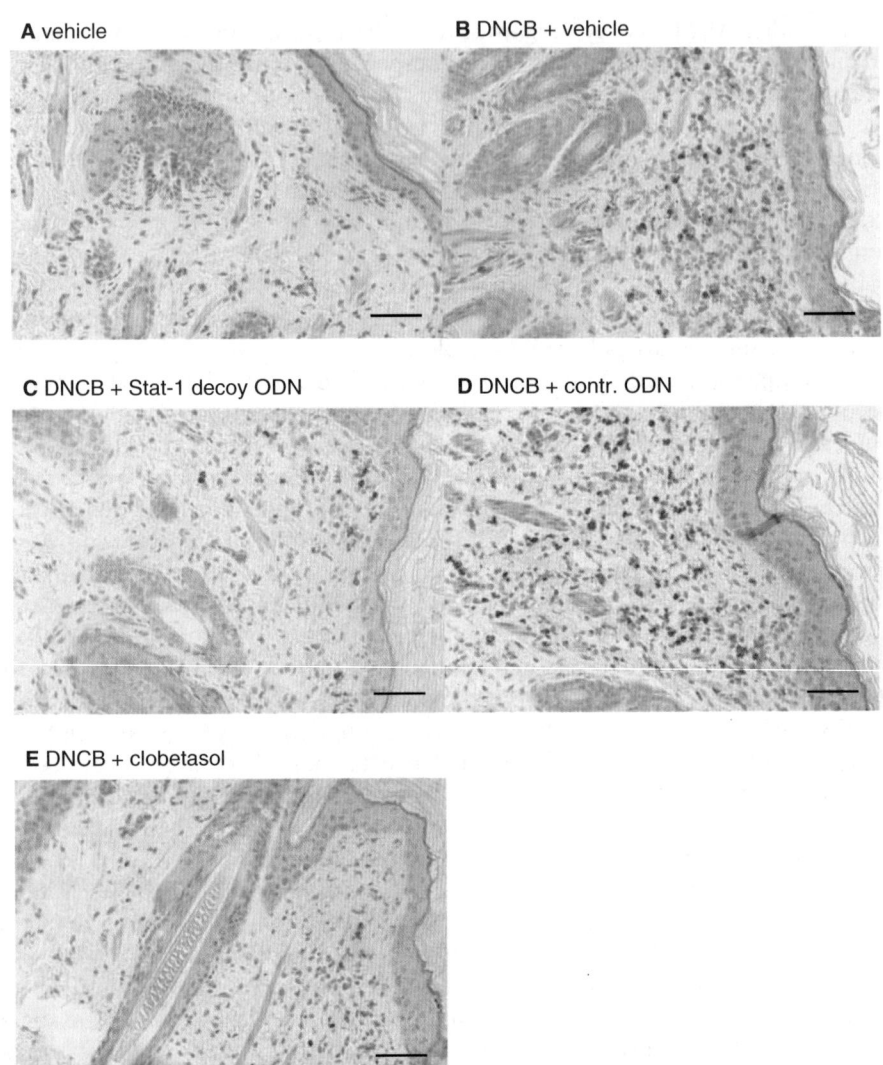

A vehicle

B DNCB + vehicle

C DNCB + Stat-1 decoy ODN

D DNCB + contr. ODN

E DNCB + clobetasol

Figure 7.9 Effects of (A) vehicle, (B) 0.5% DNCB (dinitrochlorobenzene), (C) DNCB plus 0.13% Stat-1 decoy ODN, (D) DNCB plus control ODN and (E) DNCB plus 0.25% clobetasol propionate on polymorphonuclear neutrophils (PMN) skin infiltration over 24 hours in a guinea pig ACD model. The naphthol AS D chloroacetate esterase stain was used to visualize the infiltrating PMNs (magnification 200×).

Finally, also, a Stat-6 decoy ODN was found to be effective especially against the late-phase response in a dinitrofluorobenzene-induced contact dermatitis model in mice.[34] However, the decoy ODN was rather long (28-mer) and therefore had to be administered as a haemagglutinating virus of Japan (HVJ)–liposomal formulation. The most prominent effect was observed upon subcutaneous injection at concentrations ranging from 5 to 20 μM, a topical route of administration has not been attempted and the magnitude of the anti-inflammatory effect of the Stat-6 decoy ODN turned out to be strain-specific, which suggests that it is directed against a Th2- rather than a Th1-driven immune response.

7.5.4 Rheumatoid Arthritis

Apart from arthrosis, rheumatoid arthritis is the most prevalent (approximately 1% in Caucasians) and severe chronic inflammatory disease of the joints, with a particularly high socioeconomic impact. It is thought to be an immune-mediated disease that promotes inflammation and progressive destruction of the joints. Besides the pro-inflammatory cytokines TNFα, IL-1β and IL-6, which mainly derive from macrophages, a dominant type 1 Th cell response is associated with the disease that is characterized by an imbalance of IFN-γ over IL-4.[30] While TNFα and IL-1β may signal via the transcription factor NF-κB, IL-6 and in particular IFN-γ exert their pro-inflammatory effects through activation of members of the Stat family of transcription factors, namely Stat-1. Stat-1 in turn upregulates expression, for example of the co-stimulatory molecule CD40, either by direct binding to the interferon γ-activated sequence (GAS) element in the CD40 promoter or by inducing the *de novo* synthesis of the transcription factor IFN regulatory factor-1 (IRF-1).[35,36] Interesting therapeutic targets for a decoy ODN approach, therefore, are pro-inflammatory transcription factors, such as NF-κB or Stat-1. Decoy ODNs directed against these two transcription factors have been tested in two different animal models, collagen-induced arthritis in the rat and antigen-induced arthritis in mice. Both decoy ODNs were administered by intra-articular injection. While the NF-κB decoy ODN was encapsulated into HVJ–liposomes, the Stat-1 decoy ODN was administered as naked DNA prior to the ultimate challenge with the antigen. Single treatment with the Stat-1 decoy ODN exerted a profound and dose-dependent (0.05–10 nmol, near-maximum effect at 0.25 nmol) inhibitory effect not only on joint swelling and the histopathological signs of both acute and chronic arthritis, but also on delayed-type hypersensitivity and cytokine release into the synovia. Moreover, CD40 mRNA expression in stimulated macrophages was downregulated, and these effects of the decoy ODN were mimicked by intra-articular injection of a monoclonal antibody neutralizing the CD40 ligand CD154.[30] It was concluded, therefore, that the beneficial effect of the Stat-1 decoy ODN in this model was mediated in part by affecting CD40 signalling in macrophages and/or other antigen-presenting cells. In the collagen-induced arthritis model, intra-arterial injection of the liposomal NF-κB decoy ODN formulation also afforded a marked suppression of joint destruction as well as cytokine release into the synovia.[37] When translating these findings into the situation in human rheumatoid arthritis that does not normally affect a single joint only, however, a note of caution may be appropriate with respect to the route of administration taken in the aforementioned studies. Here, a systemic delivery method is a must. Moreover, with respect to the relevance of these two models it is stressed that the cartilage destruction seen in the murine antigen-induced arthritis model more closely resembles the human disease. Recently, another decoy ODN approach targeted at the transcription factor E2F has been shown to exert a protective effect in immunodeficient severe combined immune deficiency (SCID) mice that were transplanted with human cartilage and rheumatoid arthritis tissue, probably through inhibition of synovial cell proliferation.[38]

As mentioned before, NO is a potent anti-inflammatory mediator and the primarily anti-inflammatory cytokine IL-10 is capable of upregulating eNOS expression in human endothelial cells through activation of the transcription factor Stat-3.[39] Moreover, the resulting increase in NO production blocks the CD40-mediated expression of the pro-inflammatory cytokine IL-12 in the endothelial cells. However, in endothelial cells derived from individuals that are homozygous for the T-786C SNP of the eNOS gene, IL-10 fails to upregulate eNOS expression.[15] In addition, the affected individuals had a significantly increased risk to develop rheumatoid arthritis. As with the SNP in the context of CHD, pre-incubation of C/C genotype endothelial cells with a decoy ODN directed against the C-type but not the T-type NO synthase gene promoter also restored the responsiveness of the defective gene to IL-10. The same effect could be achieved with decoy ODNs as small as 10 base pairs, provided these were directed against the C-type promoter around position -786. The molecular basis of this additional deblocking effect of the decoy ODNs is that Stat-3, the intracellular mediator of the effect of IL-10, exerts its effect on eNOS expression by binding to a generic Stat motif approximately 60 base pairs upstream of the mutation. Again, it appears that the putative inhibitory DNA-binding protein that binds with high affinity to the mutated sequence (see above) sterically hinders Stat-3 from inducing expression of the eNOS gene. As a result, the NO-synthesizing capacity of the endothelial cells in response to IL-10 is weakened, leading to an increased risk of inflammation in the joints of individuals homozygous for the T-786C SNP of the eNOS gene. As with CHD, this context provides another opportunity for employing decoy ODNs as a novel approach towards personalized medicine. Moreover, these findings point towards a common pathogenetic basis of CHD and rheumatoid arthritis in which the eNOS gene has an important modifying role. Clinically, the increased risk of patients with rheumatoid arthritis to contract CHD and vice versa is well known.[40,41]

7.5.5 Acute and Chronic Transplant Rejection

Transplant rejection is an almost ideal indication for the therapeutic application of decoy ODNs. Thus, they can be periprocedurally infused into the arterial blood vessels of the donor organ and flushed out immediately before re-establishing perfusion of the transplant in the recipient to avoid any systemic exposure. The only possible pitfall when employing naked dsDNA is to allow for a brief period of warm ischaemia so that the decoy ODNs can be taken up by the endothelial cells of the donor organ (see Section 7.3). Moreover, these cells also appear to be the most appropriate target for the decoy ODN-mediated inhibition of pro-inflammatory transcription factors, since not only acute, but also chronic, transplant rejection is closely associated with endothelial dysfunction and microcirculatory perfusion failure. Consequently, decoy ODNs directed against NF-κB, which has been implicated in endothelial cell adhesion molecule and chemokine expression, at least in rodents, were the first to be tested in animal models of allograft rejection and vasculopathy. Pressure-mediated decoy ODN administration to rat cardiac allografts[42] as well as perfusion of rat renal allografts with naked DNA[43] or by using a contrast agent–ultrasound procedure[44] resulted in a significantly increased survival of the transplants in the absence of immunosuppression or reduced graft vasculopathy in immunosuppressed recipients. Similarly, periprocedural administration of a Stat-1 decoy ODN in rat small bowel allografts[45] or that of a Stat-1 or AP-1 decoy ODN in rat cardiac allografts[46] resulted in a significant amelioration of the microcirculatory and histopathological signs of acute rejection. The same effect was found in mouse cardiac allografts exposed to the Stat-1 decoy ODN (T. Stojanovic *et al.*, unpublished observation) while, in particular, the Stat-1 decoy ODN and, to a somewhat lesser extent, the AP-1 decoy ODN virtually abolished cardiac allograft vasculopathy in immunosuppressed rats (H. Hölschermann *et al.*, unpublished observation). While cardiac allograft vasculopathy can be considered as a particularly aggressive form of CHD, the powerful protective effect of both the Stat-1 decoy ODN and the AP-1 decoy ODN may be explained by their inhibitory effect on cytokine-stimulated and basal CD40 expression, respectively, in the allograft endothelial cells. Moreover, both in rodent and primate models CD40- and/or CD154-mediated interactions between graft endothelial cells and recipient Th cells and monocyte and/or macrophages have been shown to play a decisive role in the development of acute and chronic allograft rejection.[47,48]

Related to transplant rejection is ischaemia/reperfusion (I/R) injury, a condition in which decoy ODNs directed against the transcription factors NF-κB have been shown to exert beneficial therapeutic effects in rat heart,[49] kidney[50] and liver I/R injury.[51]

7.5.6 Other

Other indications in which animal experimental proof of principle studies have been performed include chronic inflammatory bowel disease, *i.e.* different mouse models of colitis in which NF-κB decoy ODNs were successfully administered intrarectally, either encapsulated in a viral envelope[52] or as naked DNA, albeit fully phosphorothioate-modified,[53] osteoporosis in rats (NF-κB decoy ODN),[54] diabetic nephropathy in rats (AP-1 decoy ODN[55] or ring-Sp-1 decoy ODN[56]), experimental autoimmune myocarditis in rats (NF-κB decoy

ODN)[57] and carrageenin-induced acute inflammation in the rat hind paw (NF-κB decoy ODN).[58]

7.6 Perspective

On the basis of the preclinical and the clinical data available to date (see below), decoy ODNs hold every promise to eventually turn out as highly effective corticosteroid-like anti-inflammatory drugs which are readily available, fast-acting and without major side effects. Even though they may interfere with the expression of some immune response-related genes as, for example, in the case of the Stat-1 decoy ODN, they do not seem to act as immunosuppressant drugs and limit an exacerbated pro-inflammatory response rather than abolishing the subjacent physiological immune response against bacteria, viruses or parasites. If the problems of an effective and cell-specific systemic drug delivery can be solved, decoy ODNs may become one of the therapeutic approaches with blockbuster potential.

7.6.1 Nucleic Acid Based Drugs in Comparison

Decoy ODNs specifically bind their target transcription factor, thus inhibiting the expression of genes which are controlled by the said transcription factor; they normally do not completely block their expression. Their distinct advantage is that they are rapidly transported into their target cells without any additives (see Section 7.3) and immediately thereafter can act on their target transcription factor. Once inside the cell, especially terminally phosphorothioated decoy ODNs tend to be rather stable and thus exert long-lasting effects, which must be considered favourable both in terms of dosing and costs of goods. Moreover, it appears that the target cells (upon local delivery) and most likely the whole organism (upon systemic administration) has an inbuilt tolerance towards these drug candidates and the mechanism by which they act. This is not normally the case with ssDNA, *i.e.* antisense ODNs, as well as with double-stranded siRNA molecules. Furthermore, most of the transcription factors addressed thus far, *i.e.* AP-1, NF-κB or Stat-1, control the expression of many pro-inflammatory genes, so that inhibition of either transcription factor will result in diminishing the expression of several target genes simultaneously. Decoy ODNs may thus be perceived as drugs that act rather non-specifically when compared to the single-target gene knockdown that is typically attained when employing an antisense ODN or siRNA approach. This relative lack of specificity, however, may explain why the Stat-1 decoy ODN, for example, exerts such a powerful corticosteroid-like anti-inflammatory effect in various animal models of inflammation without any appreciable side effects.[21,30] Though specific for their target gene product, antisense ODNs and siRNAs face the problems of most likely being less efficacious, comparatively slow in onset, depending on the turnover of their target protein and rather difficult to deliver to their target cells.

7.6.2 Clinical Development

Clinical indications in which decoy ODNs have been studied thus far comprise asthma, psoriasis and atopic dermatitis. Here, decoy ODNs directed against the transcription factors Stat-1 (asthma, psoriasis) and NF-κB (atopic dermatitis) are in extended Phase IIa trials. While the results of the Phase I trials suggest that topical application of the Stat-1 decoy ODN (inhalation, ointment), as well as that of the NF-κB decoy ODN ointment (see www.anges-mg.com), is safe and well-tolerated, the Phase IIa data for the Stat-1 decoy ODN, though on the basis of a rather limited statistical power, also demonstrate efficacy in both indications (see www.avontec.de). The same was true for topical administration of Avrina's, formerly Corgentech's, NF-κB decoy ODN drug candidate, in a Phase I/II trial in patients with atopic dermatitis (see www.anesiva.com). Another decoy ODN developed by the same company and directed against the transcription factor E2F has even completed two large-scale Phase III studies for the prevention of vein graft failure following both peripheral (PREVENT III)[59] and coronary artery bypass graft surgery (PREVENT IV),[60] but failed to reach the clinical endpoints. This failure, despite some promising Phase I and II data, most likely was not because of a lack of biological efficacy of the decoy ODN. However, not until the Phase III trials were its effects compared to those in an appropriate control group (pressure-mediated delivery of a placebo) the outcome of which was not distinguishable from that of the treatment group. One interpretation is that the pressure-mediated delivery of the ODNs itself has caused a preconditioning effect which ultimately affected the development of vein graft disease. Nonetheless, the rather disappointing outcome of these trials has meant that the development of the E2F decoy ODN for this indication has been suspended.

However, clinical development of the other decoy ODNs is progressing well and will soon reach Phase IIb status. Several other decoy ODNs for other indications, which are currently still in the research phase, will then hopefully enter the development phase. And the decision will be reached at this level as to which type of nucleic acid-based medicine is more adequate to combat certain diseases.

References

1. C. A. Stein, C. Subasinghe, K. Shinozuka and J. S. Cohen, *Nucleic Acids Res.*, 1988, **16**, 3209.
2. A. Bene, R. C. Kurten and T. C. Chambers, *Nucleic Acids Res.*, 2004, **32**, e142.
3. F. Shi and D. Hoekstra, *J. Controlled Release*, 2004, **97**, 189.

4. M. J. Mann, G. H. Gibbons, H. Hutchinson, R. S. Poston, E. G. Hoyt, R. C. Robbins and V. J. Dzau, *Proc. Natl. Acad. Sci. U. S. A.*, 1999, **96**, 6411.
5. L. Fisher, U. Soomets, V. Cortés Toro, L. Chilton, Y. Jiang, U. Langel and K. Iverfeldt, *Gene Ther.*, 2004, **11**, 1264.
6. G. De Rosa, M. C. Maiuri, F. Ungaro, D. De Stefano, F. Quaglia, M. I. La Rotonda and R. Carnuccio, *J. Gene Med.*, 2005, **7**, 771.
7. Y. Hattori, M. Sakaguchi and Y. Maitani, *Biol. Pharm. Bull.*, 2006, **29**, 1516.
8. N. Hashiya, M. Aoki, K. Tachibana, Y. Taniyama, K. Yamasaki, K. Hiraoka, H. Makino, K. Yasufumi, T. Ogihara and R. Morishita, *Biochem. Biophys. Res. Commun.*, 2004, **317**, 508.
9. R. Bekeredjian, P. A. Grayburn and R. V. Shohet, *J. Am. Coll. Cardiol.*, 2005, **45**, 329.
10. A. H. Wagner, J. Fleischer, B. C. Burckhardt, G. Burckhardt and M. Hecker, *Pflüger's Arch. Eur. J. Physiol.*, 2004, **447**(Suppl. 1), S30.
11. X. Wei, G. Dai, G. Marcucci, Z. Liu, D. Hoyt, W. Blum and K. K. Chan, *Pharm. Res.*, 2006, **23**, 1251.
12. M. Dajee, T. Muchamuel, B. Schryver, A. Oo, J. Alleman-Sposeto, C. G. De Vry, S. Prasad, D. Ruhrmund, R. Shyamsundar, D. Mutnick, K. Mai, T. Le, C. Parham, J. Zhang, L. Komuves, T. Colby, S. Hudak, L. M. McEvoy and R. O. Ehrhardt, *J. Invest. Dermatol.*, 2006, **126**, 1792.
13. M. Mann and V. J. Dzau, *J. Clin. Invest.*, 2000, **106**, 1071.
14. M. Cattaruzza, T. J. Guzik, W. Slodowski, A. Pelvan, J. Becker, M. Halle, A. B. Buchwald, K. M. Channon and M. Hecker, *Circ. Res.*, 2004, **95**, 841.
15. I. Melchers, S. Blaschke, M. Hecker and M. Cattaruzza, *Arthritis Rheum.*, 2006, **54**, 3144.
16. A.H. Brivanlou and J. E. Darnell, *Science*, 2002, **295**, 813.
17. K. Cartharius, K. Frech, K. Grote, B. Klocke, M. Haltmeier, A. Klingenhoff, M. Frisch, M. Bayerlein and T. Werner, *Bioinformatics*, 2005, **21**, 2933.
18. A.B. Buchwald, A.H. Wagner, C. Webel and M. Hecker, *J. Am. Coll. Cardiol.*, 2002, **39**, 732.
19. C. Desmet, P. Gosset, E. Henry, V. Garze, P. Faisca, N. Vos, F. Jaspar, D. Melotte, B. Lambrecht, D. Desmecht, B. Pajak, M. Moser, P. Lekeux and F. Bureau, *Am. J. Respir. Crit. Care Med.*, 2005, **172**, 671.
20. C. Desmet, P. Gosset, B. Pajak, D. Cataldo, M. Bentires-Alj, P. Lekeux and F. Bureau, *J. Immunol.*, 2004, **173**, 5766.
21. D. Quarcoo, S. Weixler, D. Groneberg, R. Joachim, B. Ahrens, A. H. Wagner, M. Hecker and E. Hamelmann, *J. Allergy Clin. Immunol.*, 2004, **114**, 288.
22. D. Sampath, M. Castro, D. C. Look and M.J. Holtzman, *J. Clin. Invest.*, 1999, **103**, 1353.
23. M.A. Gimbrone Jr, J. N. Topper, T. Nagel, K. R. Anderson and G. Garcia-Cardena, *Ann. N. Y. Acad. Sci.*, 2000, **902**, 230.

24. M. Nakayama, H. Yasue, M. Yoshimura, Y. Shimasaki, K. Kugiyama, H. Ogawa, T. Motoyama, Y. Saito, Y. Ogawa, Y. Miyamoto and K. Nakao, *Circulation*, 1999, **99**, 2864.
25. G. P. Rossi, M. Cesari, M. Zanchetta, S. Colonna, G. Maiolino, L. Pedon, M. Cavallin, P. Maiolino and A. C. Pessina, *J. Am. Coll. Cardiol.*, 2003, **41**, 930.
26. U. Kelkenberg, A. H. Wagner, J. Sarhaddar, M. Hecker and H. E. von der Leyen, *Arterioscler. Thromb. Vasc. Biol.*, 2002, **22**, 949.
27. B. Kusch, S. Waldhans, A. Sattler, A. Wagner, M. Hecker, R. Moosdorf and S. Vogt, *Thorac. Cardiovasc. Surg.*, 2006, **54**, 388.
28. A. Ehsan, M. J. Mann, G. Dell'Acqua and V. J. Dzau, *J. Thorac. Cardiovasc. Surg.*, 2001, **121**, 714.
29. T. Miyake, M. Aoki, S. Shiraya, K. Tanemoto, T. Ogihara, Y. Kaneda and R. Morishita, *J. Mol. Cell. Cardiol.*, 2006, **41**, 431.
30. M. Hückel, U. Schurigt, A. H. Wagner, R. Stöckigt, P. K. Petrow, K. Thoss, M. Gajda, S. Henzgen, M. Hecker and R. Brauer, *Arthritis Res. Ther.*, 2006, **8**, R17.
31. H. Nakamura, M. Aoki, K. Tamai, M. Oishi, T. Ogihara, Y. Kaneda and R. Morishita, *Gene Ther.*, 2002, **9**, 1221.
32. I. Isomura, K. Tsujimura and A. Morita, *J. Invest. Dermatol.*, 2006, **126**, 97.
33. R. M. Attar, J. Caamano, D. Carrasco, V. Iotsova, H. Ishikawa, R. P. Ryseck, F. Weih and R. Bravo, *Semin. Cancer. Biol.*, 1997, **8**, 93.
34. K. Sumi, H. Yokozeki, M.H. Wu, T. Satoh, Y. Kaneda, K. Takeda, S. Akira and K. Nishioka, *Gene Ther.*, 2004, **11**, 1763.
35. V. T. Nguyen and E. N. Benveniste, *J. Biol. Chem.*, 2000, **275**, 23674.
36. A. H. Wagner, M. Gebauer, B. Pollok-Kopp and M. Hecker, *Blood*, 2002, **99**, 520.
37. T. Tomita, E. Takeuchi, N. Tomita, R. Morishita, M. Kaneko, K. Yamamoto, T. Nakase, H. Seki, K. Kato, Y. Kaneda and T. Ochi, *Arthritis Rheum.*, 1999, **42**, 2532.
38. T. Tomita, Y. Kunugiza, N. Tomita, H. Takano, R. Morishita, Y. Kaneda and H. Yoshikawa, *Int. J. Mol. Med.*, 2006, **18**, 257.
39. M. Cattaruzza, W. Slodowski, M. Stojakovic, R. Krzesz and M. Hecker, *J. Biol. Chem.*, 2003, **278**, 37874.
40. K. J. Warrington, P. D. Kent, R. L. Frye, J. F. Lymp, S. L. Kopecky, J. J. Goronzy and C. M. Weyand, *Arthritis Res. Ther.*, 2005, **7**, R984.
41. I. F. Gazi, D. T. Boumpas, D. P. Mikhailidis and E. S. Ganotakis, *Clin. Exp. Rheumatol.*, 2007, **25**, 102.
42. B. T. Feeley, D. N. Miniati, A. K. Park, E. G. Hoyt and R. C. Robbins, *Transplantation*, 2000, **70**, 1560.
43. I. H. Vos, R. Govers, H. J. Grone, L. Kleij, M. Schurink, R. A. De Weger, R. Goldschmeding and T. J. Rabelink, *FASEB J.*, 2000, **14**, 815.
44. H. Azuma, N. Tomita, Y. Kaneda, H. Koike, T. Ogihara, Y. Katsuoka and R. Morishita, *Gene Ther.*, 2003, **10**, 415.

45. T. Stojanovic, L. Scheele, A.H. Wagner, P. Middel, J. Bedke, I. Lautenschlager, I. Leister, S. Panzer and M. Hecker, *Gene Ther.*, 2007, **14**, 883.
46. H. Hölschermann, T. H. Stadlbauer, A. H. Wagner, H. Fingerhuth, H. Muth, S. Rong, F. Güler, H. Tillmanns and M. Hecker, *Cardiovasc. Res.*, 2006, **71**, 527.
47. C. Y. Wang, S. P. Mazer, K. Minamoto, S. Takuma, S. Homma, M. Yellin, L. Chess, A. Fard, S. L. Kalled, M. C. Oz and D. J. Pinsky, *Circulation*, 2002, **105**, 1609.
48. R. N. Pierson, A. C. Chang, M. G. Blum, K. S. Blair, M. A. Scott, J. B. Atkinson, B. J. Collins, J. P. Zhang, D. W. Thomas, L. C. Burkly and G. G. Miller, *Transplantation*, 1999, **68**, 1800.
49. R. Morshita, T. Sugimoto, M. Aoki, I. Kida, N. Tomita, A. Moriguchi, K. Maeda, Y. Sawa, Y. Kaneda, J. Higaki and T. Ogihara, *Nat. Med.*, 1997, **3**, 894.
50. C. C. Cao, X. Q. Ding, Z. L. Ou, C.F. Liu, P. Li, L. Wang and C. F. Zhu, *Kidney Int.*, 2004, **65**, 834.
51. M. Q. Xu, X. R. Shuai, M. L. Yan, M. M. Zhang and L. N. Yan, *World J. Gastroenterol.*, 2005, **11**, 6960.
52. S. Fichtner-Feigl, I. J. Fuss, J. C. Preiss, W. Strober and A. Kitani, *J. Clin. Invest.*, 2005, **115**, 3057.
53. C. G. De Vry, S. Prasad, L. Komuves, C. Lorenzana, C. Parham, T. Le, S. Adda, J. Hoffman, N. Kahoud, R. Garlapati, R. Shyamsundar, K. Mai, J. Zhang, T. Muchamuel, M. Dajee, B. Schryver, L. M. McEvoy and R. O. Ehrhardt, *Gut*, 2007, **56**, 524.
54. H. Shimizu, H. Nakagami, I. Tsukamoto, S. Morita, Y. Kunugiza, T. Tomita, H. Yoshikawa, Y. Kaneda, T. Ogihara and R. Morishita, *Gene Ther.*, 2006, **13**, 933.
55. J. D. Ahn, R. Morishita, Y. Kaneda, H. J. Kim, Y. D. Kim, H. J. Lee, K. U. Lee, J. Y. Park, Y. H. Kim, K. K. Park, Y. C. Chang, K. H. Yoon, H. S. Kwon, K. G. Park and I. K. Lee, *Gene Ther.*, 2004, **11**, 916.
56. Y. M. Chae, K. K. Park, I. K. Lee, J. K. Kim, C. H. Kim and Y. C. Chang, *Gene Ther.*, 2006, **13**, 430.
57. O. Yokoseki, J. Suzuki, H. Kitabayashi, N. Watanabe, Y. Wada, R. Morishita, Y. Kaneda, T. Ogihara, H. Futamatsu, Y. Kobayashi and M. Isobe, *Circ. Res.*, 2001, **89**, 899.
58. F. D'Acquisto, A. Ialenti, A. Ianaro, R. Di Vaio and R. Carnuccio, *Gene Ther.*, 2000, **7**, 1731.
59. M. S. Conte, D. F. Bandyk, A. W. Clowes, G. L. Moneta, L. Seely, T. J. Lorenz, H. Namini, A. D. Hamdan, S. P. Roddy, M. Belkin, S. A. Berceli, R. J. DeMasi, R. H. Samson, S. S. Berman and PREVENT III Investigators, *J. Vasc. Res.*, 2006, **43**, 742.
60. J. H. Alexander, G. Hafley, R. A. Harrington, E. D. Peterson, T. B. Ferguson Jr, T. J. Lorenz, A. Goyal, M. Gibson, M. J. Mack, D. Gennevois, R. M. Califf, N. T. Kouchoukos and PREVENT IV Investigators, *JAMA*, 2005, **294**, 2446.

CHAPTER 8

AS1411: Development of an Anticancer Aptamer

NIGEL COURTENAY-LUCK[*,1] AND
DONALD M. MILLER[2]

[1] Antisoma, West Africa House, Hanger Lane, London, W5 3QR, UK;
[2] Brown Cancer Center, University of Louisville, 529 South Jackson Street,
Louisville, Kentucky 40202, USA

8.1 An Introduction to Aptamers

Over the past two decades, oligonucleotides (ONs) have emerged as a viable alternative to virus-mediated gene complementation, which has proved effective in the treatment of genetic diseases in clinical trials,[1,2] but is also associated with pathogenicity and immunogenicity.[3,4] Although ON technology is still at an early stage of development, the therapeutic potential of these molecules has been explored extensively in terms of their capacity to correct genetic defects (using single-stranded ONs, triplex-forming ONs, small fragment homologous replacement, chimeraplasts and spliceosome-mediated RNA trans-splicing) and to inhibit specifically the expression of a target gene. This has been done using antisense ONs, ribozymes, DNAzymes and, more recently, small-interfering RNAs (siRNAs) and RNA interference. Sequestration of pathogenic proteins using decoy ONs and aptamers[5] has also been examined, and the development of the first aptamer in a clinical oncology setting will now be discussed.

Aptamers, derived from the Latin aptus meaning 'to fit', are relatively short (12–40 base) DNA or RNA ONs (or peptides) that were first discovered as a result of research into viruses in the late 1980s.[6,7] Some aptamers assume

RSC Biomolecular Sciences
Therapeutic Oligonucleotides
Edited by Jens Kurreck
© Royal Society of Chemistry 2008

specific three-dimensional (3D) structures *in vivo*, which are able to bind with high affinity to target proteins (in a manner analogous to that of an antibody to an antigen) and to elicit a biological response. This specificity for a given protein makes it possible to isolate an aptamer for virtually any target, and underpins their tremendous potential as targeted molecular therapies.

Although aptamers have been described as having binding affinities akin to those of antibodies (in the low nanomolar to picomolar range), some of their other properties make them more attractive therapeutic agents than antibodies.[5,8] Prominent among these are stability, which facilitates long-term storage (they can be heated to 80 °C, or stored in various solvents or harsh environments, and return to their original conformation), lack of immunogenicity in humans and markedly lower variability between batches. In addition, the smaller size of aptamers may enable them to penetrate tumours more effectively and be cleared more rapidly from the blood.

In contrast to antisense technologies, in which ONs target messenger RNA (mRNA) and must therefore be delivered inside the cell, aptamers can also target extracellular proteins, which eliminates the need for intracellular transportation. Furthermore, the 3D structure of aptamers may render them less vulnerable to nuclease degradation, and they can be chemically modified (as is done extensively with drugs based on antisense ONs) to further reduce any such susceptibility. For example, a reduction in nuclease degradation of aptamers has been achieved using amino- or fluoro- modifications at the 2′ position of pyrimidines.[9,10] The biodistribution and clearance of aptamers can also be altered by the chemical addition of polyethylene glycol and other moieties, which expands their clinical utility.[11]

Other properties that make aptamers an attractive class of therapeutic compounds include the capacity to retain activity in multiple organisms, which facilitates preclinical development, and the existence of 'antidotes' – short complementary ONs that interact with aptamers specifically and impair their active conformational structure, thereby inhibiting their effect and ensuring safe, tightly controlled therapeutics.[12]

The only aptamer currently approved for use in humans is an RNA-based molecule (Macugen, pegaptanib), which is administered locally (intravitreally) to treat age-related macular degeneration by targeting vascular endothelial growth factor. There is little doubt, however, that a growing number of aptamers will soon be used in the treatment of a variety of diseases; in particular, research has focussed on their potential as antiproliferative agents in the treatment of cancer. The most advanced aptamer in the cancer field, and the first to be tested in humans, is AS1411 (formerly AGRO100).

8.2 AS1411: An Anticancer Aptamer

AS1411 is a guanosine-rich oligonucleotide (GRO) derived from GRO29A-OH (an analogue of the 29-mer GRO29A), and comprises a single-stranded DNA chain of 26 bases (5′-GGT-GGT-GGT-GGT-TGT-GGT-GGT-GGT-GG-3′)

with unmodified phosphodiester linkages.[13,14] AS1411 was discovered *via* serendipity rather than following selection using systematic evolution of ligands by exponential enrichment (SELEX) as used in the case of other ON-based aptamers. It was developed for clinical trial based on early observations that GROs possessed antiproliferative properties against cancer cells *in vitro*. These studies revealed that antiproliferative activity of certain GROs (such as GRO29A) correlated with their binding to a specific cellular protein, which was detectable in both nuclear and cytoplasmic extracts, and in proteins derived from the plasma membrane of cells. The GRO-binding protein was eventually identified as nucleolin.[15]

8.2.1 Nucleolin: An Anticancer Target

Human nucleolin is a 76.7 kDa (710 amino acid) Q3 protein, which consists of an acidic histone-like NH_2 terminus, a central domain that contains four RNA-binding domains, a COOH terminus rich in arginine–glycine–glycine (RGG) repeats, and multiple sites of phosphorylation. Nucleolin seems to be fundamental to the survival and proliferation of cells and is thought to play a multifunctional role involving many cellular processes, including transcription, packing, and transport of ribosomal RNA (rRNA), together with replication and recombination of DNA.[16] Although mainly characterized as a nucleolar protein, nucleolin also functions as a cell surface receptor, where it is associated with the actin cytoskeleton and acts as a shuttling protein between cytoplasm and nucleus.[17] Over recent years, preclinical research has led to the emergence of cell surface nucleolin as an attractive anticancer target.[18]

Nucleolin has been shown to be the cell surface molecule that is recognized by and responsible for the internalization of a tumour-homing 34-amino acid peptide (F3), which binds selectively to tumours (but not skin, heart or brain) in mice, and to angiogenic endothelial cells within the tumour vasculature.[19] Cell surface nucleolin may function in angiogenesis as an adhesion molecule to modulate cell matrix interaction and regulate migration.[20] Furthermore, the expression of nucleolin at the cell surface was found to correlate with the proliferative and metabolic activity of a breast cancer cell line.[21] Thus, the pathophysiological function of cell surface nucleolin may involve internalization of specific ligands. One of these is the heparin-binding growth factor midkine, a cytokine involved in growth regulation and differentiation;[21] midkine mRNA expression is increased in various human carcinomas. A second appears to be lactoferrin, an iron-binding glycoprotein capable of inhibiting cancer cell proliferation; lactoferrin complexed to cell surface nucleolin is internalized into MDA-MB-231 breast cancer cells.[22] Third, the antitumour (anti-angiogenic) activity of a novel glycosaminoglycan – acharin sulfate, related to heparin sulfate – has been attributed to binding to a phosphorylated form of nucleolin on the surface of cancer cells.[23]

A role for nucleolin in protecting cells against apoptosis has been identified in studies that involve the bcl-2 proto-oncogene, which encodes a protein that inhibits apoptosis.[24–26] Overexpression of bcl-2 is thought to be an important component in the development of certain haematological malignancies,[27,28] enabling cancer cells to avoid apoptosis and become resistant to chemotherapeutic agents.[29,30] The mRNA for the bcl-2 protein contains an AU-rich element (ARE) that is involved in the regulation of bcl-2 stability. Studies with the acute myelogenous leukaemia (AML) cell line HL-60 have shown that non-nuclear nucleolin acts as an ARE-binding protein and bcl-2-stabilizing factor.[25]

The potential of nucleolin as an anticancer target in this setting is illustrated by the finding that in HL-60 cells exposed to either paclitaxel or all-trans-retinoic acid, cytoplasmic nucleolin was phosphorylated, cleaved into fragments, and could no longer stabilize bcl-2 mRNA, thereby leading to apoptosis.[25,26]

8.2.2   Anticancer Mechanism of Action of AS1411

As AS1411 is known to target nucleolin and to be internalized efficiently by cancer cells, the possibility that its uptake is mediated by nucleolin was explored.[18] The cellular uptake of fluorescein isothiocyanate (FITC)-labelled AS1411 was greatly enhanced in nucleolin-overexpressing breast carcinoma cells (MCF-7) compared to the normal counterpart (MCF-10A), which expresses very low levels of nucleolin (Figure 8.1). Hence the cellular internalization of AS1411 depends on nucleolin levels, which, as described in Section 8.2.1, are higher in cancerous tissues than in normal tissues.

The primary effect on cancer cell lines of the binding of AS1411 to nucleolin and subsequent internalization of the AS1411–nucleolin complex is a gradual accumulation of cells in the S phase of the cell cycle – a process that occurs over

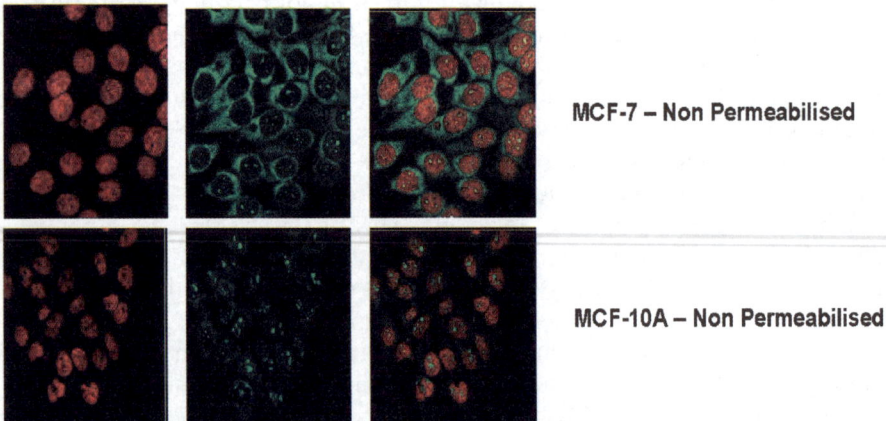

MCF-7 – Non Permeabilised

MCF-10A – Non Permeabilised

Figure 8.1   Cellular uptake of fluorescein isothiocyanate (FITC)-labelled AS1411 in nucleolin-overexpressing breast carcinoma cells (MCF-7) compared to the normal counterpart (MCF-10A). (The figure was kindly provided by Sridharan Soundararajan and Daniel J. Fernandes of the Medical University of South Carolina, Charleston, USA.)

several cell cycles.[31] This is related to an arrest of DNA replication, possibly *via* modulation of helicase activity, at time points where RNA and protein synthesis are unaffected. By contrast, cells exposed to a 15-mer control GRO, which does not contain G quartets, behave in the same way as unexposed cells; further studies have confirmed that G-quartet formation is essential for the biological activity of GROs.[14] A cell line derived from normal foreskin fibroblasts showed no major cell-cycle perturbations following exposure to GRO29A.

Recent data have shed a little more light on the anticancer mechanism of AS1411 in revealing that it forms an intracellular complex with the nuclear factor (NF-κB) essential modulator (NEMO) and nucleolin, thereby inhibiting activation of NF-κB.[32]

8.3   *In vitro* Inhibition of Growth by AS1411

AS1411 has been tested extensively in tumour cell proliferation and/or inhibition experiments *in vitro*.[8,13–15,31] Original studies showed that GRO29A at a concentration of $15\,\mu\mathrm{mol\,L}^{-1}$ induced growth inhibition of prostate (DU-145), breast (MDA-MB-231 and MCF-7) and Henrietta Lack (HeLa) cancer cell lines.[15] In most cell lines, growth inhibition was clearly evident after five days of exposure. Further studies showed that a $10\,\mu\mathrm{mol\,L}^{-1}$ concentration of either GRO29A or the unmodified GRO29AOH inhibited growth of the same panel of cell lines.[13]

In more recent studies to evaluate the biologically active concentration of AS1411 *in vitro*, various human tumour cells lines were grown in the presence of AS1411 for six days, and cytotoxicity was determined using the sulforhodamine B protein dye assay.[8] All tumour cell lines tested were sensitive to AS1411, with IC_{50} values – the concentration of AS1411 required to elicit cell death in 50% of tumour cells – typically in the low micromolar range (2–10 μM).[8] A notable finding was that 3–4 days of continuous exposure of the cancer cells to AS1411 was found to be necessary for cell viability to drop, markers of apoptosis to appear and cell death to occur.

8.4   *In vivo* Biodistribution of AS1411

Biodistribution studies have been done to establish an AS1411 dosing regimen that enables plasma concentrations to be attained *in vivo* that reflect the low micromolar levels of drug required to kill cells *in vitro*.[8] Firstly, mice bearing subcutaneous lung and renal human tumour xenografts derived from A549 and A498 cells, respectively, were injected with a single intravenous bolus of tritiated AS1411 (1, 10 and $25\,\mathrm{mg\,kg}^{-1}$). Approximately 63% of the radiolabel was recovered in the urine within hours, reflecting a rapid α-elimination phase. It is assumed that AS1411 is excreted metabolically intact, although this remains to be confirmed experimentally using a novel reversed-phase

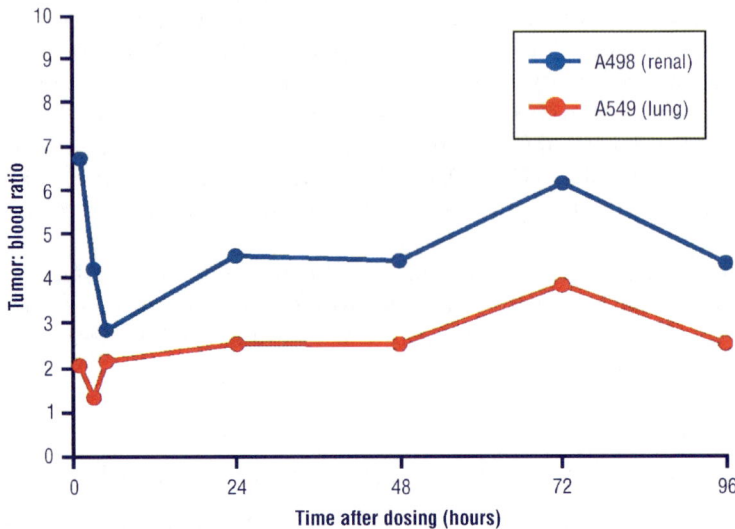

Figure 8.2   Biodistribution of tritium-labelled AS1411 after a single intravenous dose of 10 mg kg^{-1} in mice bearing subcutaneous renal and non-small-cell lung carcinoma xenografts; AS1411 levels were determined by liquid scintillation counting. (Reprinted with permission from the American Association for Cancer Research, *Mol. Cancer Ther.*, 2006, **5**, 2957–2962.)

high-pressure liquid chromatography method developed recently for the detection of AS1411 in biological fluids and tissues. The major route of elimination of AS1411 appears to be renal, as less than 1% of the injected dose of AS1411 was detected in the bile. The β-phase elimination half-life of AS1411 in plasma and whole blood was two days, and the dose-related pharmacokinetics were observed. The tumour-to-blood ratios after all three doses were 4:6 and 2:4 for the renal carcinoma and lung carcinoma xenografts, respectively (Figure 8.2). These data suggest that AS1411 is accumulating in tumour tissue, although further experimental evidence is required to confirm this observation.

The pharmacokinetics of AS1411 were also investigated in rats after a single intravenous bolus, after multiple injections on subsequent days and after continuous infusion for four days. Maximum plasma levels of AS1411 were attained shortly after dosing or the start of infusion and were found to be above the limit of quantification until about 4 hours after cessation of dosing. The extended half-life of AS1411 in plasma is in marked contrast to that of other unmodified phosphodiester ONs, which would be degraded rapidly by serum nucleases.[14]

8.5   *In vivo* Antitumour Activity of AS1411

Evidence of the efficacy of AS1411 *in vivo* was ascertained from a study in nude mice bearing established subcutaneous human tumour xenografts derived from A549 non-small-cell lung cancer cells. AS1411 was administered intravenously

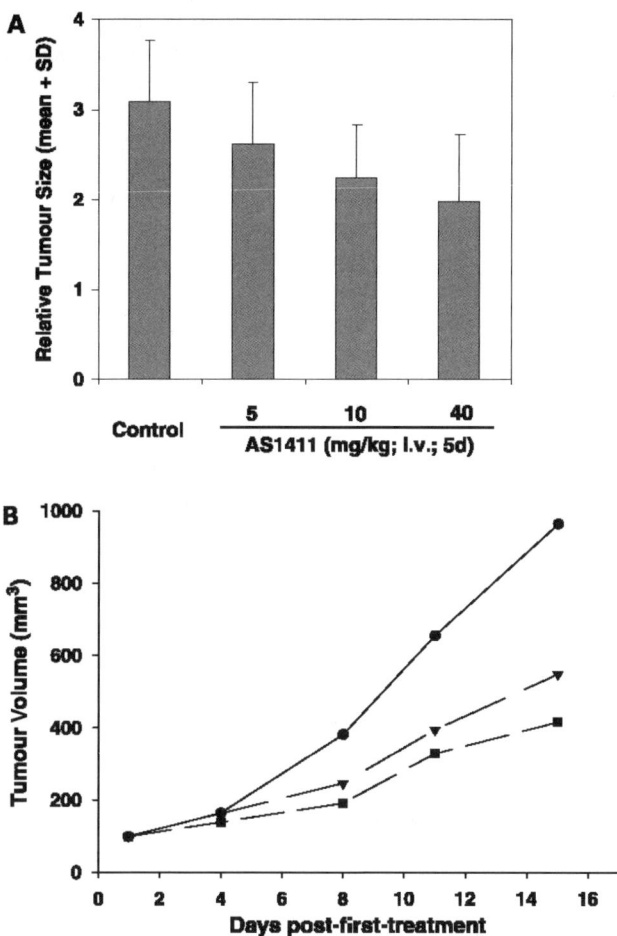

Figure 8.3 Effect of AS1411 on the growth *in vivo* of (A) A549 and (B) SKMES human lung carcinoma xenografts grown subcutaneously in nude mice. (A) Vehicle (phosphate-buffered saline) or AS1411 (5, 10 and 40 mg kg^{-1}, intravenously) was administered daily for five consecutive days; tumour volumes were measured on day 14 and expressed as tumour volume relative to the volume on the initial day of treatment. Columns, mean ($n = 8$); bars, standard deviation. (B) Vehicle (●), AS1411 5 mg kg^{-1} (▲) or AS1411 10 mg kg^{-1} (■) administered intraperitoneally on days 0, 2 and 4 from the time tumours had become established (~ 100 mm^3). (Reprinted with permission from the American Association for Cancer Research, *Mol. Cancer Ther.*, 2006, **5**, 2957–2962.)

for five consecutive days at doses of 5, 10 and 40 mg kg^{-1} and the tumour volume was determined relative to the initial volume (Figure 8.3A). Administration of AS1411 resulted in a delay of tumour growth *in vivo* and a dose response was seen, although the difference between 10 and 40 mg kg^{-1} was marginal. Antitumour effects were also observed at a dose of 10 mg kg^{-1}

administered either daily for five consecutive days, or on a Monday, Wednesday and Friday for one week, in mice bearing A498 renal cancer, SKMES lung cancer (Figure 8.3B) or MX1 breast cancer xenografts. These antitumour effects were characterized by cytostasis rather than cytotoxicity, with a resultant delay in tumour growth.[8]

As discussed in Section 8.2.2, the treatment of cells with AS1411 *in vitro* results in them being blocked in the S phase of the cell cycle. Consequently, there is a mechanistic rationale for combining AS1411 with nucleoside analogues such as gemcitabine, which also act on cells during the S phase. Gemcitabine is the first line of treatment for pancreatic cancer. The addition of AS1411 (100 mg kg^{-1} given intraperitoneally 12 hours apart on day 1) to gemcitabine (160 mg kg^{-1} given intraperitoneally every three days for four cycles) in mice bearing human tumour xenografts derived from PANC-1 pancreatic cancer cells resulted in a significant gain in antitumour activity ($p = 0.028$ on day 23 post-treatment).[8] Further studies are ongoing to define the optimum use of AS1411 in combination with existing approved chemotherapy.

8.6   Preclinical Toxicology of AS1411

The acute toxicity of AS1411 has been evaluated in rats (using a single bolus intravenous dose of up to 100 mg kg^{-1}) and in dogs (using a 96 hour continuous intravenous infusion at doses up to 10 mg kg^{-1} per day). In neither species was any significant toxicity observed, as determined by general clinical observations, clinical pathology evaluation or gross examination of tissues at necroscopy. Further evaluation including evaluation of chronic toxicity is ongoing.

8.7   Clinical Studies with AS1411

Based on the original promising preclinical properties of GROs (growth inhibition of cancer cells *in vitro* at low micromolar concentrations), and subsequent preclinical data that indicated specific potential for AS1411 against a wide range of solid tumours (antitumour activity *in vivo* in mice bearing human tumour xenografts) and a lack of significant toxicity in rats and dogs, AS1411 was selected for a Phase I, open label, non-randomized, dose escalation clinical study in 30 patients with advanced solid tumours refractory to conventional and/or standard treatments. After 17 patients had been enrolled, study entry was restricted to patients with renal cell carcinoma (RCC) or non-small-cell lung cancer, both to provide more homogeneous data in patient populations and to investigate further promising activity findings observed in the first three RCC patients. AS1411 was given by continuous intravenous infusion for 96 or 168 hours. Patients were enrolled in cohorts of three, with the first cohort receiving AS1411 1 mg kg^{-1} per day. If no patient in a cohort experienced dose-limiting toxicity, the dose for the next cohort was escalated using a standard dose-escalation scheme, to a maximum dose of 40 mg kg^{-1} per day.

The first aim of the study was to establish the maximum tolerated dose and dose-limiting toxicity of AS1411 treatment. However, the maximum tolerated dose could not be determined because dose escalation was limited by the maximum dose to be tested (40 mg kg^{-1} per day), and no dose-limiting toxicity occurred before or at this dose. Doses up to 40 mg kg^{-1} per day were well-tolerated with no serious toxicities related to drug administration observed.

The study also aimed to obtain preliminary evidence of any clinical and biological responses to AS1411 treatment. The results from this study have not yet been reported in full. However, initial data reported at the 31st European Society of Medical Oncology conference showed that of 12 patients, eleven showed clinical benefit, with two having objective responses.

8.8   Future Development of AS1411

AS1411 is the first nucleic acid-based aptamer to be tested in humans for the treatment of cancer, and has produced promising results in terms of antitumour activity and tolerability. At doses of up to 40 mg kg^{-1} per day administered continuously over 7 days, AS1411 seems to be devoid of significant toxicities, which reflects preclinical data in the rat and dog. Moreover, within the context of a Phase I dose-escalation trial in patients with advanced solid tumours, both multiple cases of stable disease and the reports of a partial response and a complete response in two patients with RCC clearly warrant further clinical investigation. Indeed, the clinical benefit seen in patients with RCC has led to the planning of a Phase II trial in this setting, with a view to building on the excellent safety and efficacy findings from Phase I. A Phase II study to test AS1411 in combination with cytarabine in patients with AML is underway. This trial is in patients with relapsed or refractory AML. Like RCC, AML cells have shown very high sensitivity to AS1411; the IC$_{50}$ of the acute myelogenous cell lines and patient cells is less than 3 µm. Together with the potential for synergy with cytarabine, this provides a compelling rationale for the evaluation of efficacy in this indication.

The swift movement of AS1411 from the research laboratory to the clinic is precisely the type of outcome for which researchers strive, and this aptamer may ultimately have potential against a variety of solid and blood cancers. There are recent reports of other aptamers being studied in the preclinical setting, and it can be reasonably anticipated that further anticancer aptamers will enter clinical study over the next few years.

References

1. M. Cavazzana-Calvo, S. Hacein-Bey, G. de Saint Basile, F. Gross, E. Yvon, P. Nusbaum, F. Selz, C. Hue, S. Certain, J. L. Casanova, P. Bousso, F. L. Deist and A. Fischer, *Science*, 2000, **288**, 669.

2. S. Hacein-Bey-Abina, F. Le Deist, F. Carlier, C. Bouneaud, C. Hue, J. P. De Villartay, A. J. Thrasher, N. Wulffraat, R. Sorensen, S. Dupuis-Girod, A. Fischer, E. G. Davies, W. Kuis, L. Leiva and M. Cavazzana-Calvo, *N. Engl. J. Med.*, 2002, **346**, 1185.

3. S. Hacein-Bey-Abina, C. Von Kalle, M. Schmidt, M. P. McCormack, N. Wulffraat, P. Leboulch, A. Lim, C. S. Osborne, R. Pawliuk, E. Morillon, R. Sorensen, A. Forster, P. Fraser, J. I. Cohen, G. de Saint Basile, I. Alexander, U. Wintergerst, T. Frebourg, A. Aurias, D. Stoppa-Lyonnet, S. Romana, I. Radford-Weiss, F. Gross, F. Valensi, E. Delabesse, E. Macintyre, F. Sigaux, J. Soulier, L. E. Leiva, M. Wissler, C. Prinz, T. H. Rabbitts, F. Le Deist, A. Fischer and M. Cavazzana-Calvo, *Science*, 2003, **302**, 415.

4. S. E. Raper, N. Chirmule, F. S. Lee, N. A. Wivel, A. Bagg, G. P. Gao, J. M. Wilson and M. L. Batshaw, *Mol. Genet. Metab.*, 2003, **80**, 148.

5. Y. Fichou and C. Ferec, *Trends Biotechnol.*, 2006, **24**, 563.

6. B. R. Cullen and W. C. Greene, *Cell*, 1989, **58**, 423.

7. A. D. Ellington and J. W. Szostak, *Nature*, 1990, **346**, 818.

8. C. R. Ireson and L. R. Kelland, *Mol. Cancer Ther.*, 2006, **5**, 2957.

9. D. Jellinek, L. S. Green, C. Bell, C. K. Lynott, N. Gill, C. Vargeese, G. Kirschenheuter, D. P. McGee, P. Abesinghe, W. A. Pieken, R. Shapiro, D. B. Riflcin, D. Moscatelli and N. Janjic, *Biochemistry*, 1995, **34**, 11363.

10. M. C. Willis, B. D. Collins, T. Zhang, L. S. Green, D. P. Sebesta, C. Bell, E. Kellogg, S. C. Gill, A. Magallanez, S. Knauer, R. A. Bendele, P. S. Gill and N. Janjic, *Bioconjugate Chem.*, 1998, **9**, 573.

11. C. E. Tucker, L. S. Chen, M. B. Judkins, J. A. Farmer, S. C. Gill and D. W. Drolet, *J. Chromatogr. B. Biomed. Sci. Appl.*, 1999, **732**, 203.

12. S. M. Nimjee, J. R. Keys, G. A. Pitoc, G. Quick, C. P. Rusconi and B. A. Sullenger, *Mol. Ther.*, 2006, **14**, 408.

13. V. Dapic, P. J. Bates, J. O. Trent, A. Rodger, S. D. Thomas and D. M. Miller, *Biochemistry*, 2002, **41**, 3676.

14. V. Dapic, V. Abdomerovic, R. Marrington, J. Peberdy, A. Rodger, J. O. Trent and P. J. Bates, *Nucleic Acids Res.*, 2003, **31**, 2097.

15. P. J. Bates, J. B. Kahlon, S. D. Thomas, J. O. Trent and D. M. Miller, *J. Biol. Chem.*, 1999, **274**, 26369.

16. M. Srivastava and H. B. Pollard, *FASEB J.*, 1999, **13**, 1911.

17. A. G. Hovanessian, F. Puvion-Dutilleul, S. Nisole, J. Svab, E. Perret, J. S. Deng and B. Krust, *Exp. Cell Res.*, 2000, **261**, 312.

18. P. Bates, S. Jueliger, A. Girvan, Y. Teng, L. Casson, S. Thomas, Y. Mi, X. Xu, S. Barve, D. Miller and J. Trent, 18th European Organization for Research and Treatment of Cancer, National Cancer Institute and American Association for Cancer Research Symposium on Molecular Targets and Cancer Therapeutics, Philadelphia, USA, 2005.

19. S. Christian, J. Pilch, M. E. Akerman, K. Porkka, P. Laakkonen and E. Ruoslahti, *J. Cell Biol.*, 2003, **163**, 871.

20. Y. Huang, H. Shi, H. Zhou, X. Song, S. Yuan and Y. Luo, *Blood*, 2006, **107**, 3564.

21. E. A. Said, B. Krust, S. Nisole, J. Svab, J. P. Briand and A. G. Hovanessian, *J. Biol. Chem.*, 2002, **277**, 37492.
22. D. Legrand, K. Vigie, E. A. Said, E. Elass, M. Masson, M. C. Slomianny, M. Carpentier, J. P. Briand, J. Mazurier and A. G. Hovanessian, *Eur. J. Biochem.*, 2004, **271**, 303.
23. E. J. Joo, G. B. ten Dam, T. H. van Kuppevelt, T. Toida, R. J. Linhardt and Y. S. Kim, *Glycobiology*, 2005, **15**, 1.
24. Y. Mi, S. D. Thomas, X. Xu, L. K. Casson, D. M. Miller and P. J. Bates, *J. Biol. Chem.*, 2003, **278**, 8572.
25. T. K. Sengupta, S. Bandyopadhyay, D. J. Fernandes and E. K. Spicer, *J. Biol. Chem.*, 2004, **279**, 10855.
26. Y. Otake, T. K. Sengupta, S. Bandyopadhyay, E. K. Spicer and D. J. Fernandes, *Mol. Pharmacol.*, 2005, **67**, 319.
27. M. L. Cleary, S. D. Smith and J. Sklar, *Cell*, 1986, **47**, 19.
28. K. G. Steube, A. Jadau, D. Teepe and H. G. Drexler, *Leukemia*, 1995, **9**, 1841.
29. J. C. Reed, *Adv. Pharmacol.*, 1997, **41**, 501.
30. G. Gao and Q. P. Dou, *Mol. Pharmacol.*, 2000, **58**, 1001.
31. X. Xu, F. Hamhouyia, S. D. Thomas, T. J. Burke, A. C. Girvan, W. G. McGregor, J. O. Trent, D. M. Miller and P. J. Bates, *J. Biol. Chem.*, 2001, **276**, 43221.
32. A. C. Girvan, Y. Teng, L. K. Casson, S. D. Thomas, S. Juliger, M. W. Ball, J. B. Klein, W. M. Pierce Jr., S. S. Barve and P. J. Bates, *Mol. Cancer Ther.*, 2006, **5**, 1790.

CHAPTER 9

Spiegelmer NOX-E36 for Renal Diseases

DIRK EULBERG,[1] WERNER PURSCHKE,[1]
HANS-JOACHIM ANDERS,[2] NORMA SELVE[1] AND
SVEN KLUSSMANN[*,1]

[1] NOXXON Pharma AG, Max-Dohrn-Strasse 8-10, D-10589 Berlin
Germany; [2] Nephrological Center, Medical Policlinic,
Ludwig-Maximilians-University, Munich, Germany

9.1 Spiegelmers: Functional Mirror-Image Oligonucleotides

To understand what a Spiegelmer is, it is useful to first introduce the concept of aptamers. An aptamer is a nucleic acid structure that can bind to a target molecule in a manner similar to an antibody recognizing an antigen.[1] Based on the SELEX (systematic evolution of ligands by exponential enrichment) process,[2] aptamers can be identified from huge combinatorial nucleic acids libraries of more than 10^{15} different sequences. The sequences in the library are comprised of a central randomized region flanked by fixed sequences that permit amplification by polymerase chain reaction (PCR). The library size depends on the length of the random portion of the molecule, which creates a diverse universe of molecules ready for screening. Specific aptamers are selected by contacting the target with the library and eluting the bound aptamers. After amplification by PCR, these binding molecules are re-selected, eluted and amplified again. By repeating this procedure multiple times, the molecules with greatest affinity and specificity are selected.

RSC Biomolecular Sciences
Therapeutic Oligonucleotides
Edited by Jens Kurreck
© Royal Society of Chemistry 2008

Aptamers have binding characteristics similar to those of peptides or antibodies, with affinities in the low nanomolar to picomolar range. However, there is one major drawback to aptamers as useful therapeutic products: as natural nucleic acid polymers they are prone to rapid degradation by nucleases that are present in all tissues in the body. To overcome this problem, methods of creating nuclease-resistant molecules that retain the binding capabilities of aptamers needed to be discovered. A highly suitable answer to this problem are the Spiegelmers.[3,4] Spiegelmers are biostable aptamers which have all of the diversity characteristics of aptamers, but possess a structure that prevents enzymatic degradation.

9.1.1 Identification of Spiegelmer Products

The name 'Spiegelmer' is derived from the German '*Spiegel*', meaning mirror, and reflects the molecule's mirror image configuration. This reversed configuration is the key to Spiegelmers' biostability and is achieved by the use of non-natural L-nucleotides rather than the natural D-nucleotides for their synthesis. The L-form cannot be recognized by nucleases so that the molecule remains intact in the biological environment. Methods have been developed that allow us to identify L-oligonucleotide (L-ON) binders that possess outstanding affinities and functionalities.

The identification of a Spiegelmer is based on the reciprocal specificities of mirror-image binder pairs (see Figure 9.1). If an aptamer binds a specific configuration of a target, the mirror image of the aptamer will identically bind the mirror image of the target. NOXXON carries out the process of aptamer selection against the mirror-image target and isolates a highly specific aptamer against this mirror image. The corresponding mirror-image nucleic acid (L-ON) of this aptamer, the Spiegelmer, will now bind to the natural target with the same binding characteristics (Figure 9.2). In addition, this Spiegelmer is resistant to nuclease degradation.

Thus, Spiegelmers possess the high-affinity binding characteristics of the best aptamers and antibodies, while defying enzymatic degradation that severely limits the utility of aptamers. Figure 9.3 illustrates the increased stability of Spiegelmers compared to aptamers. Spiegelmers are stable in human plasma for over 60 hours at 37 °C (right panel, Figure 9.3), while non-modified RNA aptamers are degraded in seconds under the same conditions (left panel, Figure 9.3).

9.2 MCP-1 and Inflammatory Kidney Diseases

9.2.1 The Target: MCP-1 (CCL-2)

The pro-inflammatory chemokine monocyte chemotactic protein-1 [MCP-1; systematic name, CCL2; alternative names MCAF (monocyte chemoattracting and activating factor); JE; SMC-CF (smooth muscle cell-colony stimulating

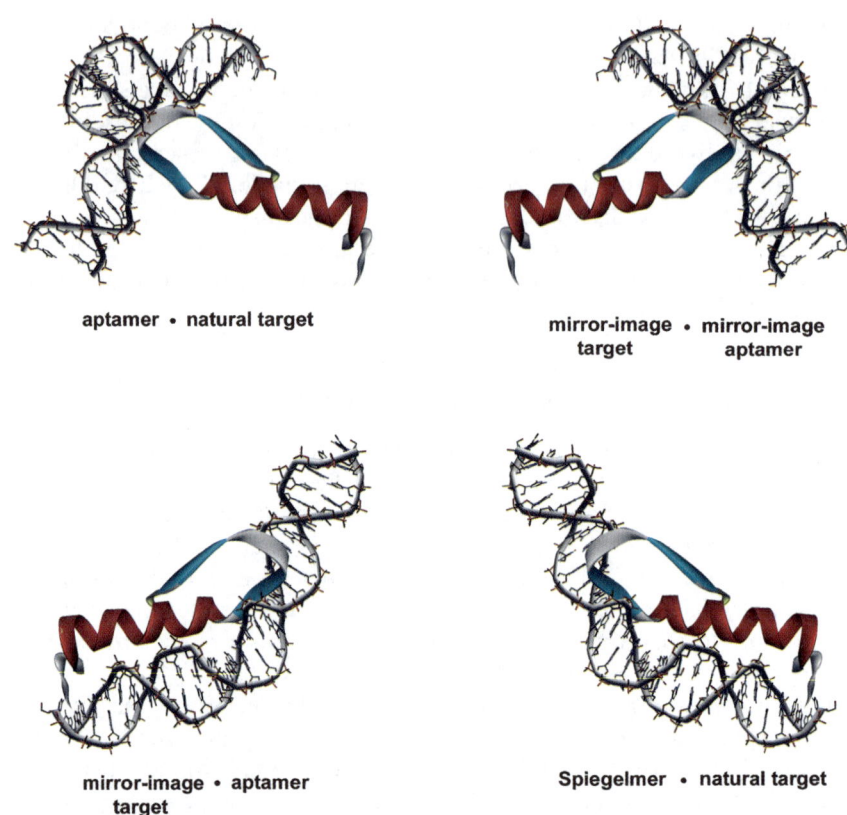

aptamer • natural target

mirror-image • mirror-image
    target   aptamer

mirror-image • aptamer
    target

Spiegelmer • natural target

Figure 9.1 Reciprocal specificities of aptamers (D-RNA) and Spiegelmers (L-RNA). (Reprinted with permission from ed. C Schmuck C. and H Wennemers, *Highlights in Bioorganic Chemistry: Methods and Applications*, 2004, Wiley VCH.)

factor); HC-11; UniProtKB/Swiss-Prot entry P13500] is a small secreted, heparin-binding protein that attracts and activates immune and non-immune cells. Human MCP-1 was characterized by three research groups independently in the late 1980s.[5–7] There are additional MCP family members in both humans (MCP-2, -3, -4) and mice (MCP-2, -3, -5). The other human MCPs are approximately 70% homologous with human MCP-1.

The structure of MCP-1 has been solved by nuclear magnetic resonance (NMR)[8] and X-ray analysis.[9] The MCP-1 monomer has the typical chemokine fold in which the N-terminal cysteines are followed by a long loop that leads into three antiparallel β-pleated sheets in a Greek key motif. The protein terminates in an α-helix that overlies the three β-sheets (Figure 9.4).[9]

MCP-1 is a potent attractor of monocytes and/or macrophages, basophils, activated T-cells and natural killer (NK) cells. A wide variety of cell types express MCP-1 upon induction with activators such as growth factors, cytokines and bacterial endotoxins [lipopolysaccharide (LPS)]. Rather unusual in

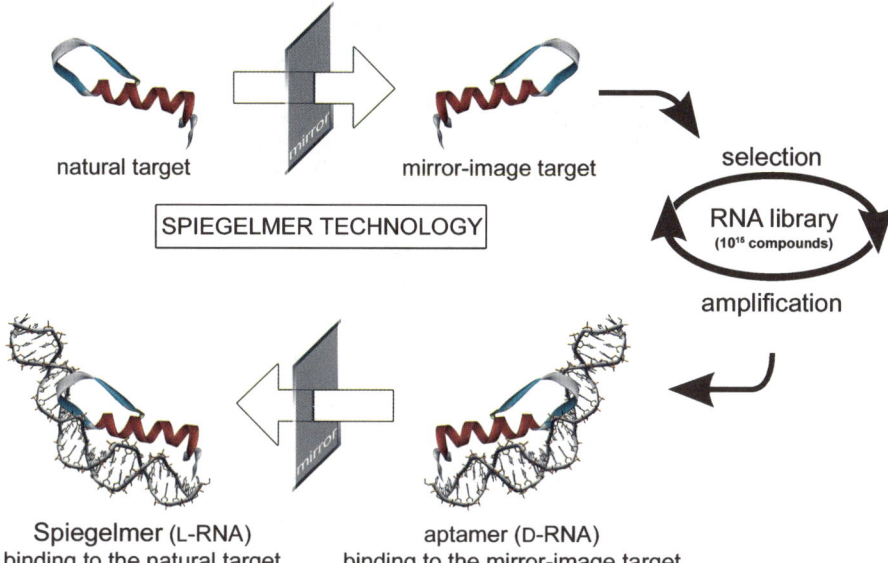

Figure 9.2 Spiegelmer identification process.

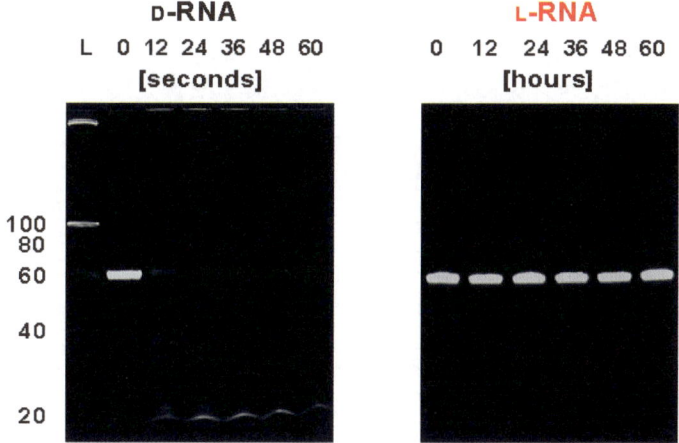

Figure 9.3 Stability of Spiegelmers (L-RNA) in human serum (*in vitro*). (Reprinted with permission from the Nature Publishing Group, *Nat. Biotechnol.*, 1996, **14**, 1112–1115.)

the promiscuous chemokine system, MCP-1 is highly specific in its receptor usage, binding only to the chemokine receptor CCR2 with high affinity. Like all chemokine receptors, CCR2 is a G-protein-coupled receptor.[10]

MCP-1 is involved in monocyte recruitment into inflamed tissues (Figure 9.5). Here, resident macrophages release chemokines and cytokines (step 1) like tumour necrosis factor (TNF) and interleukin-1β (IL-1β) and others, which

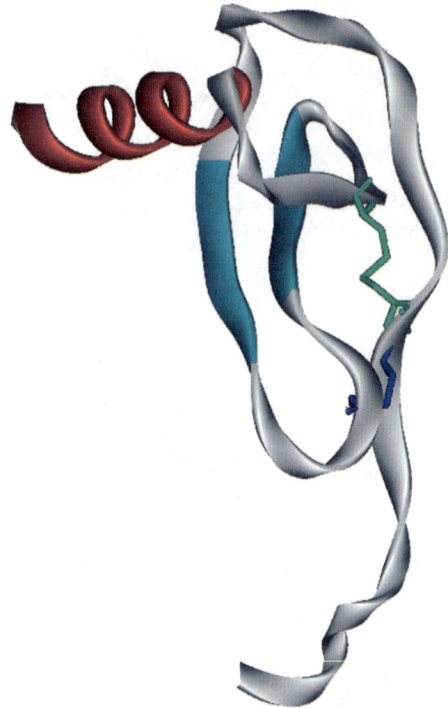

Figure 9.4   Crystal structure of MCP-1, adapted from Protein Data Base (1DOK).[9]

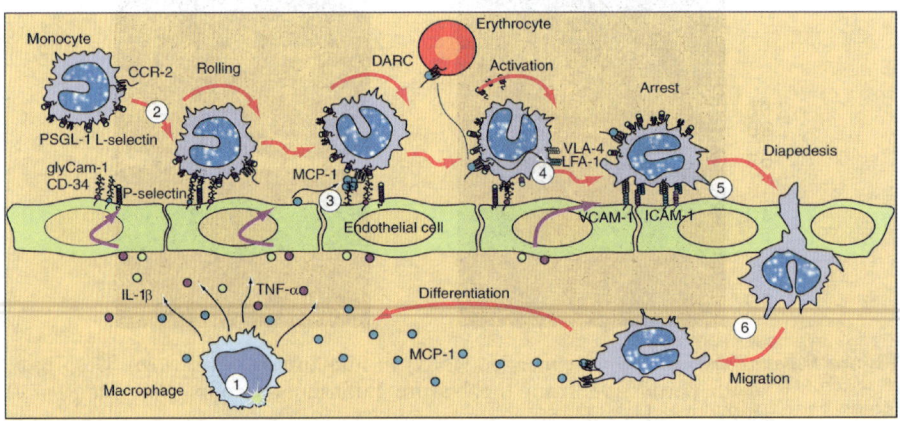

Figure 9.5   MCP-1 involvement in monocyte recruitment. (Reprinted with permission
from Ashley Publications, *Expert Opin. Ther. Targets*, 2003, **7**, 35–48.)[10]

activate endothelial cells to express a battery of adhesion molecules. The
resulting 'sticky' endothelium causes monocytes in the blood vessel to roll along
the surface (step 2). Here, the monocytes encounter MCP-1 on the endothelial
surface, which binds to CCR2 on monocytes (step 3) and activates them (step 4).

This finally leads to firm arrest (step 5), spreading of monocytes along the endothelium and transmigration into the surrounding tissue (step 6), where the monocytes differentiate into macrophages.

9.2.1.1 Species Homology

Although the overall three-dimensional structure of MCP-1 from different mammalian species is maintained, the amino acid sequence is not particularly well conserved (Figure 9.6). Within the first 76 amino acids, the sequence alignment shows 55% overall sequence similarity between human and murine MCP-1. Apart from the amino acid sequence, murine MCP-1 (also called JE) also differs from human MCP-1 in molecular size (125 aa) and the extent of

a)

Homo sapiens	QPDAINAPVTCCYNFTNRKISVQRLASYRRITSSKCPKEAVIFKTIVAKEICA
Pongo pygmaeus	QPDAINAPVTCCYNFTNRKISVQRLASYRRITSSKCPKEAVIFKTIVAKEICA
Macaca mulatta	QPDAINAPVTCCYNFTNRKISVQRLASYRRITSSKCPKEAVIFKTIVAKEICA
Sus scrofa	QPDAINSPVTCCYTLTSKKISMQRLMSYRRVTSSKCPKEAVIFKTIAGKEICA
Equus caballus	QPDAINSPVTCCYTFTGKKISSQRLGSYKRVTSSKCPKEAVIFKTILAKEICA
Canis familiaris	QPDAIISPVTCCYTLTNKKISIQRLASYKRVTSSKCPKEAVIFKTVLNKEICA
Bos taurus	QPDAINSQVACCYTFNSKKISMQRLMNYRRVTSSKCPKEAVIFKTILGKEICA
Oryctolagus cuniculus	QPDAVNSPVTCCYTFTNKTISVRRLMSYRRINSTKCPKEAVIFMTKLAKGICA
Mus musculus	QPDAVNAPLTCCYSFTSKMIPMSRLESYKRITSSRCPKEAVVFVTKLKREVCA
Rattus norvegicus	QPDAVNAPLTCCYSFTGKMIPMSRLENYKRITSSRCPKEAVVFVTKLKREICA
Cavia porcellus	QPDGVNTP-TCCYIFN-KQIPLKRVKGYERITSSRCPQEAVIFRTLKNKEVCA

Homo sapiens	DPKQKWVQDSMDHLDKQTQTPKT
Pongo pygmaeus	DPKQKWVQDSMDHLDKQTQTIKT
Macaca mulatta	DPKQKWVQDSMDHLDKQTQTPKP
Sus scrofa	EPKQKWVQDSISHLDKKNQTPKP
Equus caballus	DPEQKWVQDAVKQLDKKAQTPKP
Canis familiaris	DPKQKWVQDSMAHLDKKSQTQTAKP
Bos taurus	DPKQKWVQDSINYLNKKNQTPKP
Oryctolagus cuniculus	DPKQKWVQDAIANLDKKMQTPKTLTSYSTTQEHTTNLSSTRTPSTTTSL
Mus musculus	DPKKEWVQTYIKNLDRNQMRSEPTTLFKTASALRSSAPLNVKLTRKSEANASTTFSTTTSSTSVGVTSVTVN
Rattus norvegicus	DPNKEWVQKYIRKLDQNQVRSETTVFYKIASTLRTSAPLNVNLTHKSEANASTLFSTTTSSTSVEVTSMTEN
Cavia porcellus	DPTQKWVQDYIAKLDQRTQQKQNSTAPQTSKPLNIRFTTQDPKNRS

b)

Pongo pygmaeus (Orang Utan)	99 %
Macaca mulatta (Rhesus monkey)	97 %
Sus scrofa (Pig)	79 %
Equus caballus (Horse)	78 %
Canis familiaris (Dog)	76 %
Oryctolagus cuniculus (Rabbit)	75 %
Bos taurus (Bovine)	72 %
Mus musculus (Mouse)	55 %
Rattus norvegicus (Rat)	55 %
Cavia porcellus (Guinea pig)	55 %

Figure 9.6 (a) Clustal W alignment of MCP-1 between different mammalian species, and (b) percentage of identical positions in comparison to *homo sapiens* (amino acids 1–76 only).

glycosylation. Murine MCP-1 contains a 49-amino acid C-terminal domain that is not present in human MCP-1 and is not required for *in vitro* bioactivity.

9.2.1.2 Pathophysiology of MCP-1

The expression pattern of MCP-1 suggests that it plays important roles in human diseases characterized by mononuclear cell infiltration. Such cell infiltration is present in many inflammatory and autoimmune diseases. A large body of evidence argues in favour of a unique role for the MCP-1–CCR2 axis in monocyte chemoattraction and thus chronic inflammation:

- MCP-1- or CCR2-deficient mice show a markedly reduced macrophage chemotactic response while otherwise appearing normal.[11–14]
- Despite functional redundancy with other chemokines *in vitro*, loss of MCP-1 effector function alone is sufficient to impair monocytic trafficking in several inflammatory models.[15–22]
- MCP-1 levels are elevated in many inflammatory diseases. In fact, elevated levels of MCP-1 are associated with many diseases with and without an obvious inflammatory component (Table 9.1).

Table 9.1 Diseases associated with elevated levels of MCP-1.

Disease	Reference(s)
Rheumatoid arthritis (RA)	44, 45
Renal disease	29, 32, 47, 35, 33, 46, 48
Systemic lupus erythematosus (SLE)	49–51
Multiple sclerosis (MS)	52–54
Atherosclerosis	55–57
Restenosis	58
Amyotrophic lateral sclerosis	59
Allergy and asthma	17, 60, 61
Cancer (breast, gastric, bladder, ovarian, colorectal carcinoma; lymphoblastic leukaemia; hamartomous tumours)	62–68
Diabetic retinopathy	69–71
Uveitis	72
Psoriasis	73
Atopic dermatitis	74
Endometriosis	75
Polymyositis and dermatomyositis	76
Obesity	77, 78
Myocarditis	79
Alzheimer's disease	80
Chronic liver disease	81
Peyronie's disease	82
Behçet's disease	83
Preterm labour and/or delivery	84

Therapeutic intervention with anti-MCP-1 agents – or CCR2 antagonists – would affect the excess inflammatory monocyte trafficking, but may spare basal trafficking of phagocytes, thereby avoiding general immunosuppression and increased risk of infections.[10]

Several arguments support the validity of targeting the ligand MCP-1 instead of the receptor CCR2:

- Chemokine antagonists can act in *two* ways – an anti-MCP-1 Spiegelmer may not only inhibit CCR2 signalling by preventing the MCP-1–CCR2 interaction, but also by destruction of the MCP-1 gradient formation on endothelial surfaces (prevention of CCL2:glycosaminoglycan interaction on the endothelial surface).
- An anti-MCP-1 antagonist may be capable of stripping off the danger signal MCP-1 from the endothelial surface; MCP-1 is dissipated in the bloodstream and is thus innocuous. A CCR2 antagonist, in contrast, is efficacious only as long as it is bound to CCR2; the pro-inflammatory MCP-1 signal on the endothelial surface remains untouched.
- CCR2 may not be the only MCP-1 receptor in the kidney. The fact that renal tubular epithelial cells (TECs) are susceptible to MCP-1-specific inflammatory activation despite the lack of CCR2 on these cells is experimental evidence that the MCP-1 receptor on human TECs is different from that of CCR2.[23]
- CCR2 may play an important role in defence against viruses. Evidence exists that $CCR2^{-/-}$ mice have impaired ability to clear viral infections from the brain, whereas $MCP-1^{-/-}$ mice did not show increased morbidity or mortality.[24]

9.2.2 Inflammatory Kidney Diseases

End-stage renal disease requiring renal replacement is a major health problem with an estimated prevalence of 1 million and an incidence of 220 000 new cases each year. The number of patients on dialysis in the USA is expected to rise from 300 000 in the year 2000 to up to 500 000 in 2010 based on the rising prevalence of diabetes and hypertension. Therefore, new strategies that reduce the incidence of end-stage renal failure are required.

Based on the increasing knowledge of the molecular mechanisms of the inflammatory process and the interplay of locally secreted mediators of inflammation, new targets for the therapy of kidney diseases have been identified.[25,26] One of these targets, for which robust data on expression and interventional studies with specific antagonists in appropriate animal models exist, is MCP-1. The role of this protein for immune-cell recruitment to sites of inflammation seems to be non-redundant. Infiltration of immune cells to the kidney is thought to be a major mechanism of structural renal damage and decline of renal function in the development of various forms of kidney disease.

9.2.2.1   Expression of MCP-1 in Inflamed Kidneys

All types of renal cells can express chemokines, including MCP-1 upon stimulation *in vitro*;[26] a long list of stimuli trigger MCP-1 expression *in vitro* including cytokines, oxygen radicals, immune complexes and lipid mediators.

In healthy kidneys of rats and mice, MCP-1 is not expressed, but it is readily upregulated during the course of acute and chronic models of renal inflammation, including immune complex glomerulonephritis, rapid progressive glomerulonephritis, proliferative glomerulonephritis, diabetic nephropathy, obstructive nephropathy or acute tubular necrosis.[26,27] The expression data for MCP-1 in rodents correlate well with the respective expression found in human renal biopsies.[28–30] Furthermore, renal expression in human kidneys is associated with disease activity and declines with the appropriate therapy-induced disease remission.[31]

9.2.2.2   Role of MCP-1 in Inflammatory Renal Diseases

Glomerular mononuclear cell infiltration is associated with the development of a diffuse glomerulosclerosis in patients with diabetic nephropathy. MCP-1 plays an important role in the recruitment and accumulation of monocytes and lymphocytes within the glomerulus.[32,33]

Locally produced MCP-1 seems to be particularly involved in the initiation and progression of tubulointerstitial damage, as documented in experiments using transgenic mice with nephrotoxic serum-induced nephritis (NSN). MCP-1 was mainly detected in vascular endothelial cells, TECs and infiltrated mononuclear cells in the interstitial lesions. The MCP-1-mediated activation of TECs is consistent with the notion that MCP-1 contributes to tubulointerstitial inflammation, a hallmark of progressive renal disease.[34,35]

9.2.2.3   Main Indication: Lupus Nephritis

Systemic lupus erythematosus (SLE) is a heterogeneous autoimmune inflammatory disease that can affect multiple organs in the body. In 30–40% of individuals with SLE, the kidney is the primary target organ. Glomerular inflammation with invasion of acute and chronic inflammatory cells is the hallmark of destruction in this organ, and evidence exists that MCP-1 may play a major role in attracting these cells to the kidney. Through this mechanism renal function is impaired, evident by rising serum creatinine levels, diminished creatinine clearance, loss of proteins in the urine and the presence of blood cells in the urine sediment. Therefore, stabilization and improvement of renal function are measurable and relevant clinical efficacy parameters recommended in the respective draft guidance for industry.[36]

A significant correlation between glomerular expression of MCP-1 and the histological activity index (AI), as well as between tubulointerstitial expression of MCP-1, and the histological chronicity index (CI) was demonstrated very recently on biopsies of SLE patients.[37]

9.3 Identification and Characterization of NOX-E36

9.3.1 Path of Selection for NOX-E36

The nucleotide sequence NOX-E36 was determined by screening a library of synthetic RNAs for binding to the human protein sequence of MCP-1 in the non-natural D-configuration. The D-protein carried a biotin at the NH_2-terminus which facilitated its immobilization and the ligands binding to it. The RNA molecules consisted of 73 nucleotides comprising a central 30-nucleotide region with random sequence. Six nanomoles of RNA were used in the initial selection round, which corresponds to approximately 4×10^{15} individual molecules. Both, the target bio-D-MCP-1 and the design of the combinatorial RNA library are indicated in Figure 9.7.

After eight rounds of *in vitro* selection, the RNA was reverse transcribed and PCR amplified. The resulting dsDNA was ligated into a cloning vector and transformed into *Escherichia coli*. Plasmids from 24 bacterial clones were prepared and the nucleotide sequences of the inserts were determined (Figure 9.8). All inserts had a length of 79 nucleotides and have thus acquired six additional nucleotides during *in vitro* selection. As the process relies on enzymatic nucleic acid amplification steps, mutant RNA molecules may have been generated. If such mutants exhibit a favourable phenotype – which means high-affinity binding to the target molecule of interest – they may eventually succeed in the selection.

From the most frequently represented molecules 180-A4 and 180-D1, truncated versions with lengths of 58 and 43 nucleotides, respectively, were synthesized. A comparison of their binding ability to D-MCP-1 revealed that both

```
bio-QPDAINAPVTCCYNFTNRKISVQRLASYRRITSSKCPKEAVIFKTIVAKEICADPKQKWVQDSMDHLDKQTQTPKT

                                   -------
                                  |       |
     5'-GGAGCUCAGAC-UGGCACGCUG--   |       |
       ||||||  |||       ||||    N30 |
     3'-CACCUUGG--CUGUUGAACACGAC--   |       |
                                  |       |
                                   -------
```

Figure 9.7 Amino acid sequence of the target molecule human biotinylated D-MCP-1 and nucleotide sequence of the used RNA library DE4-4.30.

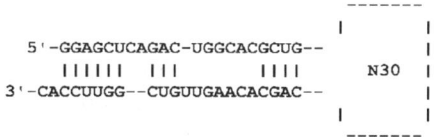

Seq.ID		Frequency
180-A4	GGAGCUCAGACUGGCACGCACGUCCCUCACCGGUGCAAGUGAAGCCGUGGCUCUGCGCAGCACAAGUUGUCGGUUCCAC	38 %
180-D1	GGAGCUCAGACUGGCACGCACGUCCCUCACCGGUGCAAGUGAAGCCGUGGCUCUGCGUAGCACAAGUUGUCGGUUCCAC	25 %
180-B1	GGAGCUCAGACUGGCACGCACGUCCCUCACCGGUGCAAGUGAAGCCGCGGCUCUGCGUAGCACAAGUUGUCGGUUCCAC	8 %
180-B3	GGAGCUCAGACUGGCACGCACGUCCCUCACCGGUGCAAGUGAAGCCGUAGCUCGCGUAGCACAAGUUGUCGGUUCCAC	4 %
180-F1	GGAGCUCAGACUGGCACGCACAUCCCUCACCGGUGCAAGUGAAGCCGUGGCUCUGCGUAGCACAAGUUGUCGGUUCCAC	4 %
180-A2	GGAGCUCAGACUGGCACGCACGUCCCUCACCGGUGUAAGUGAAGCUGUGGCUCUGCGCAGCACAAGUUGUCGGUUCCAC	4 %
180-G2	GGAGCUCAGACUGGCACGCACGUCCCUCACCGGUGCAAGUGAAGCCGUGGCUAUGCGCUGCACAAGUUGUCGGUUCCAC	4 %
180-C2	GGAGCUCAGACUGGCACGCACGUCCCUCACCGGUGCAAGUGAAGCAGUGGCUCUGCACAGCACAAGUUGUCGGUUCCAC	4 %
180-G3	GGAGCUCAGACUGGCACGCUGUGCGGAGUGCGAAACUCCCCGAGACUGCUUUGAAGCAGGCAUAAGUUGUCGGUUCCAC	8 %

Figure 9.8 Nucleotide sequences resulting from the *in vitro* selection.

Seq. ID		nt	KD
180-D1-001	UGGCACGCACGUCCCUCACCGGUGCAAGUGAAGCCGUGGCUCUGCCGUAGCACAAGUUG	58	++
180-D1-002	ACGCACGUCCCUCACCGGUGCAAGUGAAGCCGUGGCUCUGCGU	43	++
180-D1-011	GCACGUCCCUCACCGGUGCAAGUGAAGCCGUGGCUCUGCGU	41	++
180-D1-012	ACGCACGUCCCUCACCGGUGCAAGUGAAGCCGUGGCUCUGC	41	+
180-D1-018	GCACGUCCCUCACCGGUGCAAGUGAAGCCGUGGCUCUGC	39	+
180-D1-034	CGCACGUCCCUCACCGGUGCAAGUGAAGCCGUGGCUCUGCGU	42	++
180-D1-035	CGCACGUCCCUCACCGGUGCAAGUGAAGCCGUGGCUCUGCG	41	++
180-D1-036	GCACGUCCCUCACCGGUGCAAGUGAAGCCGUGGCUCUGCG	40	++

Figure 9.9   Truncations of 180-D1. nt, length in nucleotides; K_D, dissociation constant as determined by pull-down assays; $++$, $\leq 2\,\mathrm{nM}$; $+$, $\leq 10\,\mathrm{nM}$.

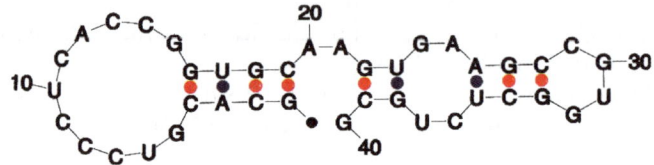

Figure 9.10   *mfold* secondary structure prediction for NOX-E36.

180-D1 variants (Figure 9.9) showed slightly better dissociation constants. Efforts towards further truncation of the selected nucleic acids were therefore concentrated on 180-D1. Several shorter sequences were prepared and tested in binding and cell culture assays, finally resulting in 180-D1-036, which still displayed a comparable affinity compared to 180-D1 (Figure 9.9). 180-D1-036 was defined as the final candidate and renamed to NOX-E36; a secondary structure (minimum free-energy conformation), as predicted by *mfold*[38] is depicted in Figure 9.10.

9.3.1.1   Biophysical Properties of NOX-E36

For the aptamer D-NOX-E36 and its (mirror-image protein) target D-MCP-1, dissociation constants of $890 \pm 65\,\mathrm{pM}$ at $37\,^\circ\mathrm{C}$ and of $146 \pm 13\,\mathrm{pM}$ at room temperature were determined in direct pull-down assays (Figure 9.11).

Via Biacore analysis, the K_D of the Spiegelmer NOX-E36 to the natural protein (L-)MCP-1 was determined to be approximately $900\,\mathrm{pM}$ ($k_a = 2.2 \times 10^5\,\mathrm{M}^{-1}\,\mathrm{s}^{-1}$; $k_d = 1.93 \times 10^{-4}\,\mathrm{s}^{-1}$; Figure 9.12).

9.3.1.2   Biophysical Properties of NOX-E36-5′-Polyethyleneglycol (PEG)

NOX-E36 was PEGylated at its 5′-end to improve its pharmacokinetic properties through an increase in its hydrodynamic volume.[39] Biacore analysis of the 5′-PEGylated NOX-E36 variant allowed the dissociation constant to be determined. The K_D of NOX-E36-5′-PEG was determined to be $2.4 \pm 0.4\,\mathrm{nM}$ ($k_a = 9.31 \pm 0.48 \times 10^4\,\mathrm{M}^{-1}\,\mathrm{s}^{-1}$; $k_d = 2.20 \pm 0.26 \times 10^{-4}\,\mathrm{s}^{-1}$; Figure 9.13).

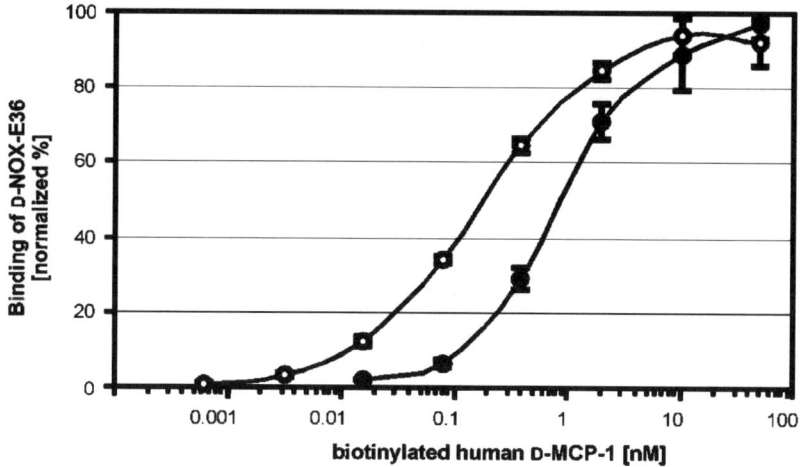

Figure 9.11 Direct pull-down assay with D-NOX-E36; full circles, 37 °C; hollow circles, RT.

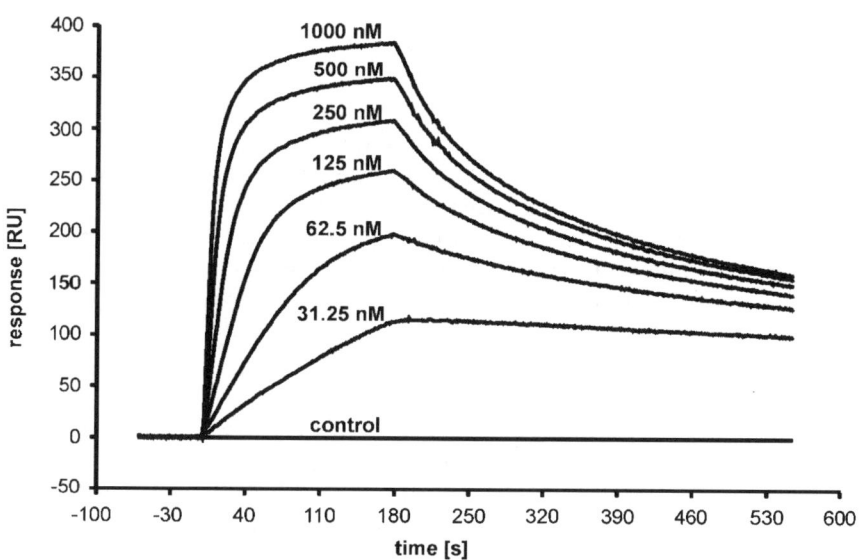

Figure 9.12 Kinetics of NOX-E36 binding to human (L-)MCP-1.

9.3.2 *In vitro* Profiling of NOX-E36

Two different cell-based assays were employed to functionally characterize the MCP-1 binding Spiegelmers. The Ca^{2+}-release assay represents a fast, reproducible and robust quantitative assay for screening purposes. The more sensitive and predictive chemotaxis assay was performed as follow-up and to

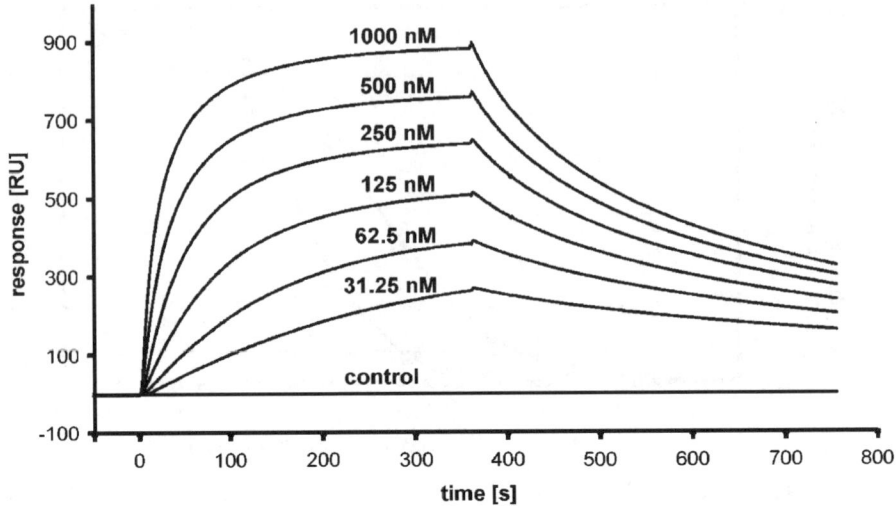

Figure 9.13   Kinetics of NOX-E36-5′-PEG binding to human (L)-MCP-1.

characterize and select the final candidate. This assay is of high relevance for the intended clinical indications, as it reflects a crucial aspect of all inflammatory processes.

In the chemotaxis assay, the final candidate NOX-E36 revealed an IC_{50} of 500 pM and 800 pM at 37 °C in its non-PEGylated and PEGylated forms, respectively. In the Ca^{2+}-release assay, IC_{50} values of 3 and 4 nM were obtained. In both assays, the determined IC_{50} values are at the lower limit of detection, as stimuli used for chemotaxis and Ca^{2+}-release measurements were 0.5 and 3 nM, respectively.

9.3.2.1   Calcium-Release Assay

To test the inhibitory actions of NOX-E36, THP-1 acute monocytic leukaemia cells which express CCR2 and other chemokine receptors were used in a Ca^{2+}-release assay. Stimulation of THP-1 cells with MCP-1 resulted in a dose-dependent increase of the cytoplasmic calcium concentration $[Ca^{2+}]$. Plotting the difference between the maximal and the baseline signals, a dose–response curve for MCP-1 was obtained that indicated a half effective concentration (EC_{50}) of about 3 nM (data not shown).

To demonstrate the efficacy of NOX-E36, THP-1 cells were stimulated with 3 nM MCP-1 and preincubated with various amounts of Spiegelmer. NOX-E36 inhibits MCP-1-induced Ca^{2+}-release with an IC_{50} of about 3 nM (Figure 9.14).

An arbitrary L-RNA sequence consisting of 40 nucleotides (UAAG-GAAACUCGGUCUGAUGCGGUAGCGCUGUGCAGAGCU) did not show any inhibition up to the highest concentration (3 μM) tested (data not shown).

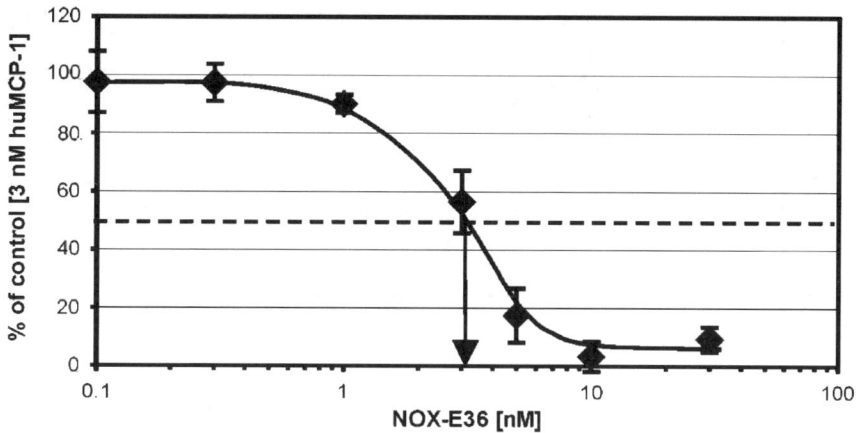

Figure 9.14   Inhibition of MCP-1 induced Ca^{2+}-release in THP-1 cells.

9.3.2.2   Chemotaxis Assay

As a second assay for analysis of the inhibitory potency of NOX-E36, a chemotaxis assay was established. In this assay, THP-1 cells are allowed to migrate through a porous membrane towards a solution containing MCP-1 or MCP-1 supplemented with Spiegelmer, respectively.

MCP-1 stimulates migration of THP-1 cells in a dose dependent manner, with maximal stimulation at about 1 nM and a reduced stimulation at higher concentrations (data not shown). This confirms the typical bell-shaped dose–response curve reported in the literature.[40] For the following experiments on the inhibition of chemotaxis by Spiegelmers an MCP-1 concentration of 0.5 nM was used.

When cells are allowed to migrate towards a solution containing MCP-1 supplemented with increasing concentrations of NOX-E36, a dose-dependent inhibition was observed, with an IC_{50} of about 0.5 nM; complete inhibition was accomplished at approximately 3 nM (Figure 9.15). An arbitrary L-RNA sequence consisting of 40 nucleotides (UAAGAAACUCGGUCUGAUGCG-GUAGCGCUGUGCAGAGCU) did not show any inhibition up to the highest concentration tested (1 μM, data not shown).

9.3.2.3   In vitro Activity of NOX-E36-5'-PEG

In both cell assays, the non-PEGylated NOX–E36 and its PEGylated form NOX–E36-5'-PEG displayed similar inhibitory activity (Table 9.2).

9.3.2.4   Specificity of NOX-E36

The specificity of NOX-E36 binding was tested with regard to human, monkey (cynomolgus), porcine, canine, rabbit, rat and mouse MCP-1 by surface

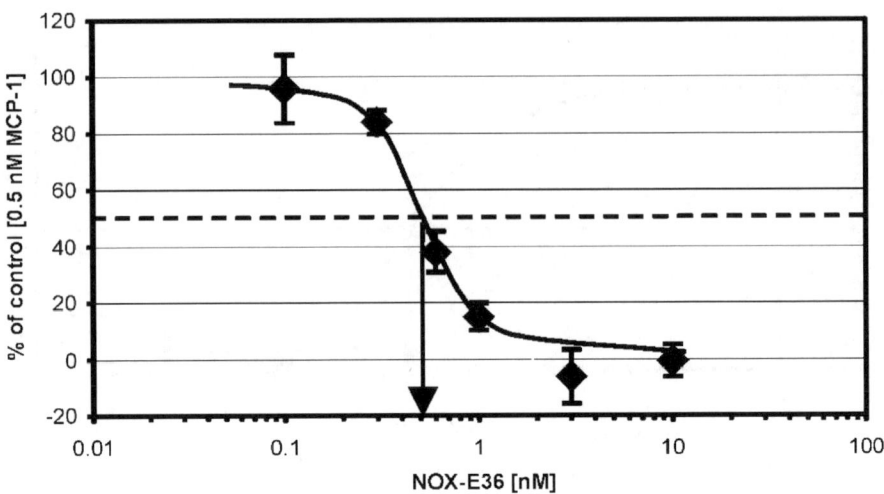

Figure 9.15 Inhibition of MCP-1-induced chemotaxis in THP-1 cells.

Table 9.2 IC$_{50}$ values of NOX-E36 and NOX-E36-5'-PEG.

Type of assay	Spiegelmer	IC$_{50}$ (nM)	MCP-1 stimulus (nM)
Ca^{2+}-release	NOX-E36	3.0	3.0
Ca^{2+}-release	NOX-E36-5'-PEG	4.0	3.0
Chemotaxis	NOX-E36	0.5	0.5
Chemotaxis	NOX-E36-5'-PEG	0.8	0.5

plasmon resonance analysis. Kinetic analyses revealed that NOX-E36 binds to human, monkey, porcine and canine MCP-1 with comparable dissociation constants (K_D) of 0.89–1.2 nM, whereas MCP-1 from mouse, rat and rabbit were not recognized (Figure 9.16; Table 9.3).

9.3.2.5 Selectivity of NOX-E36

Selectivity of NOX-E36 was also assessed by surface plasmon resonance analyses. Binding of a panel of human chemokines from all four chemokine subgroups (CC, CXC, CX$_3$C and XC) to covalently immobilized NOX-E36 was determined (Table 9.4). Specific high-affinity binding to immobilized NOX-E36 could only be detected for CCL2–MCP-1, CCL8–MCP-2, CCL11–eotaxin, CCL3–MIP-1α and CXCL7–NAP-2. The finding that MCP-2 and eotaxin are bound by NOX-E36 is not surprising because of the relatively high homology of MCP-1 to these chemokines (62% and 70%, respectively). The unexpected finding of binding to CCL3–MIP-1α and CXCL7–NAP-2, however, did not translate to a measurable efficacy in cell culture assays (data not shown).

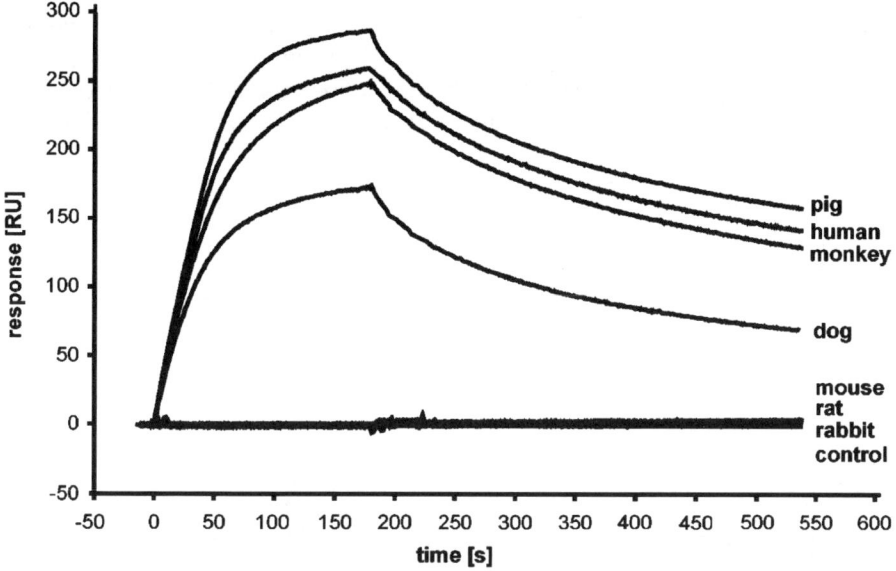

Figure 9.16 NOX-E36 binding to MCP-1 from different species.

Table 9.3 Specificity of NOX-E36 (- means no binding).

MCP-1, species	NOX-E36 K_D (nM)
Monkey	0.90
Porcine	0.89
Canine	1.2
Rabbit	–
Rat	–
Mouse	–

Finally, the kinetic parameters of interaction between NOX-E36 and CCL2–MCP-1, CCL8–MCP-2, CCL11–eotaxin, CCL3–MIP-1α, CXCL7–NAP-2, CCL7–MCP-3 and CCL13–MCP-4 were determined in the 'inverted' system. Here, the chemokines were immobilized and free NOX-E36 was injected. Kinetic data are summarized in Table 9.5.

9.3.3 *In vivo* Efficacy of an Anti-MCP-1 Spiegelmer

The most relevant model for lupus nephritis is established in rodents using MRL$^{lpr/lpr}$ mice which develop symptoms that are similar to human SLE. The relatively low homology of *ca.* 55% between human and murine MCP-1 (Figure 9.6) means, however, that NOX-E36 does not show any binding to murine MCP-1. To be able to perform efficacy studies in rodents, it was necessary to identify a murine-specific Spiegelmer mNOX-E36.

Table 9.4 Selectivity of NOX-E36 binding as determined by Biacore analysis. +, specific binding < 10 nM; (+), specific binding > 1 μM; −, no interaction measurable; −*, unspecific polyanion (control Spiegelmer or dextran matrix) binding > 250 nM; −**, unspecific polyanion (control Spiegelmer or dextran matrix) binding > 10 μM.

Chemokine/synonym	Binding	Chemokine/synonym	Binding
CCL1/I-309	−	CXCL1/GROα	−
CCL2/MCP-1	+	CXCL2/GROβ	−
CCL3/MIP-1α	+	CXCL3/GROγ	−
CCL4/MIP-1β	−	CXCL4/PF4	−**
CCL5/RANTES	−*	CXCL5/ENA-78	−
CCL7/MCP-3	−	CXCL6/GCP-2	−**
CCL8/MCP-2	+	CXCL7/NAP-2	+
CCL11/eotaxin	+	CXCL8/IL-8	−
CCL13/MCP-4	(+)	CXCL9/MIG	−*
CCL14/HCC-1	−	CXCL10/IP-10	−
		CXCL11/I-TAC	−**
CX₃CL1/fractalkine	−	CXCL12α/SDF-1α	−
XCL1/lymphotactin	−	CXCL12β/SDF-1β	−

Table 9.5 Kinetic analysis of NOX-36 interaction with different chemokines.

	Association rate k_a ($M^{-1}s^{-1}$)	Dissociation rate k_d (s^{-1})	Dissociation constant K_D (nM)
CCL2–MCP-1	$1.8 \pm 0.4 \times 10^5$	$1.9 \pm 0.1 \times 10^{-4}$	1.1 ± 0.2
CCL33–MIP-1α	$1.6 \pm 0.3 \times 10^5$	$6.4 \pm 1.1 \times 10^{-4}$	4.1 ± 1.3
CCL7–MCP-3	−	−	−
CCL8–MCP-2	$2.0 \pm 0.7 \times 10^5$	$6.7 \pm 2.0 \times 10^{-4}$	4.2 ± 2.5
CCL11–eotaxin	$1.6 \pm 0.4 \times 10^5$	$1.1 \pm 0.6 \times 10^{-3}$	7.7 ± 5.2
CCL13–MCP-4	−	−	> 1000
CXCL7–NAP-2	$1.8 \pm 0.5 \times 10^5$	$4.1 \pm 0.4 \times 10^{-4}$	2.5 ± 0.9

9.3.3.1   Murine-Specific Spiegelmer: mNOX-E36

Employing truncated murine D-MCP-1 (Δ 77–125) and *in vitro* selection techniques (see Section 9.1.1), high-affinity RNA ONs were identified. The best binding molecule was truncated to 50 nucleotides and named D-mNOX-E36. Profiling was done analogously to the human-specific candidate NOX-E36. Pull-down assays revealed that D-mNOX-E36 binds to murine MCP-1 (Δ 77–125), displaying a dissociation constant of approximately 200 pM (data not shown); Biacore kinetic analysis showed a K_D of 200–300 pM (Figure 9.17).

9.3.3.2   Functional in vitro Profile of Murine-Specific mNOX-E36

THP-1 cells respond to murine MCP-1 similarly to human MCP-1. Thus, the same assays described for the characterisation of NOX-E36 under Sections 9.3.2.1

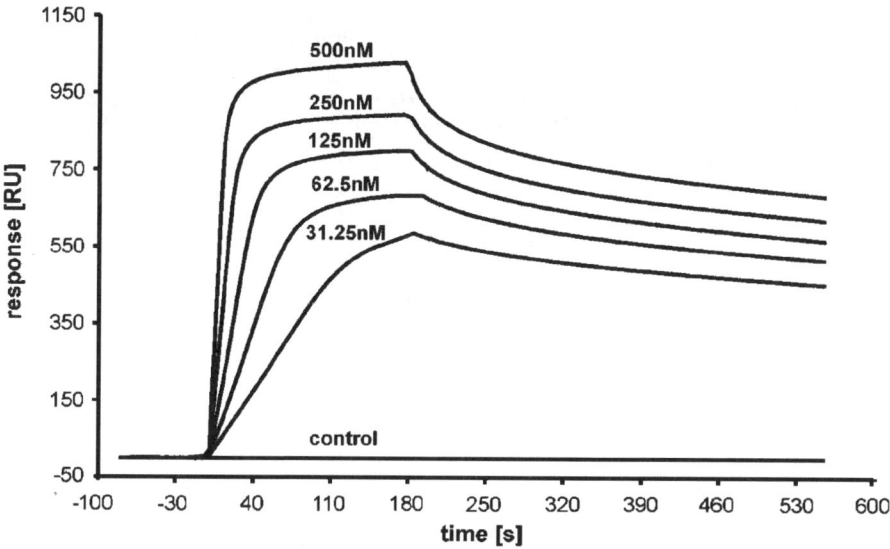

Figure 9.17 Kinetics of D-mNOX-E36 binding to murine MCP-1 (Δ 77–125).

Table 9.6 IC_{50} values of mNOX-E36 and mNOX-E36-3'-PEG.

Type of assay	Spiegelmer	IC_{50} (nM)	Stimulus
Ca^{2+}-release	mNOX-E36	7	5 nM mMCP-1
Ca^{2+}-release	mNOX-E36-3'-PEG	8	5 nM mMCP-1
Chemotaxis	mNOX-E36	5	0.5 nM mMCP-1
Chemotaxis	mNOX-E36-3'-PEG	3	0.5 nM mMCP-1

and 9.3.2.2 could be used for the characterization of mNOX-E36. In the chemotaxis assay mNOX-E36 displayed an IC_{50} of 5 nM and 3 nM in its non-PEGylated and PEGylated forms, respectively; in the Ca^{2+}-release assay, IC_{50} values of 7 and 8 nM were obtained. The data are summarized in Table 9.6.

9.3.3.3 MRL$^{lpr/lpr}$ Mouse ('Lupus Mouse') – Chronic Progressive Renal Inflammation

Mice of the MRL$^{lpr/lpr}$ strain spontaneously develop systemic autoimmunity characterized by skin lesions, enlargement of lymph nodes and progressive inflammatory lung and kidney disease similar to human SLE. Female MRL$^{lpr/lpr}$ mice develop lupus nephritis-like symptoms at 12 weeks and die from end-stage renal disease at 22–26 weeks of age. This allows mortality studies with therapeutic interventions that interfere with renal inflammation. Throughout progression of renal disease, MCP-1 is expressed in increasing amounts in the glomerular and tubulointerstitial compartment in the kidneys of MRL$^{lpr/lpr}$ mice.[41] MCP-1 knockout mice with a MRL$^{lpr/lpr}$ background had less renal

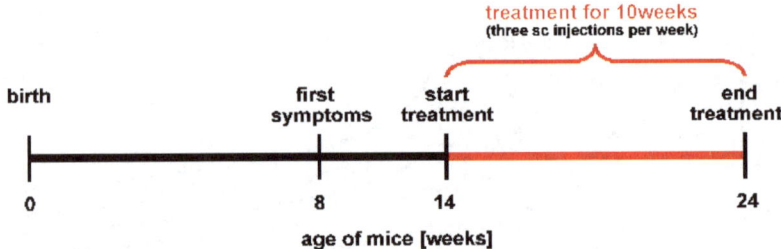

Figure 9.18   Treatment scheme of murine lupus therapy study.

damage, which was associated with improved survival compared with MRL$^{lpr/lpr}$ control mice.[42]

In this study, mNOX-E36 and its PEGylated derivative mNOX-E36-3′-PEG were administered subcutaneously three times per week every second or third day; the doses were 0.9 μmol kg^{-1} for the PEGylated and 1.5 μmol kg^{-1} for the non-PEGylated Spiegelmer. The treatment lasted 10 weeks and started at an age of 14 weeks when lupus symptoms are manifest, reflecting the situation in the clinic (see treatment scheme in Figure 9.18). In total, five experimental groups with 12 animals each were investigated: two groups with active substances (PEGylated and non-PEGylated Spiegelmer), and three control groups (vehicle, non-functional PEGylated Spiegelmer 'PoC–PEG', and non-functional non-PEGylated Spiegelmer 'PoC').

After termination of the study, the animals were sacrificed, kidneys and other organs were sampled and histological kidney sections were prepared. Parameters that were recorded included survival, count of infiltrated macrophages in the kidney tissue, count of T cells and a histopathological scoring of lesions concerning tubuli and interstitium.

9.3.3.4   mNOX-E36-3′-PEG Improves Kidney Disease of MRL$^{lpr/lpr}$ Mice

Both mNOX-E36 and mNOX-E36-3′-PEG improved the indices for activity and chronicity, as described for human lupus nephritis,[43] as well as the aforementioned markers of renal inflammation and the glomerular filtration rate (GFR; *ca.* 300 μl min^{-1} for healthy murine kidneys). Non-PEGylated mNOX-E36 was less effective on the CI and interstitial macrophage and T cell counts (Table 9.7).

Advanced chronic kidney disease was further illustrated by tubular atrophy and confluent areas of interstitial fibrosis in vehicle-, PoC- and PoC–PEG-treated mice (Figure 9.19). In fact, all these histomorphological markers of lupus nephritis were associated with significant mortality. Quantification of these changes by calculating the percentage of high power field and expressing it as means ± SEM (*, $p < 0.05$ respective PoC) revealed that mNOX-E36 and mNOX-E36-3′-PEG reduced interstitial volume, tubular dilation and tubular cell damage, all being markers of the severity and prognosis of chronic kidney disease (data not shown).

Table 9.7 Renal function and histological findings in MRL$^{lpr/lpr}$ mice (hpf, high-power field).

Paramter	Vehicle	PoC	mNOX-E36	PoC–PEG	mNOX-E36-3'-PEG
GFR (μl min)	179 ± 41	200 ± 49	245 ± 69	157 ± 74	293 ± 72
Activity index (score)	17.4 ± 4.9	17.8 ± 4.2	10.3 ± 5.0*	17.4 ± 2.7	9.4 ± 4.2*
Chronicity index (score)	6.0 ± 2.0	7.2 ± 2.6	2.6 ± 2.5	5.4 ± 1.0	1.6 ± 1.8*
Mac-2 + (cells/glom)	13.4 ± 2.0	12.6 ± 0.9	8.5 ± 2.3*	13.6 ± 2.3	8.2 ± 3.5*
Mac-2 + (cells/hpf)	20.3 ± 8.1	20.6 ± 6.7	10.8 ± 5.1	19.3 ± 3.7	7.7 ± 4.0*
CD3 + (cells/hpf)	44.6 ± 14.7	39.4 ± 7.5	23.8 ± 10.2	36.2 ± 3.1	19.0 ± 8.0*

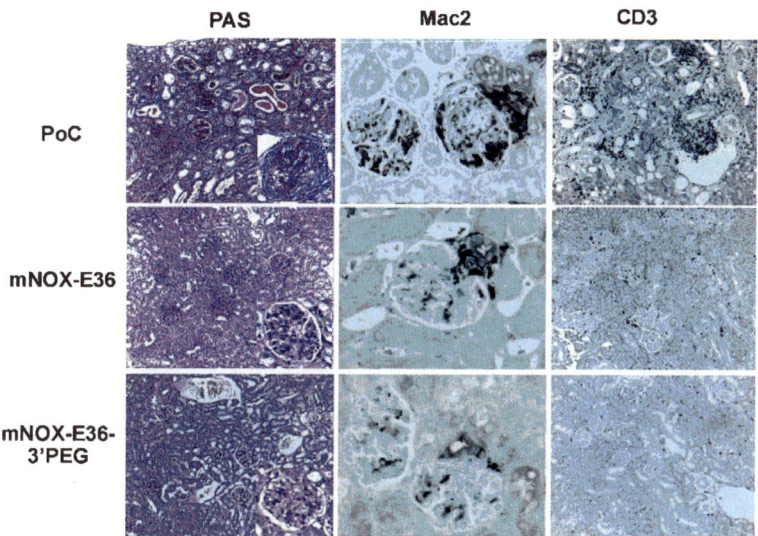

Figure 9.19 Renal histopathology in MRL$^{lpr/lpr}$ mice. Renal sections of 24-week-old animals stained with periodic acid Schiff (PAS), antibodies for Mac2 (macrophages) and CD3 (T cells) as indicated. Images are representative for 7–12 mice in each group (original magnification PAS, 100 ×; PAS inserts, 400 ×; Mac2, 400 ×; CD3, 100 ×) (Reprinted with permission from the American Society of Nephrology, *J. Am. Soc. Nephrol.*, 2007, **18**, 2350–2358.)

9.3.3.5 mNOX-E36 Treatment Improves Condition of Skin and Eyes

The manifestations of systemic lupus include the skin and typically occur at the facial or neck area in 60–80 % of mice. In the groups that had been treated with active Spiegelmer these parameters were reduced to 28% affected animals (data not shown). Treated mice exhibited less alopecia at the right and left side of the

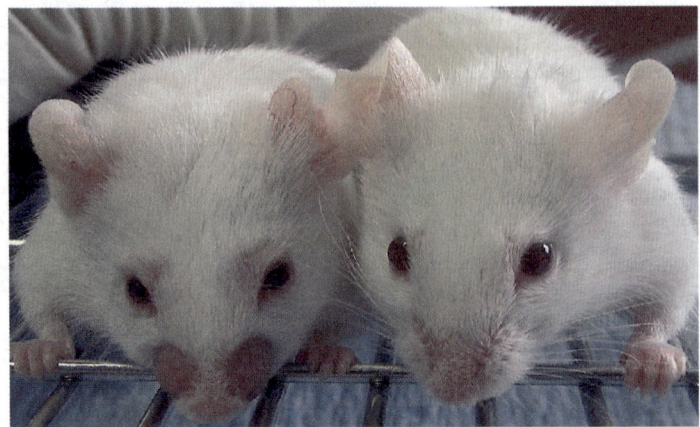

Figure 9.20 Skin- and eye-related systemic effects of mNOX-E36 treatment in MRL$^{lpr/lpr}$ mice (Reprinted with permission from the American Society of Nephrology, *J. Am. Soc. Nephrol.*, 2007, **18**, 2350–2358.)

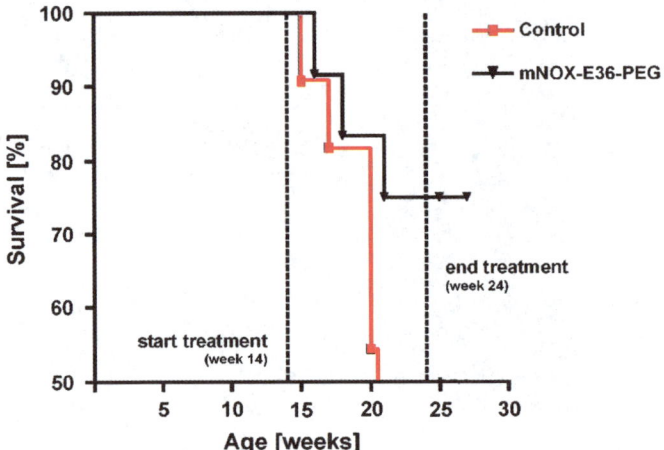

Figure 9.21 Survival of MRL$^{lpr/lpr}$ mice with experimental lupus as calculated by Kaplan-Meier analysis.

snout and had not lost the vibrissae; in addition, neither a two-sided enophthalmos nor the conjunctivitis could be observed in any of the treated animals (Figure 9.20).

9.3.3.6   mNOX-E36-3′-PEG Improves Survival of MRL$^{lpr/lpr}$ Mice

With end-stage renal disease, untreated MRL$^{lpr/lpr}$ mice start to die at around 22 weeks of age. In this study, the 50% survival for vehicle and non-functional Spiegelmer (PoC) was at around 21–23 weeks, whereas treatment with mNOX-E36-3′-PEG prolonged 50% survival to >30 weeks (Figure 9.21).

9.4 Summary and Outlook

Employing Spiegelmer technology, highly potent mirror-image ONs (Spiegelmers) to the protein monocyte chemotactic protein-1 (MCP-1) were generated. MCP-1 is a pro-inflammatory chemokine that is overexpressed in many chronic inflammatory disorders. A large body of evidence exists that inflammatory processes in the kidney are particularly enhanced by this chemokine. NOX-E36 and its murine-specific counterpart mNOX-E36 are MCP-1 inhibitors that exert their functionality through preventing MCP-1 binding to its receptor (CCR2). The inhibitory concentrations (IC_{50}) in relevant cell assays are in the low or even sub-nanomolar range. Treatment of MRL$^{lpr/lpr}$ mice with mNOX-E36-3'-PEG had beneficial effects on clinically relevant (primary and secondary) endpoints, parameters and disease scores with half-maximal efficacy doses of about $0.9\,\mu mol\,kg^{-1}$ and lower. Thus, mouse model data demonstrate that Spiegelmer-based interference with the MCP-1–CCR2 axis leads to improvements in several parameters of kidney inflammation. Further profiling of NOX-E36 and/or mNOX-E36 is ongoing to evaluate the suitability of this approach for treating human kidney diseases.

Acknowledgements

The authors thank Klaus Buchner for providing the cell culture data, Christian Maasch for performing the Biacore analyses, Stefan Vonhoff for synthesis of oligonucleotides, Christian Mihm for help with the figures and Jenny Fischer for critically reading the manuscript.

References

1. A. D. Ellington and J. W. Szostak, *Nature*, 1990, **346**, 818.
2. C. Tuerk and L. Gold, *Science*, 1990, **249**, 505.
3. S. Klussmann, A. Nolte, R. Bald, V. A. Erdmann and J. P. Furste, *Nat. Biotechnol.*, 1996, **14**, 1112.
4. A. Vater and S. Klussmann, *Curr. Opin. Drug. Discov. Devel.*, 2003, **6**, 253.
5. K. Matsushima, K. Morishita, T. Yoshimura, S. Lavu, Y. Kobayashi, W. Lew, E. Appella, H. F. Kung, E. J. Leonard and J. J. Oppenheim, *J. Exp. Med.*, 1988, **167**, 1883.
6. B. J. Rollins, P. Stier, T. Ernst and G. G. Wong, *Mol. Cell Biol.*, 1989, **9**, 4687.
7. T. Yoshimura, E. A. Robinson, S. Tanaka, E. Appella and E. J. Leonard, *J. Immunol.*, 1989, **142**, 1956.
8. T. M. Handel and P. J. Domaille, *Biochemistry*, 1996, **35**, 6569.
9. J. Lubkowski, G. Bujacz, L. Boque, P. J. Domaille, T. M. Handel and A. Wlodawer, *Nat. Struct. Biol.*, 1997, **4**, 64.
10. J. Dawson, W. Miltz, A. K. Mir and C. Wiessner, *Expert Opin. Ther. Targets*, 2003, **7**, 35.

11. L. Boring, J. Gosling, S. W. Chensue, S. L. Kunkel, R. V. Farese Jr., H. E. Broxmeyer and I. F. Charo, *J. Clin. Invest.*, 1997, **100**, 2552.
12. T. Kurihara, G. Warr, J. Loy and R. Bravo, *J. Exp. Med.*, 1997, **186**, 1757.
13. W. A. Kuziel, S. J. Morgan, T. C. Dawson, S. Griffin, O. Smithies, K. Ley and N. Maeda, *Proc. Natl. Acad. Sci. U. S. A.*, 1997, **94**, 12053.
14. B. Lu, B. J. Rutledge, L. Gu, J. Fiorillo, N. W. Lukacs, S. L. Kunkel, R. North, C. Gerard and B. J. Rollins, *J. Exp. Med.*, 1998, **187**, 601.
15. C. M. Lloyd, A. W. Minto, M. E. Dorf, A. Proudfoot, T. N. Wells, D. J. Salant and J. C. Gutierrez-Ramos, *J. Exp. Med.*, 1997, **185**, 1371.
16. H. Ogata, M. Takeya, T. Yoshimura, K. Takagi and K. Takahashi, *J. Pathol.*, 1997, **182**, 106.
17. J. A. Gonzalo, C. M. Lloyd, D. Wen, J. P. Albar, T. N. Wells, A. Proudfoot, A. C. Martinez, M. Dorf, T. Bjerke, A. J. Coyle and J. C. Gutierrez-Ramos, *J. Exp. Med.*, 1998, **188**, 157.
18. K. J. Kennedy, R. M. Strieter, S. L. Kunkel, N. W. Lukacs and W. J. Karpus, *J. Neuroimmunol.*, 1998, **92**, 98.
19. J. M. Galasso, Y. Liu, J. Szaflarski, J. S. Warren and F. S. Silverstein, *Neuroscience*, 2000, **101**, 737.
20. K. Egashira, Q. Zhao, C. Kataoka, K. Ohtani, M. Usui, I. F. Charo, K. Nishida, S. Inoue, M. Katoh, T. Ichiki and A. Takeshita, *Circ. Res.*, 2002, **90**, 1167.
21. K. Furuichi, T. Wada, Y. Iwata, K. Kitagawa, K. Kobayashi, H. Hashimoto, Y. Ishiwata, N. Tomosugi, N. Mukaida, K. Matsushima, K. Egashira and H. Yokoyama, *J. Am. Soc. Nephrol.*, 2003, **14**, 1066.
22. S. Kitamoto and K. Egashira, *Expert Rev. Cardiovasc. Ther.*, 2003, **1**, 393.
23. C. Viedt, R. Dechend, J. Fei, G. M. Hansch, J. Kreuzer and S. R. Orth, *J. Am. Soc. Nephrol.*, 2002, **13**, 1534.
24. K. S. Held, B. P. Chen, W. A. Kuziel, B. J. Rollins and T. E. Lane, *Virology*, 2004, **329**, 251.
25. S. R. Holdsworth, A. R. Kitching and P. G. Tipping, *Curr. Opin. Nephrol. Hypertens.*, 2000, **9**, 505.
26. S. Segerer, P. J. Nelson and D. Schlondorff, *J. Am. Soc. Nephrol.*, 2000, **11**, 152.
27. H. J. Anders, V. Vielhauer and D. Schlondorff, *Kidney Int.*, 2003, **63**, 401.
28. B. H. Rovin, M. Rumancik, L. Tan and J. Dickerson, *Lab. Invest.*, 1994, **71**, 536.
29. P. Cockwell, A. J. Howie, D. Adu and C. O. Savage, *Kidney Int.*, 1998, **54**, 827.
30. T. Wada, K. Furuichi, C. Segawa-Takaeda, M. Shimizu, N. Sakai, S. I. Takeda, K. Takasawa, H. Kida, K. I. Kobayashi, N. Mukaida, Y. Ohmoto, K. Matsushima and H. Yokoyama, *Kidney Int.*, 1999, **56**, 995.
31. B. Amann, R. Tinzmann and B. Angelkort, *Diabetes Care*, 2003, **26**, 2421.

32. N. Banba, T. Nakamura, M. Matsumura, H. Kuroda, Y. Hattori and K. Kasai, *Kidney Int.*, 2000, **58**, 684.
33. T. Morii, H. Fujita, T. Narita, T. Shimotomai, H. Fujishima, N. Yoshioka, H. Imai, M. Kakei and S. Ito, *J. Diabetes Complications*, 2003, **17**, 11.
34. T. Wada, H. Yokoyama, K. Matsushima and K. Kobayashi, *Int. Immunopharmacol.*, 2001, **1**, 637.
35. C. Viedt and S. R. Orth, *Nephrol. Dial. Transplant.*, 2002, **17**, 2043.
36. CDER, March 2005. Draft guidance 6496 (http:\\www.fda.gov/cder/guidance/6496dft.pdf).
37. R. W. Chan, F. M. Lai, E. K. Li, L. S. Tam, K. M. Chow, K. B. Lai, P. K. Li and C. C. Szeto, *Ann. Rheum. Dis.*, 2007, **66**, 886.
38. M. Zuker, *Nucleic Acids Res.*, 2003, **31**, 3406.
39. B. Wlotzka, S. Leva, B. Eschgfaller, J. Burmeister, F. Kleinjung, C. Kaduk, P. Muhn, H. Hess-Stumpp and S. Klussmann, *Proc. Natl. Acad. Sci. U. S. A.*, 2002, **99**, 8898.
40. K. Kito and K. Nishida, *Exp. Cell Res.*, 2002, **281**, 157.
41. G. Perez de Lema, H. Maier, E. Nieto, V. Vielhauer, B. Luckow, F. Mampaso and D. Schlondorff, *J. Am. Soc. Nephrol.*, 2001, **12**, 1369.
42. G. H. Tesch, A. Maifert, A. Schwarting, B. J. Rollins and V. R. Kelley, *J. Exp. Med.*, 1999, **190**, 1813.
43. H. A. Austin 3rd, L. R. Muenz, K. M. Joyce, T. T. Antonovych and J. E. Balow, *Kidney Int.*, 1984, **25**, 689.
44. A. E. Koch, S. L. Kunkel, L. A. Harlow, B. Johnson, H. L. Evanoff, G. K. Haines, M. D. Burdick, R. M. Pope and R. M. Strieter, *J. Clin. Invest.*, 1992, **90**, 772.
45. B. J. Rollins, *Mol. Med. Today*, 1996, **2**, 198.
46. T. Wada, H. Yokoyama, K. Furuichi, K. I. Kobayashi, K. Harada, M. Naruto, S. B. Su, M. Akiyama, N. Mukaida and K. Matsushima, *FASEB J.*, 1996, **10**, 1418.
47. F. Chiarelli, F. Cipollone, A. Mohn, M. Marini, A. Iezzi, M. Fazia, S. Tumini, D. De Cesare, M. Pomilio, S. D. Pierdomenico, M. Di Gioacchino, F. Cuccurullo and A. Mezzetti, *Diabetes Care*, 2002, **25**, 1829.
48. S. Hagiwara, Y. Makita, L. Gu, M. Tanimoto, M. Zhang, S. Nakamura, S. Kaneko, T. Itoh, T. Gohda, S. Horikoshi and Y. Tomino, *Nephrol. Dial. Transplant.*, 2006, **21**, 605.
49. L. Lit, C. Wong, L. Tam, E. Li and C. Lam, *Ann. Rheum. Dis.*, 2006, **65**, 209.
50. G. Liu, J. He, S. Dou, S. Gupta, M. Rusckowski and D. J. Hnatowich, *Eur. J. Nucl. Med. Mol. Imaging*, 2005, **32**, 1115.
51. B. H. Rovin, H. Song, D. J. Birmingham, L. A. Hebert, C. Y. Yu and H. N. Nagaraja, *J. Am. Soc. Nephrol.*, 2005, **16**, 467.
52. C. McManus, J. W. Berman, F. M. Brett, H. Staunton, M. Farrell and C. F. Brosnan, *J. Neuroimmunol.*, 1998, **86**, 20.
53. T. L. Sorensen, F. Sellebjerg, C. V. Jensen, R. M. Strieter and R. M. Ransohoff, *Eur. J. Neurol.*, 2001, **8**, 665.

54. D. J. Mahad, S. J. Howell and M. N. Woodroofe, *J. Neurol. Neurosurg. Psychiatry*, 2002, **72**, 498.
55. N. A. Nelken, S. R. Coughlin, D. Gordon and J. N. Wilcox, *J. Clin. Invest.*, 1991, **88**, 1121.
56. S. Yla-Herttuala, B. A. Lipton, M. E. Rosenfeld, T. Sarkioja, T. Yoshimura, E. J. Leonard, J. L. Witztum and D. Steinberg, *Proc. Natl. Acad. Sci. U. S. A.*, 1991, **88**, 5252.
57. L. Boring, J. Gosling, M. Cleary and I. F. Charo, *Nature*, 1998, **394**, 894.
58. E. Economou, D. Tousoulis, A. Katinioti, C. Stefanadis, A. Trikas, C. Pitsavos, C. Tentolouris, M. G. Toutouza and P. Toutouzas, *Int. J. Cardiol.*, 2001, **80**, 55.
59. P. Baron, S. Bussini, V. Cardin, M. Corbo, G. Conti, D. Galimberti, E. Scarpini, N. Bresolin, S. B. Wharton, P. J. Shaw and V. Silani, *Muscle Nerve*, 2005, **32**, 541.
60. R. Alam, J. York, M. Boyars, S. Stafford, J. A. Grant, J. Lee, P. Forsythe, T. Sim and N. Ida, *Am. J. Respir. Crit. Care Med.*, 1996, **153**, 1398.
61. S. T. Holgate, K. S. Bodey, A. Janezic, A. J. Frew, A. P. Kaplan and L. M. Teran, *Am. J. Respir. Crit. Care Med.*, 1997, **156**, 1377.
62. B. Amann, F. G. Perabo, A. Wirger, H. Hugenschmidt and W. Schultze-Seemann, *Br. J. Urol.*, 1998, **82**, 118.
63. L. Hefler, C. Tempfer, G. Heinze, K. Mayerhofer, G. Breitenecker, S. Leodolter, A. Reinthaller and C. Kainz, *Br. J. Cancer*, 1999, **81**, 855.
64. M. Ohta, Y. Kitadai, S. Tanaka, M. Yoshihara, W. Yasui, N. Mukaida, K. Haruma and K. Chayama, *Int. J. Oncol.*, 2003, **22**, 773.
65. G. M. Gordillo, D. Onat, M. Stockinger, S. Roy, M. Atalay, F. M. Beck and C. K. Sen, *Am. J. Physiol. Cell Physiol.*, 2004, **287**, C866.
66. P. K. Baier, S. Eggstein, G. Wolff-Vorbeck, U. Baumgartner and U. T. Hopt, *Anticancer Res.*, 2005, **25**, 3581.
67. M. Mestdagt, M. Polette, G. Buttice, A. Noel, A. Ueda, J. M. Foidart and C. Gilles, *Int. J. Cancer*, 2006, **118**, 35.
68. A. Eisenkraft, I. Keidan, B. Bielorai, N. Keller, A. Toren and G. Paret, *Leuk. Res.*, 2006, **30**, 1259.
69. C. Capeans, M. V. De Rojas, S. Lojo and M. S. Salorio, *Retina*, 1998, **18**, 546.
70. C. Hernandez, R. M. Segura, A. Fonollosa, E. Carrasco, G. Francisco and R. Simo, *Diabet. Med.*, 2005, **22**, 719.
71. A. D. Meleth, E. Agron, C. C. Chan, G. F. Reed, K. Arora, G. Byrnes, K. G. Csaky, F. L. Ferris 3rd and E. Y. Chew, *Invest. Ophthalmol. Vis. Sci.*, 2005, **46**, 4295.
72. N. Tuaillon, F. Shen de, R. B. Berger, B. Lu, B. J. Rollins and C. C. Chan, *Invest. Ophthalmol. Vis. Sci.*, 2002, **43**, 1493.
73. C. Vestergaard, H. Just, J. Baumgartner Nielsen, K. Thestrup-Pedersen and M. Deleuran, *Acta Derm. Venereol.*, 2004, **84**, 353.
74. Y. Kaburagi, Y. Shimada, T. Nagaoka, M. Hasegawa, K. Takehara and S. Sato, *Arch. Dermatol. Res.*, 2001, **293**, 350.
75. C. Jolicoeur, A. Lemay and A. Akoum, *Am. J. Reprod. Immunol.*, 2001, **45**, 86.

76. J. L. De Bleecker, B. De Paepe, I. E. Vanwalleghem and J. M. Schroder, *Neurology*, 2002, **58**, 1779.
77. I. Dahlman, M. Kaaman, T. Olsson, G. D. Tan, A. S. Bickerton, K. Wahlen, J. Andersson, E. A. Nordstrom, L. Blomqvist, A. Sjogren, M. Forsgren, A. Attersand and P. Arner, *J. Clin. Endocrinol. Metab.*, 2005, **90**, 5834.
78. A. E. Malavazos, E. Cereda, L. Morricone, C. Coman, M. M. Corsi and B. Ambrosi, *Eur. J. Endocrinol.*, 2005, **153**, 871.
79. K. Fuse, M. Kodama, H. Hanawa, Y. Okura, M. Ito, T. Shiono, S. Maruyama, S. Hirono, K. Kato, K. Watanabe and Y. Aizawa, *Clin. Exp. Immunol.*, 2001, **124**, 346.
80. D. Galimberti, C. Fenoglio, C. Lovati, E. Venturelli, I. Guidi, B. Corra, D. Scalabrini, F. Clerici, C. Mariani, N. Bresolin and E. Scarpini, *Neurobiol. Aging*, 2006, **27**, 1763.
81. J. Marsillach, N. Bertran, J. Camps, N. Ferre, F. Riu, M. Tous, B. Coll, C. Alonso-Villaverde and J. Joven, *Clin. Biochem.*, 2005, **38**, 1138.
82. G. T. Lin, Z. Wang, B. C. Liu, T. F. Lue and C. S. Lin, *Asian J. Androl.*, 2005, **7**, 237.
83. E. Bozoglu, A. Dinc, H. Erdem, S. Pay, I. Simsek and I. H. Kocar, *Clin. Exp. Rheumatol.*, 2005, **23**, S42.
84. M. S. Esplin, R. Romero, T. Chaiworapongsa, Y. M. Kim, S. Edwin, R. Gomez, M. Mazor and E. Y. Adashi, *J. Matern. Fetal. Neonatal. Med.*, 2005, **17**, 365.

CHAPTER 10

Strategies for the Delivery of Oligonucleotides in vivo

CHRISTIAN REINSCH, EVGENIOS SIEPI, ANDREAS DIECKMANN AND STEFFEN PANZNER*

Novosom AG, Weinbergweg 22, 06120 Halle (Saale), Germany

10.1 Introduction

Over the past two decades, oligonucleotide (ON)-based approaches to inhibit the expression of target genes using, *e.g.* antisense oligonucleotides (ASOs), small interfering RNAs (siRNAs), ribozymes, decoys, or aptamers have emerged as valuable tools for functional genomics, gene target validation, and therapeutic purposes. However, the relatively large molecular size of ON molecules, their hydrophilic properties, and their extremely short half-lives in serum are challenging the development of ON-based therapeutics for *in vivo* applications. Despite encouraging progress in stabilizing ONs chemically, major obstacles, such as poor bioavailability and inefficient intracellular targeting, have still to be overcome.

In this chapter, we discuss issues involved with the systemic delivery of ASOs and siRNAs. We evaluate delivery strategies based on specific examples, analyze the impact of chemical modifications with respect to *in vivo* stability and cellular uptake, and take a close look at the impact of carrier systems on biodistribution and intracellular delivery.

RSC Biomolecular Sciences
Therapeutic Oligonucleotides
Edited by Jens Kurreck
© Royal Society of Chemistry 2008

10.2   Degradation and Excretion: Progress with Chemically Modified Oligonucleotides

10.2.1   Antisense Oligonucleotides

From the moment of systemic entry, ONs have to withstand degradation from nucleases and urinary excretion. In the past, chemical modifications of ON molecules have helped considerably in these respects. For example, replacement of the phosphodiester linkages in the backbone of the DNA by phosphorothioate linkages has improved resistance of the ASO towards nucleases and increased the half-life in human serum by approximately 10-fold.[1,2] The reason is that, in contrast to unmodified ONs, phosphorothiolated oligonucleotides (PTOs) are not readily recognized as substrates by nucleases, and are mainly protected from rapid urinary excretion by binding to serum proteins such as albumin and alpha-2 macroglobulin. As a result, PTOs accumulate mainly in the cortex and medulla of the kidneys, the liver, the lymph nodes, and the spleen.[3,4] PTOs still bind to target messenger RNAs (mRNAs) and mediate cleavage by RNAse H.[5,6] Therefore, modification of the phosphate linkages in the DNA backbone has emerged as a key for antisense drug development and resulted in a 'first generation' of therapeutically active ASO drugs.[7]

While PTO chemistry solved the most pressing issues, a number of drawbacks remained or newly appeared, including (1) binding to certain proteins may cause cellular toxicity; (2) intravenous (i.v.) bolus injections of concentrated PTOs activate the complement cascades; (3) PTOs exhibit a reduced binding affinity towards complementary RNA molecules in comparison to their unmodified counterparts; (4) PTOs are still extensively digested in plasma and tissues by exonucleases.[8,9,10] Thus, 'second generation' ASOs, with alkyl modifications at the 2' position of the ribose, emerged, which have greatly helped to solve some of the problems associated with PTOs. ASOs that contain 2'-*O*-methyl (2'-OMe) or 2'-*O*-methoxyethyl (2'-MOE) modifications exhibit less toxicity, higher affinity towards the complementary RNA, and higher nuclease resistance.[11,12,13] However, these modifications must be restricted to the ends of the ON molecule, leaving the central strand unmodified to activate RNAse H cleavage.[14] These so-called gapmers, with at least five nucleotides between the modified wings, represent the current state-of-the-art in ASOs for clinical use.

Interestingly, 2'-MOE modified ONs offer sufficient nuclease resistance, even in the absence of a thiolated backbone. Thus, omitting the phosphorothioate modification has been suggested to reduce non-specific protein binding of ONs. However, non-phosphorothioated, 2'-MOE-modified ONs exhibited enhanced renal clearance.[15]

Locked nucleic acids (LNAs) represent another class of nucleic acid analogs that are currently under development. The 'lock' is a methylene bridge that connects the 2'-oxygen with the 4'-carbon of the ribose molecule.[16] Chimeric DNA–LNA gapmers exhibit increased stability against nucleases and unprecedented binding affinity towards complementary DNA or RNA.[17,18]

While LNA-modified ASOs have the potential to improve the potency of antisense therapeutics, they may, however, impose a higher risk of hepatoxicity.[19]

Other modified ASOs, such as peptide nucleic acids (PNAs), 2'-deoxy-2'-fluoro-β-D-arabino nucleic acids (FANA), and morpholino oligonucleotides (MF), are also currently being investigated for therapeutic use, but are still in an early state of development. For details about ON modifications, see Chapters 1, 3, 4 and 5 and J. Kurreck 2003.[20]

10.2.2 Small Interfering RNAs (siRNAs)

The use of siRNAs to suppress gene expression for research or therapeutic purposes has moved from basic science into the realm of applied molecular biology and molecular medicine.[21] siRNA molecules are typically used as 19–21-mers and optionally comprise of 3'- or 5'-overhangs of two to four nucleotides in length.[22,23]

Although double-stranded RNA molecules are more resistant towards nuclease attack than single-stranded ASOs, unmodified siRNAs are nevertheless rapidly degraded after systemic administration. A variety of chemical modifications, originally developed to stabilize ASOs, can also be used for siRNA molecules. However, careful selection of the type of modification, the limited number of modified nucleotides, and precise positioning of the modifications on the RNA strands are important to retain potency of the siRNA molecules. For a comprehensive overview, see Zhang *et al.*,[24] and Behlke.[25]

Pharmaceutical companies have developed different chemical modification strategies: SR Pharma (now Silence Therapeutics) is using alternating 2'-OMe modifications in blunt-ended 19-mers,[26] Sirna Therapeutics Inc. has developed complex and extensive modification patterns, while Alnylam Pharmaceuticals and RXi Pharmaceuticals have modified their siRNA molecules only slightly.[27,28,29]

10.3 Unassisted Cellular Uptake of Oligonucleotides

Cellular internalization of free ONs has frequently been observed. Binding of ASOs to a cell membrane receptor can trigger their endocytotic or pinocytotic uptake,[30,31,32] and a number of research groups have investigated uptake of ligand–ASO conjugates.[33,34,35] In addition, Li and co-workers reported the transport of unassisted ONs through anion channels in cultured bovine adrenal cells.[36] Cellular uptake is, however, only the first barrier to be overcome. Following uptake, ONs are initially trapped in the endosomal–lysosomal compartment and only a fraction are released into the cytoplasm intact. Therefore, unassisted ON-mediated gene silencing is ineffective, even at high ON concentrations.[37,38,39] Nevertheless, a great number of reports support the pharmacological activity of chemically stabilized, but otherwise unassisted, ASOs *in vivo*.

Following systemic injection, thiolated ASOs distribute mainly to kidney cortex, liver, spleen, and lymph nodes.[40] In addition, accumulation and anti-sense activity have been observed in adipose tissue.[41] At the sub-tissue level, a dose-dependent compartmentalized uptake was observed in rat liver. ASO dosages equal to or lower than $10\,mg\,kg^{-1}$ resulted in a selective uptake in Kupffer cells and endothelial cells, and dosages of $25\,mg\,kg^{-1}$ or more were needed to penetrate into the liver parenchyma and achieve localization in hepatocytes.[42] Pharmacological responses in rodents match these results, and the ASO-mediated knockdown of hepatocellular targets, such as ApoB-100 and DGAT2, have been achieved with dosages of $10–50\,mg\,kg^{-1}$.[43,44]

Second-generation ASOs, such as 2′-MOE gapmer ASOs, exhibit a re-markable longevity in target tissues. Depending on the dosage and the organ, tissue elimination half-lives of 11–19 days have been reported.[45,46] In clinical settings, this feature enables convenient, intermittent dosing regimens while keeping maintenance dosages at low levels.

siRNA molecules cannot passively penetrate cell membranes because, in comparison to ASOs, siRNAs carry approximately twice the charge and are approximately twice the size. Although some reports claim entry of siRNA molecules into cells of the target tissue following i.v. injection, the vast majority of reports confirm that naked siRNA is not active *in vivo*.[47] Moreover, naked siRNA is used by many researchers as a negative control, which fails to produce silencing effects after i.v. injection. Hydrodynamic injections have enabled ex-perimental work *in vivo* using siRNA. This harsh and painful administration procedure lacks, however, any clinical relevance.[48,49]

Important progress for the *in vivo* application of siRNAs was made in 2004 by Soutschek and co-workers, who attached a modified cholesterol molecule at the 3′-end of the sense strand of a partially phosphorothioated and 2′-OMe-modified siRNA.[28] Following i.v. administration of this hybrid molecule, ApoB-100 levels were reduced in the liver and jejunum on both the mRNA and protein levels. Most convincingly, ApoB-100 mRNA was cleaved within the sequence targeted by the siRNA. In addition, the reduction of ApoB-100 levels was paralleled by a reduction in cholesterol, high-density lipoprotein, low-density lipoprotein, and chylomicron plasma levels. Despite these im-pressive results, the dosage of chemically modified siRNAs was relatively high, at $50\,mg\,kg^{-1}$.

The extended persistence of 2′-MOE-modified ASOs in tissues and the im-proved potency of these molecules in contrast to first-generation ASOs have lowered the dosages needed for a therapeutic effect substantially, and even made convenient subcutaneous (s.c.) injections a reality for selected indications. However, the need for improvement has been widely recognized. In clinical studies, ASOs accessed sites of inflammation but clinical endpoints have so far not been met.[50] In addition, recruitment and uptake of ASOs in cancer is poor and requires high dosages.[51,52,53] While ASOs are therapeutically active at sufficiently high doses, the hurdles for the delivery of siRNA are much higher and are not outweighed by the higher potency of siRNA molecules.

Therefore, assisted delivery of siRNA molecules into the cytosol is so far a necessary condition for systemic application.

To date, significant progress has been made in the construction of delivery systems that enable cytosolic delivery or nuclear uptake of ONs without affecting cellular integrity. Section 10.4 reviews some of the most advanced delivery vectors for ONs and highlights common principles.

10.4   Carriers for Oligonucleotide Delivery

10.4.1   Cationic Carriers

The most common feature of all ON carriers is a positive surface charge, which facilitates rapid complex formation with negatively charged ONs, resulting in high weight ratios between cargo and vector. In addition, complexes with a cationic net charge are readily adsorbed onto cells, leading to a high local ON concentration at the cell surface supporting internalization. However, aggregate formation and unspecific adsorption onto endothelia can lead to a blockage of capillaries, which represents a major challenge to these carrier systems for systemic applications.

10.4.1.1   CLP-2-Based Liposomes

Large cationic liposomal ASO formulations, consisting of dimethyldioctadecyl-ammonium bromide (DDAB) and phosphatidyl cholines (PCs), aggregate and require sonication prior to use.[54] This problem was circumvented by replacing DDAB with the bivalent cationic lipid cardiolipin-2 (CLP-2).[55] CLP-2 is a cationic version of cardiolipin, in which the phosphate head group is replaced by an ammonium group. It may also be regarded as a cross-linked dioleoyl-dimethyl-ammonium-propane (DOTAP) molecule. An ASO against c-raf, formulated in such a cationic vector, inhibited tumor growth in mouse models of human breast, ovarian, and prostate cancers.[56] NeoPhectin-AT is a commercially available formulation consisting of CLP-2, cholesterol, and L-alpha-dioleoyl phosphatidylethanolamine (DOPE). This formulation has been reported to facilitate the delivery of siRNA against c-raf. Treatment with c-raf siRNA–NeoPhectin at a dosage of $7.5\,mg\,kg^{-1}$ twice a day for five days resulted in tumor growth inhibition by 49%. A combined treatment of c-raf siRNA–NeoPhectin and the chemotherapeutic agent docetaxel (Taxotere®) resulted in tumor growth inhibition by 89%.[57] CLP-2-based liposomal ASO formulations were reported to have better efficacy and tolerability than commercially available DOTAP-based formulations, but dose-limiting toxic effects were reported at a lipid dose of approximately $100\,mg\,kg^{-1}$.[56]

10.4.1.2   Atuplex Technology

Atugen's (now Silence Therapeutics) product AtuFECT01 (Figure 10.1A) is a cationic lipid with three positive charges per head group and is used in

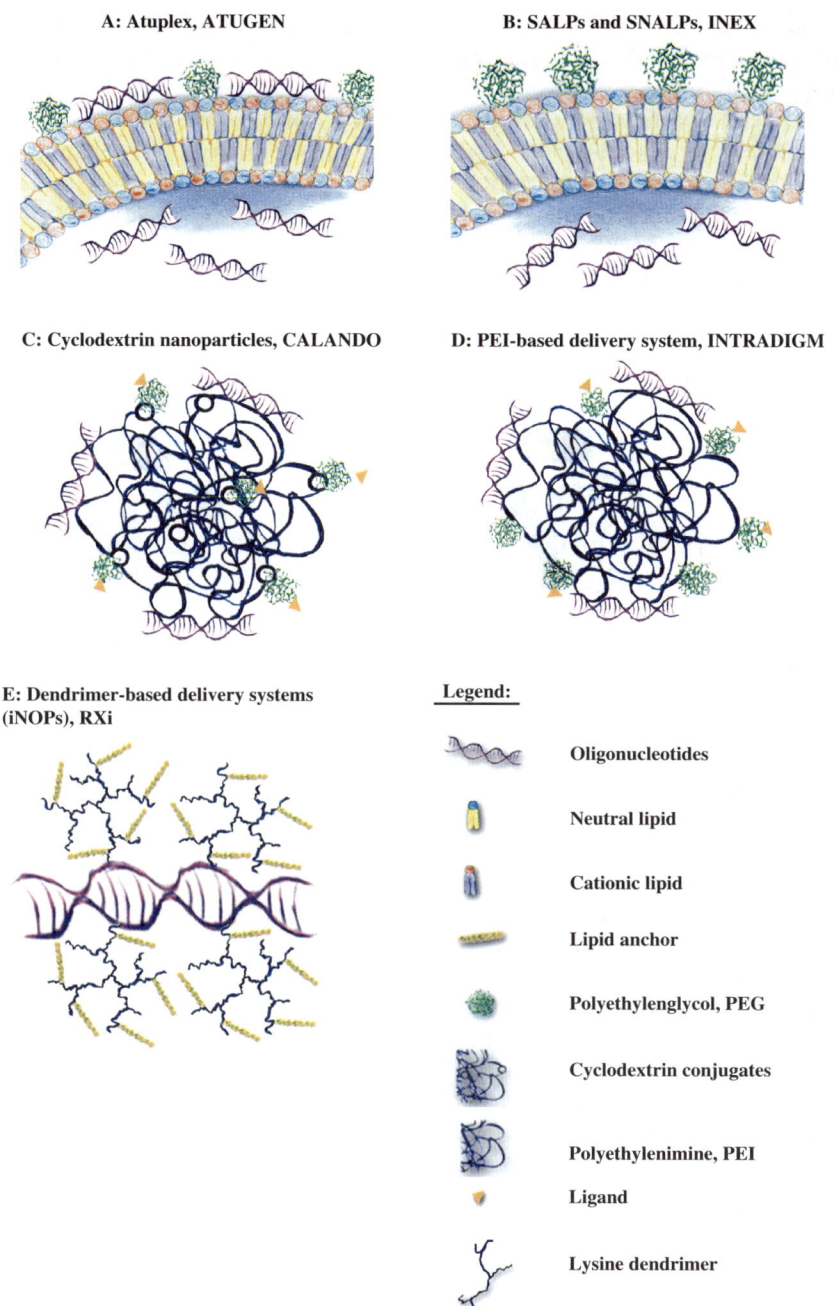

Figure 10.1   Schematic representation of delivery agents for oligonucleotides and siRNAs. For further details, see the text.

combination with the neutral, fusogenic phospholipid 1,2-diphytanoyl-sn-glycero-3-phosphoethanolamine (DPhyPE) and 1,2-distearoyl-sn-glycero-3-phospho-ethanolamine–polyethyleneglycol (DSPE–PEG) at 1 mol%. The low degree of PEGylation does not inhibit siRNA binding or cellular uptake, and such lipoplexes maintain a strong cationic surface charge after combination with the (negatively charged) siRNA. Fluorescence-labelled siRNA molecules formulated with such lipoplexes appear in all major organs within five minutes of injection. However, lipoplexes are rapidly adsorbed onto the endothelial lining of blood vessels, and penetration beyond this cell layer has not been observed.[26]

Functional delivery of siRNA molecules to endothelial cells has been demonstrated by repeated administration of complexed siRNA targeting CD31 and Tie2, respectively. A dosage of approximately 2 mg kg^{-1} resulted in approximately 60% downregulation of the target gene expression in the endothelia of the lung, heart, and liver.[58]

10.4.1.3 Stable Antisense-Lipid Particles and Stable Nucleic Acid–Lipid Particles

Stable antisense-lipid particles (SALPs) and stable nucleic acid–lipid particles (SNALPs) (Figure 10.1B) are PEGylated, cationic lipid carriers, which solve the problem of aggregation with cargo molecules or serum components. However, SALPs potentiate the immune-stimulatory presentation of ONs, even in the absence of classic immune-stimulatory motifs such as CpG.[59,60] Inex Pharmaceuticals Corp. (now Tekmira Pharmaceuticals Corp.) took advantage of this effect and developed an immune-stimulatory product (INX-0167) from this platform.

In addition to the immune-stimulatory effect, the PEGylation of the carrier also impaired its ability to transfect cells.[61] The use of transient PEGylation was therefore a major improvement for the technology, and this new generation of particles is now known as SNALPs. In these particles, PEG–lipid conjugations with rather short membrane anchors exhibit sufficient membrane residence during production and storage, but redistribute in the presence of a sink, such as lipoproteins or cellular membranes. Consequently, PEG–lipids eventually leave liposomal membranes after i.v. injection and expose their positive surface charge.[62] Further details concerning SNALP technology are outlined in Chapter 11.

Scientists at Sirna Therapeutics Inc. and Protiva Biotherapeutics Inc. reported the use of SNALPs for the delivery of siRNA and reported inhibition of HepB antigen production in a murine model for hepatitis B virus (HBV) infection.[27] Interestingly, this work re-emphasized the need for chemical modifications of the siRNA to avoid immune stimulation by the combination of ONs and cationic, PEGylated lipid carriers. Only the chemically modified, not the unmodified, version of the HepB Ag siRNA escaped the immune radar and did not trigger substantial cytokine release.

Consequently, in an approach to silence ApoB-100 in mice and cynomolgus monkeys, the chosen siRNA was specifically modified to prevent stimulation of the immune system. Efficient and persistent knockdown of ApoB-100 was achieved in the liver of mice and monkeys at dosages of $1–3 \, \text{mg kg}^{-1}$. The substantial knockdown reported in this work, together with tissue distribution data, revealed a complete penetration of this massive organ with liposomal carriers. However, the formulation exhibited a relatively narrow therapeutic index in cynomolgus monkeys. Liver enzymes, in particular aspartate aminotransferase (AST), were increased in serum after a single injection at a dosage of $2.5 \, \text{mg kg}^{-1}$.[63]

10.4.1.4 Cyclodextrin Nanoparticles

Besides liposomes, a different cationic nanoparticle has also been investigated as a carrier for siRNA. Calando Pharmaceuticals is developing cationic cyclodextrin conjugates (Figure 10.1C), which consist of individual cyclo-dextrin units connected by cationic linkers, resulting in a cationic copolymer. The cyclodextrin units can bind hydrophobic moieties, such as the cycloalkane adamantane. Thus, PEG-conjugated adamantane was used for a non-covalent PEGylation of the cationic copolymer. However, extensive PEGylation is known to interfere with the cellular uptake of siRNA. Thus, particles were eventually decorated with transferrin as a ligand to aid the cellular uptake process. In mice, a dosage of $2.5 \, \text{mg kg}^{-1}$ siRNA was well-tolerated and resulted in short term inhibition of a tumor-associated luciferase or EFBP2 gene expression.[64] In more recent reports, the carrier system has also been shown to be safe in cynomolgus monkeys. In a dose-escalation study, dosages up to $27 \, \text{mg kg}^{-1}$ siRNA did not result in fatalities. Side effects included kidney and liver toxicity, but the effects were moderate. Elevated serum alanine aminotransferase (ALT) levels were reported at a dosage of $9 \, \text{mg kg}^{-1}$.[65] Heidel and co-workers also reported antibody generation against transferrin as the targeting ligand, which might represent a monkey-specific immune response against the human-derived transferrin that was used in this experiment.[65] If the antibody generation was, however, directed against the conjugated transferrin–carrier complex, the potential of generating antibody-mediated toxicities in human trials might pose a severe risk, since transferrin is the essential iron transporter of the body.

10.4.1.5 Polyethylenimine-Based Delivery System

Intradigm Corp. uses polyethylenimine (PEI, Figure 10.1D), a cationic polymer, as a very well-characterized starting material for building gene therapy vectors with a PEI core protected by PEG shielding. To the surface of the particles, ligands for targeting and enhancement of cell uptake are added.[66] The polymer forms tight complexes with siRNA, and the combined material was shown to distribute to tumor endothelia, which express the corresponding receptor for the targeting Arg–Gly–Asp (RGD) peptides. However, substantial amounts of the carrier were also found in the liver and lung, indicating unspecific binding of the cationic nanoparticles to endothelia.

10.4.1.6   Dendrimer-Based Delivery System

RXi Pharmaceuticals is using a dendrimer-based delivery system (Figure 10.1E), called iNOPs, originally developed by Rana and co-workers.[29] The particles consist of a cationic, fourth generation dendrimer core made from lysine that is selectively functionalized with lipid groups. On average, one siRNA molecule is complexed with one or more carriers. Since full protection of the siRNA is not achieved, chemically modified siRNA molecules were used for studies in mice. At dosages as low as $1 \, \mathrm{mg \, kg^{-1}}$, substantial knockdown of the hepatocyte target ApoB-100 has been reported.[29] However, an increased dose response was not observed for higher doses, and accumulation of some material in the lung may indicate aggregate formation and limited tissue penetration, both typical for carriers with a net positive surface charge.

10.4.2   Anionic and Neutral Carriers

A few anionic and neutral delivery systems have also been developed. These vectors have a low toxicity and exhibit a relatively long circulation lifetime. However, efficient ON encapsulation and cellular uptake represent major obstacles to these delivery vectors.

10.4.2.1   Dioleoyl Phosphatidylcholine Liposomes

The delivery profile of large neutral liposomes consisting of 100% dioleoyl phosphatidylcholine (DOPC) was investigated in the labs of Sood and co-workers.[67,68] DOPC liposomes containing siRNA against EphA2 or focal adhesion kinase (FAK) were injected intraperitoneally at dosages of $150 \, \mathrm{\mu g \, kg^{-1}}$ twice a week for three weeks into orthotopic tumor-bearing mice. Tumor growth could be inhibited, and treatment was synergistic in combination with the chemotherapeutic agents paclitaxel (Taxol®) or docetaxel (Taxotere®). In biodistribution studies with labelled siRNA encapsulated into DOPC liposomes, the label was found in tumor tissue and accumulated at significant levels in the liver, kidneys, lung, and heart. High background fluorescence meant that accumulation was questionable in the spleen, brain, and pancreas. Large, multilamellar DOPC liposomes were prepared as a lyophilized powder from organic solvent, and encapsulated approximately 40% of the siRNA. While the initial results are encouraging, the upscaling of production might bear technical challenges, since large liposomes cannot be sterile-filtered and need aseptic production. Reducing the particle size might interfere with the efficacy observed in the current reports and also reduces the encapsulation efficacy dramatically, as the example in Section 10.4.2.2 shows.

10.4.2.2 Egg-PC–Cholesterol Liposomes

Neutral, stable liposomes can be prepared from egg phosphatidylcholine (PC) and cholesterol. Klimuk and colleagues loaded such liposomes with an ASO-targeting intracellular adhesion molecule-1 (ICAM-1) and tested the efficacy of the formulation in a murine inflammation model.[69] Entrapment in liposomes considerably enhanced the circulation half-life and ASO targeting to the inflammation site, which were regarded as the principal reasons for the enhanced anti-inflammatory activity of the liposomal ICAM-1 ASO observed. Despite this promising result, relatively high dosages of 25–50 mg kg^{-1} were needed to obtain a pharmacological effect, and the neutral liposomes used in this study had rather low encapsulation yields of approximately 15% because of a lack of electrostatic attraction.

10.4.3 Summary of Common Building Principles for Oligonucleotide Carriers

Carriers for ONs are either cationic or neutral and/or anionic particles. Efficient ion pair formation between ONs and cationic carriers has led to an overwhelming prevalence of these carrier systems with a net positive surface charge. Cationic carriers also bind to cells by electrostatic interaction and generate steep concentration gradients at cell membranes. The latter fact, however, also has a disadvantage: cellular attraction provides encouraging results *in vitro*, where cells are growing as two-dimensional layers on plastic supports, but *in vivo* such rapid but unspecific binding of cationic carriers onto cell surfaces results in unspecific adsorption of the materials onto endothelia throughout the body, which challenges the target tissue specificity of the vector. At higher dosages, the application of some cationic carrier systems may cause fatalities through aggregate formation and the blockage of capillaries. It may well be that local blockages of microvessels, together with drug-specific effects, both contributed to the observed toxicities in such settings. NeoPharm Inc. reported toxic effects at a carrier lipid dosage of 100 mg kg^{-1}, and even less tolerability of a commercial cationic liposome preparation consisting of dioleoyloxypropyl-N,N,N-trimethylammonium (DOTAP) and DOPE.[57]

PEGylation was therefore an important step in the evolution of cationic carrier systems. It reduces the surface charge density while retaining the ability of a carrier to bind or encapsulate the ON cargo. Even if carriers are sparsely PEGylated and retain a strong cationic surface charge, this modification results in significant improvements in tolerability. These sparsely PEGylated carriers remain cationic in nature and bind readily to endothelia without deep tissue penetration and with no apparent organ specificity.[26,58] A lack of tissue penetration may explain the efficient, but incomplete, target gene knockdown, as reported by Baigude and co-workers.[29] Unfortunately, in that study, data about the surface charge of the complexes were not reported.

Extensive PEGylation shields the cationic surface charge almost completely and prohibits electrostatic binding to the endothelia, but it also interferes with the cellular uptake process.[58,66] This 'PEGylation dilemma' has been addressed in two different ways: (1) by using weakly bound, diffusible PEG–lipids for transient protection of the carrier or (2) by insertion of ligands that provide homing and internalization in specific cell types. Schiffelers and co-workers clearly re-established the activity of otherwise inactive PEGylated poly-ethylenimine formulations after decoration with RGD peptides at the tip of the PEG chain.[66] Likewise, PEGylated cyclodextrin-based polymer formulations gain activity when a small percentage of transferrin-modified PEG molecules are blended into the surface modification mix.[64] Transient PEGylation is achieved when a PEG molecule is grafted onto a lipid with a short membrane anchor, *e.g.* a C14 chain. Cationic liposome formulations that contain these diffusible PEG molecules, the so-called SNALPs, have successfully been used to deliver siRNA in murine hepatitis models.[27]

10.5 Summary

Chemical modifications and the use of carrier systems are the two main strategies used to solve problems associated with the functional delivery of ONs *in vivo*. Chemical modifications can increase the resistance of ONs against nuclease degradation and impede their renal excretion. Chemical modifications can also increase the potency and tissue residence times of ONs. All of these effects have facilitated *in vivo* use, particularly of single-stranded ASO molecules, even in the absence of a delivery system. The number of organs or tissues that can be reached is, however, limited, and insufficient delivery to more distal sites, such as tumors and sites of inflammation, is currently hindering the development of ASO inhibitors for such indications.

The hurdles for the systemic delivery of functional siRNA into target cells are much higher and not outweighed by their higher potency. Therefore, assisted delivery of siRNA molecules is still a necessary condition for siRNA therapeutics. Carrier systems specifically address the challenges involved with the transit of ONs into cells, and are expected to unleash the full potential of ON-based therapeutics. Most carriers have a cationic surface charge that promotes complex formation with nucleic acids. These cationic complexes exhibit no organ specificity and adhere readily to endothelia, which limits their ability to penetrate into organs. A common strategy to overcome such limitations is the PEGylation of the carrier, which, however, impedes the cellular uptake process. The use of a transient, diffusible PEG shielding or decoration with ligands has helped to combine the beneficial biodistribution properties with efficient cellular uptake.

The field of ON delivery is highly dynamic and developing rapidly. The authors believe that, within the next few years, systems will evolve that can effectively encapsulate and protect ONs, penetrate from systemic circulation into specific tissues, and transfect target cells with limited side effects.

Acknowledgement

We thank Dr Ludger Ickenstein for critically reviewing the manuscript and Ute Vinzens for artistically designing the figures.

References

1. J. M. Campbell, T. A. Bacon and E. Wickstrom, *J. Biochem. Biophys. Methods*, 1990, **20**, 259.
2. M. I. Phillips and Y. C. Zhang, *Methods Enzymol.*, 2000, **313**, 46.
3. D. A. Brown, S.-H. Kang, S. M. Gryaznov, L. De Dionisio, O. Heidenreich, S. Sullivan, X. Xu and M. I. Neerenberg, *J. Biol. Chem.*, 1994, **43**, 26801.
4. A. A. Levin, *Biochim. Biophys. Acta*, 1999, **1489**, 69.
5. C. A. Stein, C. Subasinghe, K. Shinozuka and J. S. Cohen, *Nucleic Acids Res.*, 1988, **16**, 3209.
6. P. J. Furdon, Z. Dominski and R. Kole, *Nucleic Acids Res.*, 1989, **17**, 9193.
7. F. Eckstein, *Antisense Nucleic Acid Drug Dev.*, 2000, **10**, 117.
8. S. P. Henry, P. C. Giclas, J. Leeds, M. Pangburn, C. Auletta, A. A. Levin and D. J. Kornbrust, *J. Pharmacol. Exp. Ther.*, 1997, **281**, 810.
9. S. T. Crooke, *Methods Enzymol.*, 2000, **313**, 3.
10. J. Temsamani, A. Roskey, C. Chaix and S. Agrawal, *Antisense Nucleic Acid Drug Dev.*, 1997, **7**, 159.
11. K.-H. Altmann, N. M. Dean, D. Fabbro, S. M. Freier, T. Geiger, R. Haener, D. Huesken, P. Martin and B. P. Monia, *Chimia*, 1996, **50**, 168.
12. S. M. Freier and K.-H. Altmann, *Nucleic Acids Res.*, 1997, **25**, 4429.
13. L. L. Cummins, S. R. Owens, L. M. Risen, E. A. Lesnik, S. M. Freier, D. McGee, C. J. Guinosso and P. D. Cook, *Nucleic Acids Res.*, 1995, **23**, 2019.
14. B. P. Monia, E. A. Lesnik, C. Gonzalez, W. F. Lima, D. McGee, C. J. Guinosso, A. M. Kawasaki, P. D. Cook and S. M. Freier, *J. Biol. Chem.*, 1993, **268**, 14514.
15. R. S. Geary, R. Z. Yu and A. A. Levin, *Curr. Opin. Invest. Drugs*, 2001, **2**, 562.
16. D. A. Braasch, S. Jensen, Y. Liu, K. Kaur, K. Arar, M. A. White and D. R. Corey, *Biochemistry*, 2003, **42**, 7967.
17. J. Kurreck, E. Wyszko, C. Gillen and V. A. Erdmann, *Nucleic Acids Res.*, 2002, **30**, 1911.
18. H. Ørum and J. Wengel, *Curr. Opin. Mol. Ther.*, 2001, **3**, 239.
19. E. E. Swayze, A. M. Siwkowski, E. V. Wancewicz, M. T. Migawa, T. K. Wyrzykiewicz, G. Hung, B. P. Monia and C. F. Bennett, *Nucleic Acids Res.*, 2007, **35**, 687.
20. J. Kurreck, *Eur. J. Biochem.*, 2003, **270**, 1628.
21. G. J. Hannon and J. J. Rossi, *Nature*, 2004, **431**, 371.
22. J.-Y. Yu, S. L. DeRuiter and D. L. Turner, *Proc. Natl. Acad. Sci. U. S. A.*, 2002, **99**, 6047.
23. T. Tuschl, *Nat. Biotechnol.*, 2002, **20**, 446.

24. H. Y. Zhang, Q. Du, C. Wahlestedt and Z. Liang, *Curr. Top. Med. Chem.*, 2006, **6**, 893.

25. M. Behlke, *Mol. Ther.*, 2006, **13**, 644.

26. A. Santel, M. Aleku, O. Keil, J. Endruschat, V. Esche, B. Durieux, K. Löffler, M. Fechtner, T. Röhl, G. Fisch, S. Dames, W. Arnold, K. Giese, A. Klippel and J. Kaufmann, *Gene Ther.*, 2006, **13**, 1360.

27. D. V. Morrissey, J. A. Lockridge, L. Shaw, K. Blanchard, K. Jensen, W. Breen, K. Hartsough, L. Machemer, S. Radka, V. Jadhav, N. Vaish, S. Zinnen, C. Vargeese, K. Bowman, C. S. Shaffer, L. B. Jeffs, A. Judge, I. MacLachlan and B. Polisky, *Nat. Biotechnol.*, 2005, **23**, 1002.

28. J. Soutschek, A. Akinc, B. Bramlage, K. Charisse, R. Constien, M. Donoghue, S. Elbashir, A. Geick, P. Hadwiger, J. Harborth, M. John, V. Kesavan, G. Lavine, R. K. Pandey, T. Racie, K. G. Rajeev, I. Rohl, I. Toudjarska, G. Wang, S. Wuschko, D. Bumcrot, V. Koteliansky, S. Limmer, M. Manoharan and H. P. Vornlocher, *Nature*, 2004, **432**, 173.

29. H. Baigude, J. McCarroll, C.-S. Yang, P. M. Swain and T. M. Rana, *ACS Chem. Biol.*, 2007, **2**, 237.

30. L. A Yakubov, E. A. Deeva, V. F. Zarytova, E. M. Ivanova, A. S. Ryte, L. V. Yurchenko and V. V. Vlassov, *Proc. Natl. Acad. Sci. U. S. A.*, 1989, **86**, 6454.

31. E. Bonfils, C. Depierreux, P. Midoux, N. T. Thuong, M. Monsigny and A. C. Roche, *Nucleic Acids Res.*, 1992, **20**, 462.

32. W. W. Liang, X. Shi, D. Deshpande, C. J. Malanga and Y. Rojanasakul, *Biochim. Biophys. Acta.*, 1996, **1279**, 227.

33. G. Citro, D. Perrotti, C. Cucco, I. D'Agnano, A. Sacchi, G. Zupi and B. Calabretta, *Proc. Natl. Acad. Sci. U. S. A.*, 1992, **89**, 7031.

34. G. Citro, C. Szczylik, P. Ginobbi, G. Zupi and B. Calabretta, *Br. J. Cancer*, 1994, **69**, 463.

35. E. H. Feinberg and C. P. Hunter, *Science*, 2003, **301**, 1545.

36. B. Li, J. A. Hughes and M. I. Phillips, *Neurochem. Int.*, 1997, **31**, 393.

37. Q. Hu, M. B. Bally and T. D. Madden, *Nucleic Acids Res.*, 2002, **30**, 3632.

38. Y. L. Chiu, A. Ali, C. Y. Chu, H. Cao and T. M. Rana, *Chem. Biol.*, 2004, **11**, 1165.

39. Z. Zhang, T. Weinschenk, K. Guo and H. J. Schluesener, *J. Cell Biochem.*, 2006, **97**, 1217.

40. R. S. Geary, R. Z. Yu and A. A. Levin, *Curr. Opin. Investig. Drugs*, 2001, **2**, 562.

41. R. M. Crooke, *Expert Opin. Biol. Ther.*, 2005, **5**, 907.

42. M. J. Graham, S. T. Crooke, D. K. Monteith, S. R. Cooper, K. M. Lemonidis, K. K. Stecker, M. J. Martin and R. M. Crooke, *J. Pharmacol. Exp. Ther.*, 1998, **286**, 447.

43. R. M. Crooke, M. J. Graham, K. M. Lemonidis, C. P. Whipple, S. Koo and R. J. Perera, *J. Lipid Res.*, 2005, **46**, 872.

44. X. X. Yu, S. F. Murray, S. K. Pandey, S. L. Booten, D. Bao, X. Z. Song, S. Kelly, S. Chen, R. McKay, B. P. Monia and S. Bhanot, *Hepatology*, 2005, **42**, 362.

45. H. Zhang, J. Cook, J. Nickel, R. Yu, K. Stecker, K. Myers and N. M. Dean, *Nat. Biotechnol.*, 2000, **18**, 862.
46. R. S. Geary, T. A. Watanabe, L. Truong, S. Freier, E. A. Lesnik, N. B. Sioufi, H. Sasmor, M. Manoharan and A. A. Levin, *J. Pharmacol. Exp. Ther.*, 2001, **296**, 890.
47. M. S. Duxbury, E. Matros, H. Ito, M. J. Zinner, S. W. Ashley and E. E. Whang, *Ann. Surg.*, 2004, **240**, 667.
48. A. P. McCaffrey and L. Meuse, *Nature*, 2002, **418**, 38.
49. D. L. Lewis, J. E. Hagstrom, A. G. Loomis, J. A. Wolff and H. Herweijer, *Nat. Genet.*, 2002, **32**, 107.
50. A. T. Gewirtz and S. Sitaraman, *Curr. Opin. Investig. Drugs*, 2001, **2**, 1401.
51. E. Koller, S. Propp, H. Zhang, C. Zhao, X. Xiao, M. Y. Chang, S. A. Hirsch, P. J. Shepard, S. Koo, C. Murphy, R. I. Glazer and N. M. Dean, *Cancer Res.*, 2006, **66**, 2059.
52. Y. P. Hu, G. Cherton-Horvat, V. Dragowska, S. Baird, R. G. Korneluk, J. P. Durkin, L. D. Mayer and E. C. LaCasse, *Clin. Cancer Res.*, 2003, **9**, 2826.
53. Z. Zhang, M. Li, H. Wang, S. Agrawal and R. Zhang, *Proc. Natl. Acad. Sci. U. S. A.*, 2003, **100**, 11636.
54. P. C. Gokhale, C. Zhang, J. T. Newsome, J. Pei, I. Ahmad, A. Rahman, A. Dritschilo and U. N. Kasid, *Clin. Cancer Res.*, 2002, **8**, 3611.
55. K. Kasireddy, S. M. Ali, M. U. Ahmad, S. Choudhury, P. Y. Chien, S. Sheikh and I. Ahmad, *Bioorg. Chem.*, 2005, **33**, 345.
56. P. Y. Chien, J. Wang, D. Carbonaro, S. Lei, B. Miller, S. Sheikh, S. M. Ali, M. U. Ahmad and I. Ahmad, *Cancer Gene Ther.*, 2005, **12**, 321.
57. A. Pal, A. Ahmad, S. Khan, I. Sakabe, C. Zhang, U. N. Kasid and I. Ahmad, *Int. J. Oncol.*, 2005, **26**, 1087.
58. A. Santel, M. Aleku, O. Keil, J. Endruschat, V. Esche, G. Fisch, S. Dames, K. Loffler, M. Fechtner, W. Arnold, K. Giese, A. Klippel and J. Kaufmann, *Gene Ther.*, 2006, **13**, 1222.
59. B. Mui, S. G. Raney, S. C. Semple and M. J. Hope, *J. Pharmacol. Exp. Ther.*, 2001, **298**, 1185.
60. S. C. Semple, T. O. Harasym, K. A. Clow, S. M. Ansell, S. K. Klimuk and M. J. Hope, *J. Pharmacol. Exp. Ther.*, 2005, **312**, 1020.
61. L. Y. Song, Q. F. Ahkong, Q. Rong, Z. Wang, S. Ansell, M. J. Hope and B. Mui, *Biochim. Biophys. Acta*, 2002, **1558**, 1.
62. Y. P. Zhang, L. Sekirov, E. G. Saravolac, J. J. Wheeler, P. Tardi, K. Clow, E. Leng, R. Sun, P. R. Cullis and P. Scherrer, *Gene Ther.*, 1999, **6**, 1438.
63. T. S. Zimmermann, A. C. Lee, A. Akinc, B. Bramlage, D. Bumcrot, M. N. Fedoruk, J. Harborth, J. A. Heyes, L. B. Jeffs, M. John, A. D. Judge, K. Lam, K. McClintock, L. V. Nechev, L. R. Palmer, T. Racie, I. Rohl, S. Seiffert, S. Shanmugam, V. Sood, J. Soutschek, I. Toudjarska, A. J. Wheat, E. Yaworski, W. Zedalis, V. Koteliansky, M. Manoharan, H. P. Vornlocher and I. MacLachlan, *Nature*, 2006, **441**, 111.
64. S. Hu-Lieskovan, J. D. Heidel, D. W. Bartlett, M. E. Davis and T. J. Triche, *Cancer Res.*, 2005, **65**, 8984.

65. J. D. Heidel, Z. Yu, J. Y. Liu, S. M. Rele, Y. Liang, R. K. Zeidan, D. J. Kornbrust and M. E. Davis, *Proc. Natl. Acad. Sci. U. S. A.*, 2007, **104**, 5715.
66. R. M. Schiffelers, A. Ansari, J. Xu, Q. Zhou, Q. Tang, G. Storm, G. Molema, P. Y. Lu, P. V. Scaria and M. C. Woodle, *Nucleic Acids Res.*, 2004, **32**, 149.
67. C. N. Landen Jr, A. Chavez-Reyes, C. Bucana, R. Schmandt, M. T. Deavers, G. Lopez-Berestein and A. K. Sood, *Cancer Res.*, 2005, **65**, 6910.
68. J. Halder, A. A. Kamat, C. N. Landen Jr, L. Y. Han, S. K. Lutgendorf, Y. G. Lin, W. M. Merritt, N. B. Jennings, A. Chavez-Reyes, R. L. Coleman, D. M. Gershenson, R. Schmandt, S. W. Cole, G. Lopez-Berestein and A. K. Sood, *Clin. Cancer Res.*, 2006, **12**, 4916.
69. S. K. Klimuk, S. C. Semple, P. N. Nahirney, M. C. Mullen, C. F. Bennett, P. Scherrer and M. J. Hope, *J. Pharm. Exp. Ther.*, 2000, **292**, 480.

CHAPTER 11

Lipid-Mediated in vivo Delivery of Small Interfering RNAs

IAN MACLACHLAN

Protiva Biotherapeutics Inc., 100-3480 Gilmore Way, Burnaby, BC, Canada, V5G 4Y1

11.1 Lipid-Mediated Drug Delivery

It has been four years since Phil Sharp was famously quoted as saying that, "the major hurdle *right now* is **delivery, delivery, delivery**."[1] This statement continues to resonate today for the simple reason that it is still true. Although reports of naked small interfering RNAs (siRNAs) that affect RNA interference (RNAi) continue to trickle into the literature, most *in vivo* applications of siRNAs, in particular those that involve systemic administration, require the use of suitable delivery technologies. This is because canonical siRNA duplexes have poor pharmacology and require some degree of enablement to overcome their limited circulation and rapid clearance from the blood compartment, limitations that in turn limit their bioavailability at disease sites. Even when siRNA duplexes reach their target cell, they must be delivered into the cytoplasm before they can mediate RNAi. However, naked nucleic acid molecules, including siRNAs, are unable to cross the mammalian plasma membrane and are very poorly taken up by cells. For this reason, the use of delivery vehicles to enable synthetic siRNA uptake and presentation to the cytoplasmic RNAi machinery is seen as essential by researchers conducting both *in vitro* and *in vivo* studies. The most widely used vehicles for this purpose are based around positively charged (cationic) agents that complex with, or encapsulate, the negatively charged nucleic acid. These include cationic lipid formulations, for example lipoplex reagents such as Lipofectamine, and cationic polymers such

RSC Biomolecular Sciences
Therapeutic Oligonucleotides
Edited by Jens Kurreck
© Royal Society of Chemistry 2008

as polyethylenimine (PEI), both of which form positively charged complexes with siRNAs and facilitate intracellular delivery through electrostatic interactions with the membrane of the target cell. This chapter discusses lipidic carriers, in particular those that utilize lipids that are able to spontaneously self-assemble into ordered structures called liposomes, with an emphasis on those that truly encapsulate their siRNA payloads.

Discovered in 1961,[2] and subsequently investigated as model membrane systems, liposomes are artificial vesicles made up of one or more lipid bilayers encapsulating an equal number of internal aqueous compartments. They are classified according to their size and number of lipid bilayers. Multilamellar vesicles (MLVs) are formed by the aqueous hydration, and often sonication, of dried lipid films. Typically hundreds of nanometers in diameter, they are large, complex structures containing a series of concentric bilayers separated by narrow aqueous compartments. Further disruption or extrusion of MLVs results in the formation of simple unilamellar vesicles between 50–500 nm in diameter, referred to as large unilamellar vesicles (LUVs). The smallest liposomes, small unilamellar vesicles (SUVs), are vesicles smaller than 50 nm in diameter, formed by the more extensive sonication and further disruption of dried lipid films.

Liposomes have received considerable attention for their use in drug delivery. Typically, they are used as drug carriers, with the solubilized drug encapsulated in the internal aqueous space formed by the liposomal lamellae. Liposomal drug formulations can be used to overcome a drug's non-ideal properties, such as limited solubility, serum stability, circulation half-life, biodistribution and target-tissue selectivity. Experience with conventional small-molecule drugs has shown that the drugs which benefit the most from liposomal delivery are those that are chemically labile, subject to enzymatic degradation, and have an intracellular site of action.[3] This provides a clear rationale for exploiting liposomes as carriers of nucleic acid based drugs, such as antisense oligonucleotides (ASOs), ribozymes and, more recently, siRNAs. Successful application of the liposomal drug-delivery paradigm to nucleic acid based drugs results in the pharmacokinetics, biodistribution and intracellular delivery of the liposome payload being determined by the physicochemical properties of the carrier. (For example, the biodistribution of a lipid-encapsulated siRNA is independent of the siRNA sequence or chemical modification pattern.) This is only true, however, if the liposome truly encapsulates the siRNA and acts as a carrier, rather than as a mere excipient. Liposomes may be used as excipients when used to formulate hydrophobic drugs that would otherwise be difficult to administer in aqueous dosage form. Upon intravenous administration, hydrophobic drugs rapidly exchange into lipoproteins or other lipid-rich environments, resulting in comparably less controlled pharmacology. In the context of siRNA delivery, liposomes are excipients when used to enable vialing and aqueous dosing of hydrophobic lipid–nucleic acid conjugates,[4–7] applications which are not considered in this chapter.

An implicit objective when considering the use of ancillary technologies in drug development, including drug delivery technologies, is to improve the final product in such a way as to minimize the risks associated with treatment (toxicity) while maximizing the benefit to patient health (efficacy). siRNA-based drugs can be thought of as having three classes of dose-limiting toxicities. In the first case, the siRNA may be toxic as a result of exaggerated pharmacology, a direct consequence of the intentional knockdown of the siRNA target transcript. For example, antineoplastic siRNAs that act to induce apoptosis in proliferating cells may be cytotoxic when delivered to normal, healthy cells. It is only recently that exaggerated pharmacology has been a relevant concern when working with nucleic acid based drugs. In some ways the ability to elicit 'too much' of a desired effect when using siRNAs reflects the potency and accessibility of the RNAi mechanism and may ultimately bode well for the development of effective siRNA-based drugs. Clearly, effective delivery will increase the potential for exaggerated pharmacology. A second toxicological consideration relates to the chemical nature of the siRNA and its associated delivery system. In this regard, certain nucleotide chemistries, including backbone modifications commonly used to confer stability to the siRNA duplex, have their own toxicological considerations.[8] As novel chemical entities, the increasingly wide varieties of delivery vehicles and siRNA-conjugates under development also require formal evaluation. The use of delivery vehicles, including liposomes, can alter the toxicity profile of a drug through altered pharmacokinetics and biodistribution. In the case of liver-targeting formulations for the delivery of siRNA, elevations in serum transaminase levels can be a sensitive indicator of liver damage associated with an excessive chemical load into hepatocytes. The third type of toxicity is potentiated through the use of delivery technologies, including liposomes, that facilitate the unintended activation of the innate immune response by immunostimulatory siRNA. A well-designed liposomal delivery system will be capable of minimizing each of these three classes of toxicity and increasing the potency of siRNA-based drugs by optimizing delivery to target tissues. In this chapter we discuss the physical make up, methods of manufacture and pharmacological considerations specific to liposomal systems for the delivery of nucleic acid based drugs, with emphasis on those that enable systemic delivery of siRNAs.

11.2 Liposome Constituents

The earliest attempts at liposomal oligonucleotide (ON) encapsulation used passive techniques to entrap the nucleic acid in neutral liposomes.[9–11] The introduction of cationic lipid-mediated lipofection[12] resulted in a migration away from true encapsulation toward lipoplex or 'oligoplex' systems. Recent advances in formulation technology have permitted the combination of both approaches, thus reaping the benefits of using an encapsulated payload in conjunction with cationic lipids. More sophisticated systems now contain multiple lipid components, each playing a role in determining the physical and pharmacological properties of the system as a whole. Multi-component systems may contain one or more cationic lipids, neutral helper lipids or polyethyleneglycol (PEG)–lipid conjugates (PEG–lipids), each of which is discussed here.

11.2.1 Cationic Lipids

Cationic lipids play two roles in liposomal nucleic acid formulations. First, during the formulation process they promote interaction between the lipid bilayer and the negatively charged nucleic acid, yielding greatly enriched payload concentrations of nucleic acid by comparison with charge-neutral liposomes ($\sim$40% when using co-extrusion methods, and greater than 95% when using more sophisticated techniques).[13–15] Second, cationic lipids lend the particle a net positive charge, facilitating interaction with anionic cell-surface molecules (*e.g.* sulfated proteoglycans and sialic acids[16–18]) and subsequent endocytic uptake. However, the advantages of a cationic delivery system in facilitating intracellular delivery present a dilemma, as highly charged systems are promptly cleared from the blood, thereby restricting accumulation in target tissues. Conversely, particles with no surface charge are poorly internalized by cells, but display excellent circulation characteristics. This supports the concept of a modular delivery solution, an engineered system with individual lipid components fulfilling discrete functions, actively responding to the environment in a manner that best facilitates delivery and uptake by the cell. The use of titratable, ionizable cationic lipids allows for adjustments of the particle charge simply by responding to the external pH.[19] Formulation is performed at reduced pH, when the lipid system is strongly charged, and the nucleic acids are efficiently encapsulated. At the higher, physiological pH of the blood compartment, the charge on the lipids is reduced, the particles become more charge-neutral and are able to avoid opsonization by blood components.[19] More recently, the use of novel, pH titratable cationic lipids with distinct physicochemical properties regulating particle formation, cellular uptake, fusogenicity and endosomal release of nucleic acid drugs has been described.[20] The chemical and biological properties of pH titratable cationic lipids, phase transition temperature (T_c) in particular, are influenced by their degree of lipid saturation. Above the T_c, lipids adopt the more fusogenic reverse hexagonal H_{II} phase,[20–22] promoting fusion and endosomal release of the nucleic acid payload. By comparing T_c values the comparative ability of these lipids to form the fusogenic H_{II} phase can be determined. On this basis it was demonstrated that as the titratable lipid becomes less saturated, the fusogenicity of the liposomal system increases.

A correlation also exists between lipid pK_a and the degree of saturation of pH titratable cationic lipids, with saturated lipids carrying more cationic charge at physiological pH. Liposomes containing the more highly saturated lipids are therefore more effectively internalized by cells *in vitro*.[20] It might be expected that these lipids would be most effective at mediating RNAi; however, liposomes containing the more unsaturated cationic lipids DLinDMA and DLenDMA are more fusogenic, are more potent with respect to mediating endosomal release and are therefore more effective at mediating RNA interference. This clearly illustrates that cellular uptake, fusogenicity and endosomal release are discrete processes, each determined by select properties of the delivery vehicle and strongly affected by the physicochemical characteristics of the cationic lipids used.

11.2.2   Neutral Helper Lipids

As we have just discussed, cationic lipids may be designed to possess innate fusogenic properties. Until recently, however, the incorporation of a separate 'helper' lipid was generally considered necessary to confer this property to liposomal delivery systems for nucleic acids.[23-26] A fusogenic element facilitates the intracellular delivery of the nucleic acids payload by fusing with the membranes of the target cell. Fusion may occur at several junctures in delivery – at the plasma membrane, endosome or nuclear envelope. Fusion of non-encapsulated lipoplex systems with the plasma membrane is an inefficient pathway for the introduction of nucleic acids into the cytosol. Since lipoplex formulated nucleic acids are predominantly associated with the surface of the liposome, such fusion events would be expected to resolve with the nucleic acids deposited on the outer surface of the cell. On the contrary, upon fusion with either the plasma or endosomal membrane(s), encapsulated systems deliver their contents directly to the cytosol.

Lipids that form non-bilayer phases (*e.g.* H_{II} phase) promote destabilization of lipid bilayers and fusion with other membranes. The unsaturated phosphatidylethanolamine DOPE is a well-known example of one such lipid. As with fusogenic cationic lipids, a decrease in lipid saturation has been shown to increase the lipid's affinity for the H_{II} phase.[27-32] Nevertheless, not all cationic lipids require the presence of a fusogenic helper lipid, and many function adequately either alone[24,25] or in conjunction with the non-fusogenic lipid cholesterol.[33] This suggests that the lipids in question either have their own fusogenic properties, or promote a delivery mechanism that does not require membrane fusion.

When attempting to address the role of fusogenic lipids *in vivo*, it is important to distinguish the effect of fusogenic lipids on circulation behavior and delivery to disease site from their effect on intracellular delivery. Formulations of a fusogenic nature will have an increased propensity to interact with systems they encounter in the blood compartment (vascular endothelium, blood cells, lipoproteins). Systemic carriers, therefore, stand to benefit from transiently shielding their fusogenic potential with agents such as PEG.

11.2.3   PEG–Lipids

Ideally, an siRNA delivery system would be transiently shielded upon administration, yet become increasingly charged and fusogenic upon delivery to the target site. This is partially accomplished through the use of PEG–lipids. Readily incorporated in liposomal nucleic acid formulations, they assist in the formulation process by stabilizing the nascent particle during particle formation and they contribute to the shelf-life of the formulation by averting aggregation.[13] They function by forming a protective hydrophilic outer layer about the particle, thereby providing steric stabilization to the liposome. When administered *in vivo*, they curb the association of serum proteins and resulting reticuloendothelial system (RES) uptake by concealing the liposome's surface charge.[34,35]

Despite showing potential in their ability to deliver nucleic acids to disease sites, improvements in the potency of systems that contain PEG–lipids are still necessary. Nucleic acid based drugs require effective *intracellular* delivery, hence the use of the cationic and fusogenic lipids described earlier. Unfortunately, despite the benefits they confer, PEG coatings can inhibit cell association and uptake processes,[23,36,37] thereby having a negative impact on transfection efficiency. Ideally, once at the target site, PEG–lipids would dissociate from the carrier, transforming it from a stable, stealthy particle to a transfection-competent entity. Accordingly, a number of chemically labile PEG–lipids have been described.[38–44] These compounds typically possess a chemically labile bond between the lipid and PEG moieties that is cleaved under acidic conditions to remove the sterically stabilizing PEG layer. Although improved performance has been reported when using this approach, both *in vitro* and *in vivo*, two drawbacks to the strategy remain. First, the reduced pH environment within the endosome is intended to be the cleavage trigger for these acid-labile PEG–lipids. Yet PEG–lipids inhibit cellular uptake, a prerequisite to endosomal localization and therefore to PEG cleavage, and acid labile PEG–lipids limit the amount of material delivered to the endosome.[45] Second, a reduction of formulation shelf-life is expected when incorporating chemically labile lipids, relative to systems that use more stable lipid components.

An alternative to chemically labile PEG–lipids entails the use of stable, yet diffusible PEG–lipids. In the blood compartment, the PEG–lipid dissociates from the particle over time, revealing the positively charged and increasingly fusogenic lipid bilayer, and transforming the particle into a transfection-competent entity. By modulating the size (length) of the PEG-conjugate lipid anchor, and thus the dissociation rate, the pharmacology of encapsulated nucleic acids can be controlled in a predictable manner.[46–50] This approach has been successfully applied to the delivery of siRNAs in several animal models using SNALP (stable nucleic acid–lipid particle) technology,[14,15,51] and may help to resolve the two seemingly contradictory requirements imposed upon nucleic acid carriers. The carrier must be inert and persistent in the circulation to allow for accumulation at target sites. But once at the target site, the carrier must be able to interact with the target cell to facilitate intracellular delivery.

11.3   Methods of Encapsulating Nucleic Acids

To exploit fully the improved pharmacology of liposomal drug carriers, complete entrapment of the nucleic acid payload within the liposome is required.

Here it is important to distinguish between 'lipoplex' and those systems that truly encapsulate their payload. Lipoplexes are electrostatic complexes originally conceived for the delivery of plasmid DNA[12] that are generated by combining preformed cationic liposomes with nucleic acids. The result is a diverse, metastable dispersion of aggregates, effective in transfecting cells in culture, but constrained by a number of limitations when used *in vivo*. When administered intravenously, lipoplex systems are rapidly opsonized and cleared from the blood, accumulating in first-pass organs such as the lung. The cells of the innate immune system are also very effective in taking up lipoplex, a feature that contributes to their toxicities and off-target effects. In anti-tumor or anti-infective applications, these side effects may be incorrectly interpreted as 'efficacy', encouraging acceptance of false-positive results. Therefore, caution must be exercised when designing *in vivo* studies involving these and other liposomal systems in the delivery of immunostimulatory nucleic acids.

11.3.1 Passive Encapsulation

Unlike small-molecule drugs, nucleic acids are not readily packaged in preformed liposomes using pH gradients or other similar active loading techniques. This is predominantly because of the large size and hydrophilic nature of nucleic acids that conspire to prevent them from crossing intact lipid bilayers. Passive encapsulation of nucleic acids typically involves the evaporation of an organic lipid solution to form a 'lipid film'. Rehydration of the lipid film in an aqueous buffer containing the nucleic acid payload, followed by vigorous mixing and multiple cycles of freezing and thawing, results in the formation of MLV with the nucleic acid solute entrapped in the lipid bilayers. Numerous rounds of extrusion through polycarbonate or nitrocellulose filters result in the preparation of LUV-size, according to the size of the filter pores.[52] This method suffers from a number of limitations. The encapsulation efficiency is typically very low, ranging from 3 to 45%. This consequently dictates the inclusion of a post-encapsulation purification step (*e.g.* dialysis, size exclusion chromatography, or ultrafiltration) to remove non-encapsulated material. In an effort to improve the efficiency of encapsulation, an excess of lipid may be used, leading to low nucleic acid:lipid ratios and negatively impacting toxicity and cost of goods. Finally, scaling of the extrusion process is intrinsically difficult. Larger, more expensive batches require custom-built extruders to accommodate large filters. These have an increased probability of rupture during use, resulting in batch failure and even greater expense. Despite these limitations, liposome preparation by extrusion has been successfully adopted by many laboratories and has been used commercially in the preparation of liposomes encapsulating small-molecule drugs.

11.3.2 The Ethanol Drop Method of Encapsulation

The limited efficiency of passive nucleic acid encapsulation and the pharmacology of the resulting particles are partially addressed by the development of the stabilized antisense lipid particle (SALP) method of encapsulation (Figure 11.1). SALPs are prepared by dropwise addition or injection of an ethanolic lipid solution to an aqueous solution of nucleic acid (generally antisense DNA ONs).[19] By utilizing an ionizable aminolipid at an acidic pH, where the aminolipid is fully charged, highly efficient (up to 70%) encapsulation may be achieved. Subsequent adjustment of the pH allows the total charge of the system to be altered in a controlled manner. Using this methodology, ONs have been encapsulated at nucleic acid:lipid ratios as high as 0.25 (w/w).[19] At lower drug:lipid ratios, classic LUV or capped LUV are formed. At the higher nucleic acid:lipid ratios novel small MLVs (SMLVs) are the result, consisting of numerous (typically 6–9) lamellae arranged concentrically around a dense core. Ultimately, the SALP method still requires

A. Ethanol Drop (SALP)

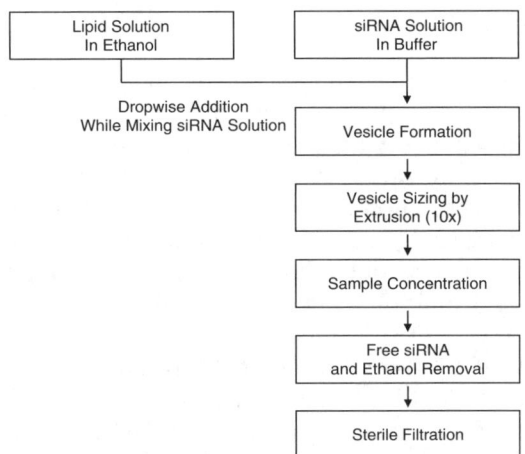

Figure 11.1 Ethanol-based methods of oligo- and polynucleotide encapsulation. (A). Ethanol drop (SALP) method of nucleic acid encapsulation. This method involves the dropwise addition of an ethanolic solution of lipid to an aqueous solution of nucleic acid, resulting in the formation of MLVs. Vesicles are then sized by extrusion through polycarbonate filters. (B). Encapsulation of nucleic acid in ethanol destabilized liposomes. A dried lipid film is rehydrated in buffer, resulting in the formation of MLVs. Multiple freeze–thaw cycles follow, and the empty vesicles are then extruded through polycarbonate filters, producing LUVs. The LUVs are then destabilized by the controlled addition of ethanol to the rapidly mixing aqueous suspension of vesicles. Nucleic acid solution is added to the destabilized liposomes in a drop-wise manner, resulting in encapsulation. (C). Ethanol dilution (SNALP) method of nucleic acid encapsulation. This method involves in-line mixing of lipids dissolved in ethanol with nucleic acid dissolved in buffer, resulting in the spontaneous formation of lipid vesicles. As the solutions are mixed, ethanol is diluted below the concentration required to maintain lipid solubility, resulting in spontaneous vesicle formation and nucleic acid encapsulation. No extrusion or resizing steps are required.

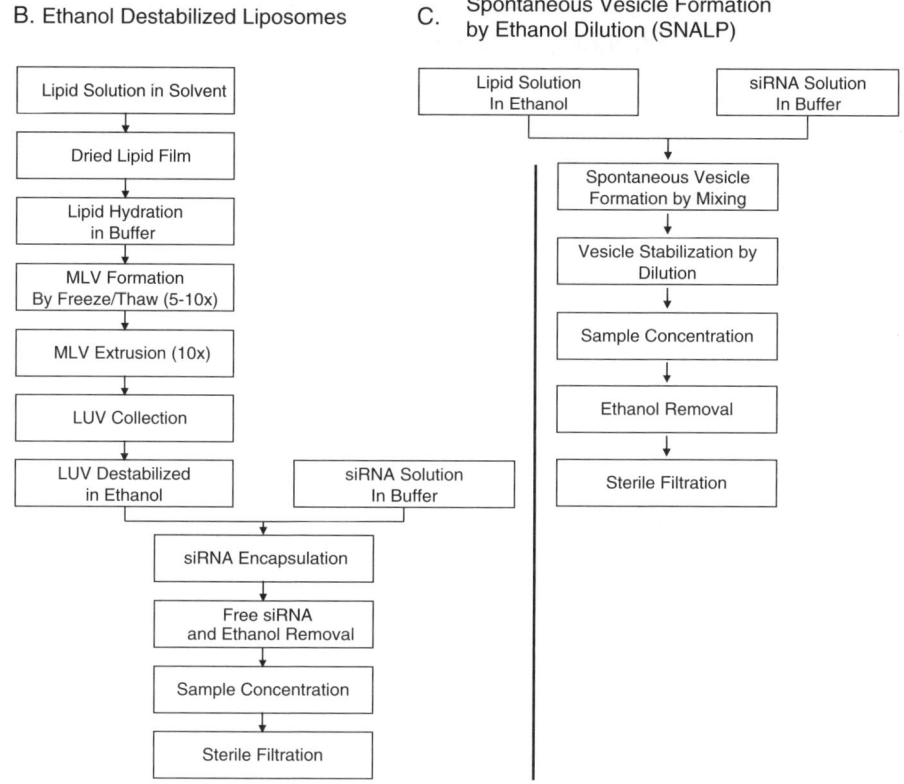

B. Ethanol Destabilized Liposomes

C. Spontaneous Vesicle Formation by Ethanol Dilution (SNALP)

Figure 11.1 (Continued).

extrusion of the formulation, thereby invoking the various disadvantages this process entails.

11.3.3 Encapsulation in Ethanol-Destabilized Liposomes

Encapsulation in ethanol-destabilized cationic liposomes requires empty liposome formation by extrusion, prior to nucleic acid addition.[53,54] Once prepared, the vesicles are destabilized by the careful addition of ethanol (to final 40% v/v) to a rapidly mixing aqueous suspension (Figure 11.1). Failure to maintain precise control of this step results in localized areas of high ethanol concentration ($>50\%$ v/v), promoting vesicular fusion and transformation into larger lipid structures. The subsequent addition of nucleic acid must also be performed carefully to avoid aggregation of the resulting particle suspension. The delicate nature of both the vesicle destabilization and nucleic acid addition, coupled with the extrusion step, are significant impediments to the adoption of this method at a clinical scale.

11.3.4 The Spontaneous Vesicle Formation by Ethanol Dilution Method of Encapsulation

The methods described above rely on extrusion steps to ensure the preparation of small, monodisperse liposomes. Expressly developed as an alternative to such techniques, the SNALP method is simple, robust, and fully scalable.[13] Originally developed to encapsulate plasmid DNA, subsequent adaptation has allowed the encapsulation of smaller nucleic payloads, such as siRNAs.

Liposomes encapsulating siRNA are formed instantaneously by combining the flow streams of an ethanolic lipid solution and an aqueous nucleic acid containing buffer (Figure 11.1). The result is an immediate dilution of ethanol concentration, below that required to maintain lipid solubility, which produces a spontaneous formation of lipid vesicles. Vesicles containing siRNA with sizes as low as 45 nm, and encapsulation efficiencies as high as 95% can routinely be prepared in this way. The term SNALP is used to differentiate from particles prepared using the SALP and ethanol-destabilized liposome methods, and denotes the broad utility of the technique, which can be applied to any charged nucleic acid species.

The ability of the SNALP method to rapidly and reproducibly prepare vesicles of the required size and encapsulation efficiency is thought to stem from the exact control of the conditions under which the lipids encounter the aqueous buffer, subsequently arrange into lipid bilayer fragments and finally form liposomes. After initial dilution, the formation of macromolecular intermediates is thought to occur, either in the form of lamellar lipid sheets or micellar formations. Electrostatic attraction then draws the anionic nucleic acid to interact with the nascent bilayer. If cationic lipid concentrations are too low, the nucleic acid will fail to associate with these fragments, promoting the formation of empty vesicles. Conversely, if the cationic lipid concentration is too high, the formation of polydisperse aggregates is favored, caused by the additional surface charge attracting excess nucleic acid. It is suggested that at optimal concentrations of cationic lipid, nucleic acid associates with the lipidic fragments in such a way as to reduce the net positive charge on the lipid surface. Ensuing interactions with additional bilayer fragments result in the formation of vesicles containing encapsulated nucleic acid. The balancing of ionic strength and cationic lipid to PEG–lipid content allows for the optimization of SNALP formulation by ethanol dilution. The method is robust, with excellent results achieved through a wide spectrum of formulation conditions.

Of the variety of methodologies available for encapsulating siRNAs into lipid-based systems, the SNALP approach, using stepwise ethanol dilution, allows for the preparation of small (<50 nm), stable, homogeneous systems with high encapsulation efficiencies (>95%) and an extensive range of nucleic acid:lipid ratios (>0.1 w/w). SNALPs exhibit the circulation lifetimes necessary to facilitate preferential accumulation at target sites, and the method is both scalable and reproducible.

11.4 Pharmacology of Liposomal Nucleic Acids

To avoid indiscriminate interaction with blood components, which may cause aggregation and/or clearance before the carrier reaches the target site, systemic delivery requires the use of a relatively charge-neutral, 'stealthy', delivery system. For systems containing large polyanionic molecules, such as siRNAs, which have a greater potential for inducing toxicity through interaction with complement and coagulation pathways,[55] this is especially important. The microcapillary beds of the first-pass organs (the lungs and the liver and the phagocytic cells of the RES) may also present obstacles to delivery. Accessing target-cell populations requires the ability to extravasate from the blood compartment to the target site. Charge-neutral carriers of suitable size can pass through the fenestrated epithelium observed in sites of clinical interest, such as tumors, sites of infection and inflammation, and in the healthy liver, and accrue *via* the 'enhanced permeation and retention' (EPR) effect[56] (also referred to as 'passive' targeting or 'disease site' targeting). To take advantage of the EPR effect (*e.g.* significant enrichment at the target site), carriers must be small (diameter on the order of 100 nm) and long-circulating (prolonged circulation lifetimes following intravenous injection in mice). siRNAs stand to benefit from the pharmaceutical enablement bestowed by encapsulation in suitably designed liposomal carriers.

11.4.1 Pharmacokinetics and Biodistribution of Liposomal siRNAs Following Systemic Administration

Following intravenous injection, the clearance properties of encapsulated siRNAs can be measured by lipid and/or nucleic acid markers. Previous experience shows that when nucleic acids are fully encapsulated in stable liposomes, they remain protected from nuclease degradation, and are cleared from the blood at a rate equal to that of the lipid components.[14,15,57] The biodistribution of intact lipid particles can be represented by monitoring radiolabeled lipid markers stably incorporated into the bilayer.[58] This principle may be applied to clearance and biodistribution studies up to 24 hours after administration, after which time most lipid markers will begin to experience some remodeling or exchange.[59]

A comparison of the clearance properties of liposomal formulations of siRNA in three different species is shown in Figure 11.2. The siRNA was liposomally encapsulated in a SNALP formulation containing the lipids 1,2-distearoyl-sn-glycero-3-phosphocholine (DSPC), cholesterol, 1,2-dilinoleyloxy-3-*N*,*N*-dimethyl-aminopropane (DLinDMA) and {3-*N*-[(q-methoxy poly(ethyleneglycol)$_{2000}$) carbamoyl]-1,2-dimyrestyloxy-propylamine} (PEG-C-DMA). The exact lipid composition was designed for efficient delivery to the liver, while avoiding accumulation in distal tissue and non-target tissues of the RES, such as the spleen. After intravenous administration to mice and guinea pigs, siRNA concentration in plasma and tissue samples was determined using the radiolabeled lipid marker ^{3}H-cholesteryl hexadecyl ether (CHE).[14,51] In cynomolgus monkeys, siRNA

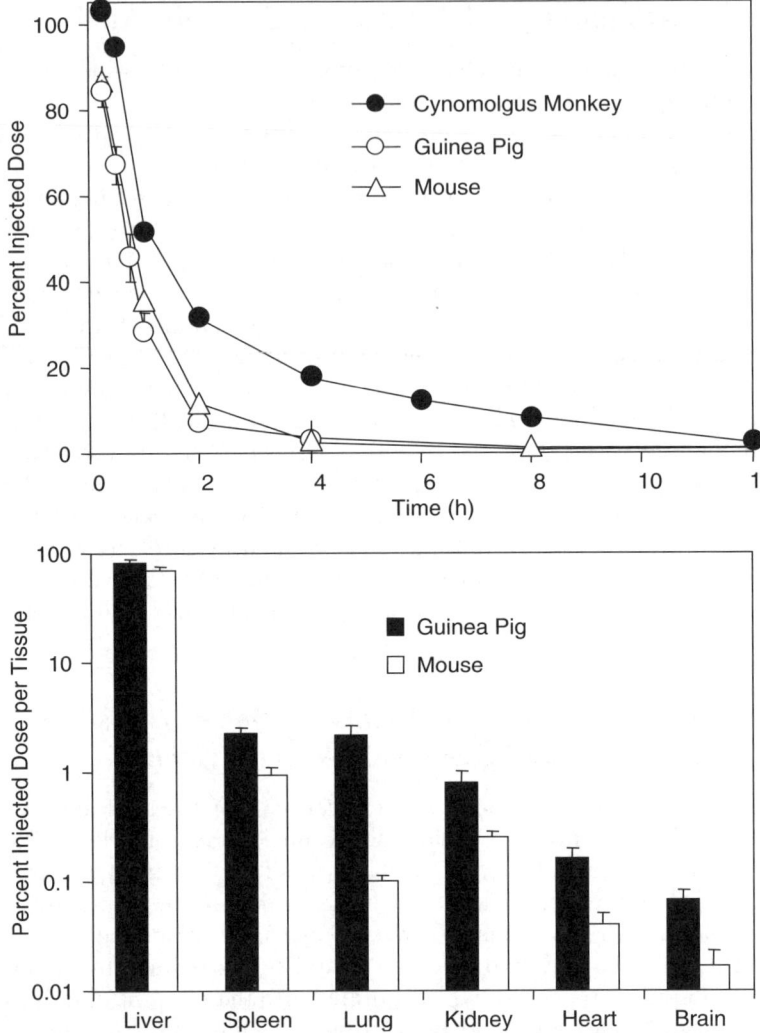

Figure 11.2   Pharmacokinetics and biodistribution of lipid-encapsulated siRNA. (A). Plasma clearance of SNALP siRNA determined in mice, guinea pigs and cynomolgus monkeys. Each animal received a single intravenous injection of SNALP-formulated siRNA. Data represent percent of the total injected dose in blood at the indicated time-points after treatment. Mouse and guinea pig data are presented as mean ± standard deviation (s.d.), $n = 5$. Cynomolgus monkey data represent the mean of two treated animals. (B). Biodistribution of SNALP siRNA was determined in mice and guinea pigs. Each animal received a single intravenous injection of ^{3}H-labeled SNALP-formulated siRNA. Data represent percent of the total injected dose in each tissue 24 hours after treatment. Data are presented as mean ± s.d., $n = 5$.

plasma clearance was determined directly by ion-exchange high-performance liquid chromatography (HPLC).[15] The plasma half-life of SNALP administered by tail vein injection in mice was found to be 38 minutes, whereas that of unprotected, unmodified phosphodiester siRNA is less than 2 minutes.[14] In guinea pigs, the plasma half-life of SNALP (administered by ear vein injection) was 39.3 minutes,[51] and in cynomolgus monkeys (bolus injection in the saphenous vein) the plasma half-life was 72 minutes. The agreement between the clearance properties in mice and in guinea pigs is striking, particularly given the different routes of administration.

Given pharmacokinetic data in a number of disparate species it is possible to perform allometric analysis whereby empirical relationships between the size, shape, surface area or life span of animals and the pharmacokinetics of drugs may be determined. In this way a drug's pharmacological behavior in one species can be used to predict its behavior in another.[60] SNALP pharmaco-kinetics scale well across species when examined using allometric correlations as a function of body weight (Figure 11.3). Unlike naked phosphorothioate oligonucleotides (PTOs), which clear more rapidly from the plasma of mice than would be predicted on this basis,[61] SNALP clearance is unaffected by the more extensive exonuclease metabolism found in mouse plasma. The allometric approach allows one to estimate certain pharmacokinetic parameters for humans. SNALP clearance is estimated to be $\sim 3900 \, \mathrm{ml \, h^{-1}}$, or $0.93 \, \mathrm{ml \, min^{-1} \, kg^{-1}}$, while the volume of distribution at a steady state is estimated to be $\sim 6690 \, \mathrm{ml}$, or $\sim 96 \, \mathrm{ml \, kg^{-1}}$. On this basis the plasma half-life in humans of the SNALP formulation used in these studies can be estimated at 71 minutes, essentially the same as that of cynomolgus monkeys.

The biodistribution of liposomal siRNA following intravenous admini-stration may be determined by similar methods. Figure 11.2 describes the accumulation of liposomal siRNA in various tissues 24 hours after administra-tion in both mice and guinea pigs. The greatest accumulation is typically observed in the liver and spleen. In this example the liver has accumulated $70.7 \pm 5.4\%$ and $83.4 \pm 6.5\%$ of the injected dose per gram in mice and guinea pigs, respectively, and the spleen has accumulated $0.94 \pm 0.15\%$ and $2.2 \pm 0.2\%$ of the injected dose per gram in mice and guinea pigs. Less than 1% of the liposomal siRNA is observed per gram in kidney, heart and brain of either species. This is worth noting, as the kidney is the prototypical target tissue associated with the toxicity of naked antisense drugs.

Although such results are not atypical for freely circulating liposomal sys-tems, biodistribution can easily be altered by effecting changes in the liposomal formulation. This is particularly true for the liver, spleen and distal disease sites, such as tumors. Manipulation of individual lipid chemistries and their relative molar ratios has profound effects on particle accumulation at distal sites. As described earlier, adjusting the size of a PEG-conjugate's lipid anchor will affect the rate at which it dissociates from the particle surface, and the resulting pharmacology of the encapsulated siRNA. Shorter PEG–lipid anchor lengths decrease SNALP circulation times and increase the rate and extent of nanoparticle accumulation in the liver of mice, while SNALP employing

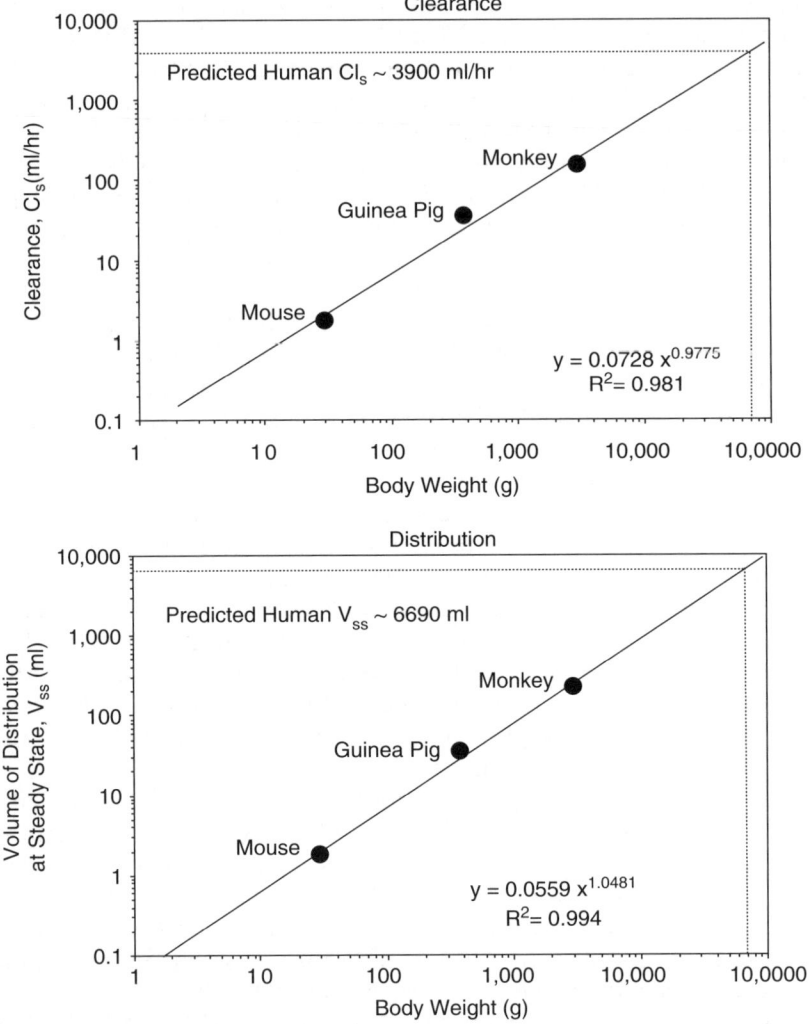

Figure 11.3 Allometric analysis of SNALP pharmacokinetic data from mouse, guinea pig and cynomolgus monkeys (Figure 11.2). On the basis of this data SNALP clearance is estimated to be $\sim 3900 \, \text{ml h}^{-1}$, or $0.93 \, \text{ml}^{-1} \, \text{min}^{-1} \, \text{kg}^{-1}$ in humans, while the volume of distribution at a steady state is estimated to be $\sim 6690 \, \text{ml}$, or $\sim 96 \, \text{ml kg}^{-1}$ in humans.

PEG–lipids with larger lipid anchors exhibit the opposite behavior with a concomitant increase in accumulation at distal sites.

The markedly increased tissue accumulation of siRNA delivered in liposomal form can be ascribed to their circulation lifetime and protection from nuclease degradation. While intravenously administered, 'naked' nucleic acids and cationic lipoplexes tend to have half-lives of minutes or less,[14,62–65] encapsulated liposomal formulations may provide plasma half-lives for intact nucleic acid of 0.5 to 60 hours.[14,15,51,57,65]

11.4.2 Toxicity of Liposomal Nucleic Acid Formulations

The basic goal of drug-delivery technology is to improve a drug's effectiveness by increasing the availability of the drug at the intended target site. However, an unwanted side effect that often accompanies the use of such technology is an increase in toxicity. Previous experience with the free drug is occasionally predictive, however changes in the target organ of toxicity are also encountered.[66,67] Intravenous administration of 'naked' antisense PTOs typically results in approximately 20% of the injected dose accumulating in the kidneys.[68] At high doses in monkeys, these concentrations manifest as toxicity, first in the form of focal tubular regeneration and then as a perturbation in N-acetylglucosamine, total protein and retinol binding protein levels at even higher doses.[69] In the same way, accretion in the liver results in hypertrophy of Kupffer cells and elevated transaminase [aspartate aminotransferase (AST) and alanine aminotransferase (ALT)] levels.[70] Since liposomal encapsulation leads to a dramatic rearrangement of siRNA biodistribution, a concomitant shift in the target organs of toxicity is expected. Indeed, since less than 1% of the total injected SNALP dose accumulates in the kidney of mice or guinea pigs,[51] little or no nephrotoxicity results, even at doses greater than those required to elicit hepatic toxicity as measured by elevations of serum transaminases.[15]

When working with liposomal systems, it is important to consider their potential to activate the complement system. Charged liposomes, in particular, are capable of binding complement proteins and activating the complement cascade.[55] The capacity for a liposomal formulation to activate complement may be examined by standard *in vitro* assays, or complementary experiments may be performed *in vivo*. Various studies have suggested that complement activation can be avoided by careful management of the liposomal surface charge. Strategies for this include the use of PEG–lipids, and adjustment of the amount or type of cationic lipid used. In this respect, PEGylated systems containing modest amounts of titratable cationic lipids, which effectively combine both approaches, appear particularly stealthy.[71]

11.4.3 Immune Stimulation

Nucleic acids can cause activation of the mammalian innate immune system, initiating the release of interferons and pro-inflammatory cytokines. Exogenous single[72] and double-stranded RNA[73] stimulate immune recognition pathways through toll-like receptors TLR7/8 and TLR3, respectively. More recently, canonical 21-mer synthetic siRNA duplexes have been reported to induce potent immune stimulation.[74–76] These immune responses are greatly potentiated by the use of delivery vehicles that are internalized *via* endosomes, the primary intracellular location of the affected TLRs.[74] A host of local and systemic inflammatory reactions can be triggered, so the consequences of immune activation can be severe, particularly in more sensitive species such as humans.[77–79] Many of the toxicities associated with the siRNA administration *in vivo* have been attributed to this response.[14,74]

The finding that immune activation by siRNA is sequence-dependent raised the intriguing possibility of designing active siRNA with negligible immunostimulatory activity by simply avoiding GU-rich domains or other immunostimulatory motifs.[74,80] This obviously limits the number of novel siRNA sequences that can be designed against a given target and requires some degree of screening because of the relatively ill-defined nature of putative RNA immunostimulatory motifs. Stabilization chemistries previously developed for ribozymes or antisense ON drugs[81] offer another approach. A range of nucleoside modifications, including many at the 2′-site [2′-OMe,[82–84] 2′-F,[82,84–86] 2′-deoxy[85] and some 'locked nucleic acid' (LNA) modifications[75,87]] appear to be compatible with the RNAi machinery. However, modification of siRNA is only tolerated in certain positional or sequence-related contexts, indiscriminate modification typically has a negative impact on efficacy.[75,83–85,87] Design of such chemically modified siRNA has, until recently, required screening to identify those that retain potent gene-silencing activity.

A more rational approach to the design of highly active, non-stimulatory siRNA molecules has been described.[88] Surprisingly, irrespective of sequence, minimal 2′-OMe modifications in one strand of a double-stranded siRNA suppress the immunostimulatory activity of the siRNA duplex. Remarkably, abrogation of the immunostimulatory activity of highly immunostimulatory sequences can be accomplished by inclusion of as few as two 2′-OMe guanosine or uridine residues, representing as little as 5% of the native 2′-OH positions in the siRNA duplex. Since complete suppression of the immune response requires only one of the RNA strands to be selectively modified, 2′-OMe modifications can be limited to the sense strand of the duplex. This strategy minimizes the potential for attenuating siRNA potency, primarily determined by the antisense 'guide' strand. Such minimally modified siRNAs retain potent gene silencing *in vivo*, without the cytokine induction, immunotoxicities or immune-mediated off-target effects caused by unmodified siRNA. This simple method permits the design of non-immunostimulatory siRNA based on native sequences with proven RNAi activity.

It is presently unclear how the introduction of 2′-OMe nucleotides into one strand of an siRNA duplex prevents immune recognition. The trans-inhibitory effect of 2′-*O*-methylation, whereby 2′-OMe-modified ssRNA annealed to unmodified immunostimulatory ssRNA generates a non-immunostimulatory duplex, is consistent with a hypothesis that involves recognition of the siRNA by its putative immune receptor, thought to be TLR-7,[75] as a double-stranded molecule. The possibility remains that a mechanism has evolved to allow for the differentiation of self- from pathogen-derived RNA, and that 2′-OMe-modified siRNA may evade recognition in this manner.

Other siRNA chemical modification strategies have the potential to influence immune stimulation. Partial reduction of immune stimulation has been achieved by using LNAs, possessing a 2'-O, 4'-C methylene bridge in the sugar ring.[75] However, sequences containing inverted deoxy abasic end caps retain immunostimulatory activity.[14] No evidence of a trans-inhibitory effect has been observed with LNA-modified duplexes. We conclude that, for reasons not currently understood, siRNA activation of the immune system may be more effectively inhibited by 2'-OMe modifications than by other stabilization chemistries. Preventing the induction of interferons and inflammatory cyto-kines with minimal 2'-OMe modification both limits the possibility of non-specific effects on gene expression and improves the tolerability of siRNA formulations. Intravenous administrations of liposomal siRNA designed by using this approach are efficacious and well-tolerated in mice.[88] Together with continued advances in delivery technology and siRNA design, the approach may be an important component in the development of synthetic siRNA therapeutics.

11.4.4 Immunogenicity

As antibody responses can severely compromise both the safety and efficacy of a drug, any potential for immunogenicity is a serious concern. The ability of liposomes to act as immunological adjuvants is well-known and is a result of their particulate nature, efficient uptake by antigen presenting cells and ability to cross-link surface receptors.[89] This behavior is exacerbated when immuno-stimulatory agents, such as 2'-deoxyribo(cytidine-phosphate-guanosine) (CpG) DNA, are incorporated into liposomes.[90,91]

Stable plasmid lipid particles (SPLPs), a liposomal system for the delivery and expression of therapeutic plasmid DNA (pDNA),[13,49] provide an example of the challenges faced when designing non-immunogenic nucleic acid carriers. Repeat administration of SPLPs containing stably integrated PEG–lipids reveals a surprisingly robust antibody response against PEG that arises from a single administration, and this greatly compromises the *in vivo* safety and efficacy of the system. This phenomenon can be greatly reduced by using alternative PEG–lipids that, after intravenous administration, dissociate more quickly from the lipid bilayer. Abrogation of the PEG antibody response allows these modified liposomes to be safely re-administered to mice, while preserving their pharmacology, including their ability to effectively deliver pDNA to distal tumor sites. Considerable cytokine induction is still associated with non-immunogenic SNALP administration, suggesting that the reduced immunogenicity is not caused by inhibition of the immunostimulatory activity of the payload. Rather, it appears the vigorous antibody responses to PEG require the close physical association of the PEG–lipid with pDNA and are driven by the specific binding and internalization of PEGylated liposomes containing stimulatory pDNA by PEG-reactive B cells.[71]

The use of non-immunostimulatory nucleic acids is a valuable adjunct approach to reducing carrier immunogenicity. The ability of CpG-free pDNA and chemically modified synthetic siRNAs to provoke cytokine responses is significantly reduced relative to their native counterparts. Unmodified synthetic siRNAs can induce powerful immune stimulation *in vivo*, driving the production of a potent anti-PEG antibody response when encapsulated in liposomes containing stably integrated C18 PEG–lipids.[71] However, using the minimal modification approach and incorporating 2'-OMe nucleotides, siRNA duplexes encapsulated in these same liposomes do not elicit an immunogenic response. The development of such strategies allows for much greater flexibility in the design of non-immunogenic liposomal systems.

Given that most unmodified siRNA species stimulate innate cytokine responses and B cell activation,[72,74,75,88,92] it should be apparent that the immunogenicity of any delivery vehicle used with immunostimulatory nucleic acids requires careful consideration. Antibody responses against surface components of liposomal systems, particularly targeting ligands and PEG molecules, should be closely monitored. However, the ability to suppress the immunogenicity of liposomal siRNA formulations by straightforward modifications to either the siRNA duplex or the lipid composition suggests multiple paths forward to facilitate the clinical development of these systems.

11.4.5 Efficacy of Liposomally Formulated Nucleic Acid Drugs

Currently, the lack of clinical experience with liposomal formulations of siRNA drugs requires us to rely on preclinical results to measure their potential (recently reviewed by Behlke[93]). A variety of models, including infectious diseases, cancers and various metabolic conditions, have been used to evaluate liposomal siRNA. However, only recently have we come to understand the extent to which non-specific effects, such as immune stimulation, may affect the results obtained in preclinical models used to gauge efficacy. The impact of immune-system activation on anti-tumor and infectious disease models is particularly troublesome. It is therefore crucial to adopt suitable experimental controls when evaluating these systems. The inclusion of a non-targeting control siRNA, with similar immunostimulatory properties to the active compounds, is a minimal requirement. With our improved understanding of chemical modification strategies that prevent immune stimulation, it is now relatively straightforward to design active siRNAs that are immunologically inert.[88]

Despite the well-documented influence that liposomal siRNA exerts on the innate immune system, some investigators fail to describe fully the immunostimulatory properties of their test article before publishing efficacy results. In our laboratory we have embarked on a retrospective analysis of many published siRNAs used in efficacy studies and strikingly, with one single exception, all were shown to be immunostimulatory. All sequences had either been previously reported as non-immunostimulatory, or their immunostimulatory properties had not been described. Possibly more disturbing still, the lone exception in our analysis was a negative control siRNA, used in at least 12 published studies to support the comparative 'efficacy' of an assortment of highly immunostimulatory siRNAs. Ideally, control siRNAs with similar immunostimulatory properties to those of the active compounds would have been used instead. A simple assay for interferon alpha and interleukin-6 (IL-6), performed between 2 and 12 hours after intravenous administration in mice or 24 hours after exposure of primary peripheral blood mononuclear cells (PBMCs), will adequately assess the immunostimulatory properties of siRNAs. Our expectation is that, when approached and interpreted with appropriate caution, future efficacy results will be substantially more convincing.

Rather than examine an exhaustive list of previously described efficacy results, we will instead describe a single example that we hope depicts the potential of liposomal formulations of siRNA drugs. The siRNA sequence under consideration targets apolipoprotein B (apoB), a vital component in the assembly and secretion of very low density lipoproteins (VLDL), precursors to LDLs. ApoB is considered 'non-druggable' with conventional small-molecule drugs, yet it is a highly relevant, genetically and clinically validated disease target. Second-generation ASOs have been used pre-clinically and clinically to target apoB, showing promising results.[94] siRNA directed against apoB covalently attached to cholesterol has successfully resulted in the knockdown of apoB message, yielding an associated reduction in total cholesterol.[95] Since the apoB transcript is regulated chiefly at the post-translational level, unlike many oncology, inflammatory or infectious disease targets, it is thought to have a degree of protection from off-target effects that might otherwise cause unintended fluctuations in apoB expression levels, making apoB a 'good' choice for proof-of-principle efficacy studies.

When administered as the naked siRNA duplex, high doses ($> 50\,\mathrm{mg\,kg^{-1}}$) of the apoB-directed siRNA sequence siApoB-1 have no *in vivo* silencing activity in mice.[95] However, a single, low dose of the same sequence encapsulated in a liposome (SNALP process) yields profound silencing of liver apoB mRNA and apoB-100 protein.[15] Corresponding treatment with SNALP-formulated mismatch siRNA (siApoB-MM) or empty SNALP vesicles yields no measurable apoB knockdown, suggesting the silencing effect is specific to the active siRNA and unrelated to the liposomal carrier or other off-target effects. Figure 11.4 illustrates the potency of siApoB-1 SNALP-mediated silencing compared to the same sequence covalently conjugated to cholesterol. The cholesterol conjugate requires doses of $100\,\mathrm{mg\,kg^{-1}}$ to achieve 50% knockdown of apoB message, while SNALP-formulated siApoB-1 requires $0.1\,\mathrm{mg\,kg^{-1}}$ to yield the same effect. This represents a 1000-fold improvement in the potency of siRNA when encapsulated, relative to the cholesterol conjugate, and, relative to the naked sequence, a potency increase substantially greater than four orders of magnitude. This degree of silencing is readily achieved in the absence of immune stimulation or other toxicities.

siRNA drugs will ultimately be used in a manner analogous to their small-molecule counterparts, including reliance on multiple-dosing regimens that have an increased potential for the development of toxicity. The purpose of a multiple-dosing regimen is to achieve a more prolonged effect than otherwise afforded by the administration of a single dose. With conventional small-molecule drugs, typically, the objective is to design a dosing regimen that achieves a steady-state concentration of drug in the target tissue. The plasma

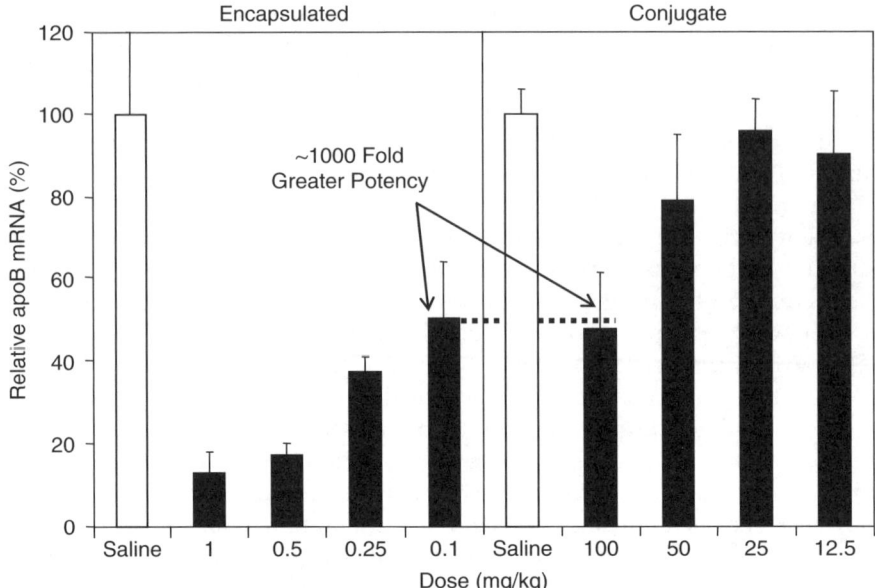

Figure 11.4 The efficacy of liposomal- (SNALP-) encapsulated siRNA compares favorably to that of chemically conjugated siRNA. The dose-dependent silencing of liver apoB mRNA after administration of either SNALP siApoB-1 (left panel) or chol-siApoB-1 (right panel) is shown. Liver apoB mRNA levels were quantified relative to glyceraldehyde-3-phosphate dehydrogenase (GAPDH) mRNA three days after intravenous administration of siRNA. Data are mean values relative to the saline treatment group $\pm$ s.d. Chol-siApoB-1 was administered at doses of 100, 50, 25 or $12.5 \, \text{mg} \, \text{kg}^{-1}$ ($n = \text{six per group}$), and SNALP siApoB-1 was administered at siRNA doses of 1, 0.5, 0.25 and $0.1 \, \text{mg} \, \text{kg}^{-1}$ ($n = \text{four per group}$).

concentration of the drug may fluctuate with each administration, achieving a maximum (peak) concentration soon after, and a minimum plasma concentration immediately prior to each dose. A well-designed dosing regimen establishes and maintains the drug concentration in the target tissue within the therapeutic range. When considering siRNA-based drugs, the relationship between the plasma siRNA concentration, the concentration of siRNA in the target tissue and the RNAi effect is less straightforward. In many cases, RNAi-mediated effects persist long after the drug is removed from the plasma.[15] An example of the efficacy associated with multiple doses of liposomal siRNA is illustrated in Figure 11.5. Mice were treated with anti-apoB siRNA formulated as SNALP for a period of 12 weeks. One group was treated with a dose of $1 \, \text{mg} \, \text{kg}^{-1}$, administered once every week, while a second group was treated with a dose of $5 \, \text{mg} \, \text{kg}^{-1}$ administered once every two weeks. Both treatments resulted in substantial inhibition of apoB expression and a concomitant reduction in total blood cholesterol. Both groups exhibited a $\sim 90\%$ reduction in apoB protein which, in the case of the $5 \, \text{mg} \, \text{kg}^{-1}$ group,

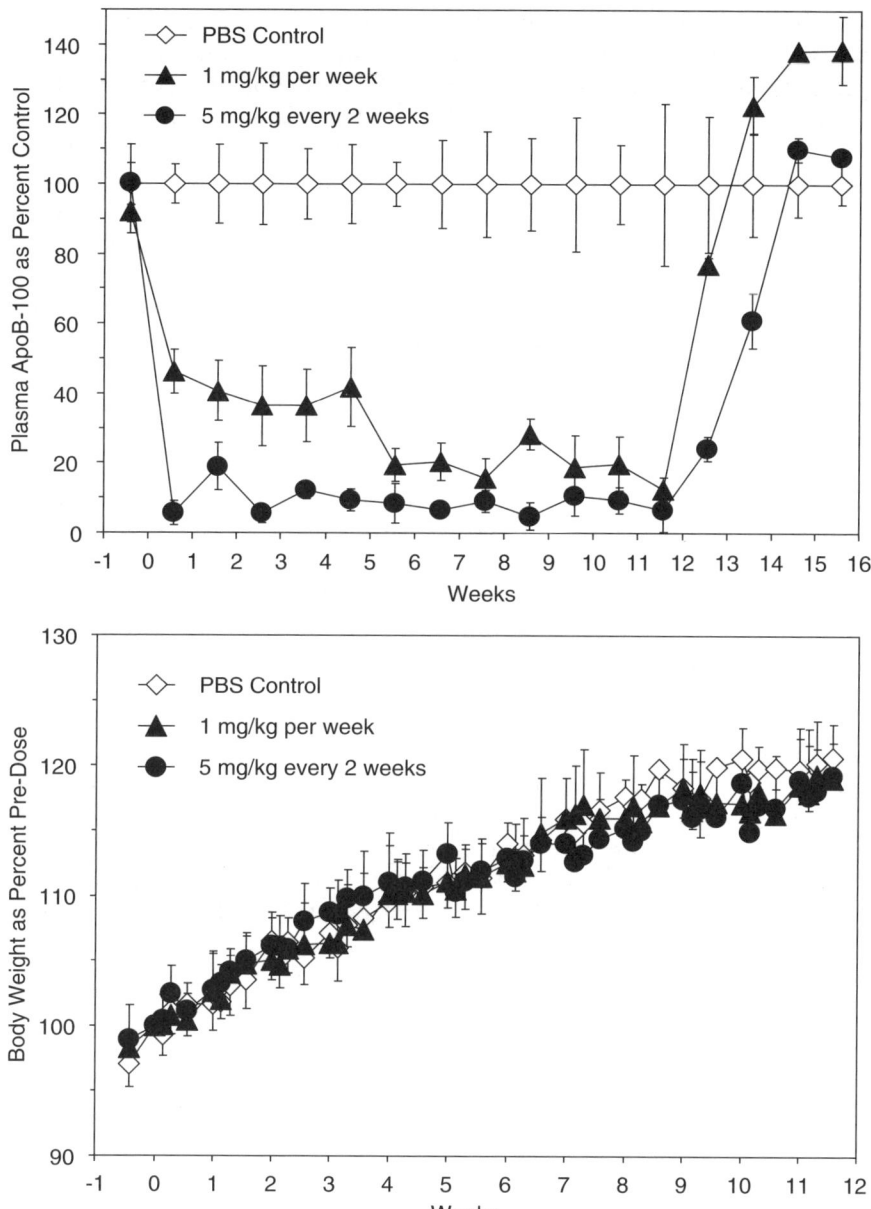

Figure 11.5 Liposomal- (SNALP-) encapsulated siRNA mediate multidosing efficacy in the absence of measurable toxicity. Mice were treated with anti-apoB siRNA formulated as SNALP for a period of 12 weeks. Mice were treated with a dose of $1\,\mathrm{mg\,kg^{-1}}$, administered once every week or $5\,\mathrm{mg\,kg^{-1}}$ administered once every two weeks. Plasma ApoB protein was reduced through the duration of the experiment in the absence of any concomitant toxicity. Data are presented as mean $\pm$ s.d., $n = 5$.

was achieved a few days after administration of the first dose while the $1 \, mg \, kg^{-1}$ group experienced a cumulative drop in apoB expression that reached its nadir a few days after the final dose. Both groups returned to normal within three weeks after treatment. Multiple doses of SNALP were well-tolerated, as exhibited by normal weight gain throughout the experiment and normal blood chemistry. Importantly, no transaminitis was observed, nor was there an adaptive immune response to the SNALP carrier, as measured by the absence of anti-PEG antibodies and their associated hypersensitivities.

In a separate, non-human primate study, another (considerably less potent) siRNA sequence was also found to potently silence apoB when formulated as SNALP.[15] The endogenous apoB message was silenced by 90% in cynomolgus monkeys treated with a single intravenous administration of $2.5 \, mg \, kg^{-1}$ SNALP-formulated siRNA. Again, this result was achieved in the absence of toxicity as measured by general tolerability, complement activation, coagulation or pro-inflammatory cytokine production. No changes were observed in the hematology parameters of SNALP-treated animals. Primates treated with ApoB SNALP exhibited no detectable changes other than a moderate, transient increase in liver enzymes in monkeys that received the highest dose. The transient transaminitis peaked 48 hours post-treatment, and was found to vary considerably between individual animals. The phenomenon was completely reversible, normalizing within six days, while apoB silencing reached a nadir after 11 days. It is important to note that this high degree of apoB silencing (90%) is unlikely to be a relevant clinical target, and that a less aggressive reduction in apoB is the presumed clinical endpoint. Furthermore, a number of opportunities for improving the therapeutic index of apoB SNALP have been examined in mice, including the use of more potent siRNA sequences, formulation refinements, and changes to the dosing regime.

The experience with apoB SNALP indicates that effective systemic delivery of siRNAs using liposomes is readily achievable. Further optimization of payloads, formulations and treatment regimens may be required, but liposomal siRNAs show considerable promise in a number of applications. It is increasingly probable that as siRNA drugs continue to make the transition from laboratory to clinic, liposomes will play an important role in their success.

References

1. E. Check, *Nature*, 2003, **425**, 10.
2. A. D. Bangham and R. W. Horne, *J. Mol. Biol.*, 1964, **8**, 660.
3. P. Tardi, E. Choice, D. Masin, T. Redelmeier, M. Bally and T. D. Madden, *Cancer Res.*, 2000, **60**, 3389.
4. R. L. Letsinger, G. R. Zhang, D. K. Sun, T. Ikeuchi and P. S. Sarin, *Proc. Natl. Acad. Sci. U. S. A.*, 1989, **86**, 6553.
5. R. G. Shea, J. C. Marsters and N. Bischofberger, *Nucleic Acids Res.*, 1990, **18**, 3777.
6. C. A. Stein, R. Pal, A. L. DeVico, G. Hoke, S. Mumbauer, O. Kinstler, M. G. Sarngadharan and R. L. Letsinger, *Biochemistry*, 1991, **30**, 2439.

7. A. M. Krieg, J. Tonkinson, S. Matson, Q. Zhao, M. Saxon, L. M. Zhang, U. Bhanja, L. Yakubov and C. A. Stein, *Proc. Natl. Acad. Sci. U. S. A.*, 1993, **90**, 1048.

8. G. Zon, *Toxicol. Lett.*, 1995, **82–83**, 419.

9. B. Bayard, L. D. Leserman, C. Bisbal and B. Lebleu, *Eur. J. Biochem.*, 1985, **151**, 319.

10. S. L. Loke, C. Stein, X. Zhang, M. Avigan, J. Cohen and L. M. Neckers, *Curr. Top. Microbiol. Immunol.*, 1988, **141**, 282.

11. P. G. Milhaud, P. Machy, B. Lebleu and L. Leserman, *Biochim. Biophys. Acta*, 1989, **987**, 15.

12. P. L. Felgner, T. R. Gadek, M. Holm, R. Roman, H. W. Chan, M. Wenz, J. P. Northrop, G. M. Ringold and M. Danielsen, *Proc. Natl. Acad. Sci. U. S. A.*, 1987, **84**, 7413.

13. L. B. Jeffs, L. R. Palmer, E. G. Ambegia, C. Giesbrecht, S. Ewanick and I. MacLachlan, *Pharm. Res.*, 2005, **22**, 362.

14. D. V. Morrissey, J. A. Lockridge, L. Shaw, K. Blanchard, K. Jensen, W. Breen, K. Hartsough, L. Machemer, S. Radka, V. Jadhav, N. Vaish, S. Zinnen, C. Vargeese, K. Bowman, C. Shaffer, L. B. Jeffs, A. D. Judge, I. MacLachlan and B. A. Polisky, *Nat. Biotechnol.*, 2005, **23**, 1002.

15. T. S. Zimmermann, A. Lee, A. Akinc, B. Bramlage, D. Bumcrot, M. N. Fedoruk, J. Harborth, J. Heyes, L. Jeffs, J. Matthias, A. D. Judge, K. Lam, K. McClintock, L. V. Nechev, L. Palmer, T. Racie, I. Röhl, S. Seiffert, S. Shanmugam, V. Sood, J. Soutschek, I. Toudjarska, A. Wheat, E. Yaworski, W. Zedalis, V. Koteliansky, M. Manoharan, H. -P. Vornlocher and I. MacLachlan, *Nature*, 2006, **441**, 111.

16. J. Zabner, A. J. Fasbender, T. Moninger, K. A. Poellinger and M. J. Welsh, *J. Biol. Chem.*, 1995, **270**, 18997.

17. K. A. Mislick and J. D. Baldeschwieler, *Proc. Natl. Acad. Sci. U. S. A.*, 1996, **93**, 12349.

18. L. C. Mounkes, W. Zhong, G. Cipres-Palacin, T. D. Heath and R. J. Debs, *J. Biol. Chem.*, 1998, **273**, 26164.

19. S. C. Semple, S. K. Klimuk, T. O. Harasym, N. Dos Santos, S. M. Ansell, K. F. Wong, N. Maurer, H. Stark, P. R. Cullis, M. J. Hope and P. Scherrer, *Biochim. Biophys. Acta*, 2001, **1510**, 152.

20. J. Heyes, L. Palmer, K. Bremner and I. MacLachlan, *J. Controlled Release*, 2005, **107**, 276.

21. I. Koltover, T. Salditt, J. O. Radler and C. R. Safinya, *Science*, 1998, **281**, 78.

22. I. M. Hafez, N. Maurer and P. Cullis, *Gene Ther.*, 2001, **8**, 1188.

23. H. Farhood, N. Serbina and L. Huang, *Biochim. Biophys. Acta*, 1995, **1235**, 289.

24. J. H. Felgner, R. Kumar, C. N. Sridhar, C. J. Wheeler, Y. J. Tsai, R. Border, P. Ramsey, M. Martin and P. L. Felgner, *J. Biol. Chem.*, 1994, **269**, 2550.

25. X. Gao and L. Huang, *Gene Ther.*, 1995, **2**, 710.

26. S. W. Hui, M. Langner, Y. L. Zhao, P. Ross, E. Hurley and K. Chan, *Biophys. J.*, 1996, **71**, 590.

27. P. R. Cullis and B. de Kruijff, *Biochim. Biophys. Acta*, 1979, **559**, 399.
28. C. J. Dekker, W. Vankessel, J. P. G. Klomp, J. Pieters and B. Dekruijff, *Chem. Phys. Lipids*, 1983, **33**, 93.
29. M. B. Sankaram, G. L. Powell and D. Marsh, *Biochim. Biophys. Acta*, 1989, **980**, 389.
30. R. M. Epand, R. F. Epand, N. Ahmed and R. Chen, *Chem. Phys. Lipids*, 1991, **57**, 75.
31. R. M. Epand, N. Fuller and R. P. Rand, *Biophys. J.*, 1996, **71**, 1806.
32. J. A. Szule, N. L. Fuller and R. P. Rand, *Biophys. J.*, 2002, **83**, 977.
33. Y. Liu, D. Liggitt, W. Zhong, G. Tu, K. Gaensler and R. Debs, *J. Biol. Chem.*, 1995, **270**, 24864.
34. A. Gabizon and D. Papahadjopoulos, *Proc. Natl. Acad. Sci. U. S. A.*, 1988, **85**, 6949.
35. J. Senior, C. Delgado, D. Fisher, C. Tilcock and G. Gregoriadis, *Biochim. Biophys. Acta*, 1991, **1062**, 77.
36. J. W. Holland, C. Hui, P. R. Cullis and T. D. Madden, *Biochemistry*, 1996, **35**, 2618.
37. Y. H. Xu and F. C. Szoka, *Biochemistry*, 1996, **35**, 5616.
38. D. Kirpotin, K. Hong, N. Mullah, D. Papahadjopoulos and S. Zalipsky, *FEBS Lett.*, 1996, **388**, 115.
39. V. A. Slepushkin, S. Simoes, P. Dazin, M. S. Newman, L. S. Guo, M. C. Pedroso de Lima and N. Duzgunes, *J. Biol. Chem.*, 1997, **272**, 2382.
40. X. Guo and F. C. Szoka, Jr., *Bioconjugate Chem.*, 2001, **12**, 291.
41. S. Simoes, V. Slepushkin, N. Duzgunes and M. C. Pedroso de Lima, *Biochim. Biophys. Acta*, 2001, **1515**, 23.
42. J. S. Choi, J. A. MacKay and F. C. Szoka, *Bioconjugate Chem.*, 2003, **14**, 420.
43. J. Shin, P. Shum and D. H. Thompson, *J. Controlled Release*, 2003, **91**, 187.
44. C. Masson, M. Garinot, N. Mignet, B. Wetzer, P. Mailhe, D. Scherman and M. Bessodes, *J. Controlled Release*, 2004, **99**, 423.
45. S. Mishra, P. Webster and M. E. Davis, *Eur. J. Cell Biol.*, 2004, **83**, 97.
46. J. J. Wheeler, L. Palmer, M. Ossanlou, I. MacLachlan, R. W. Graham, Y. P. Zhang, M. J. Hope, P. Scherrer and P. R. Cullis, *Gene Ther.*, 1999, **6**, 271.
47. Q. Hu, C. R. Shew, M. B. Bally and T. D. Madden, *Biochim. Biophys. Acta*, 2001, **1514**, 1.
48. D. B. Fenske, I. MacLachlan and P. R. Cullis, *Methods Enzymol.*, 2002, **346**, 36.
49. E. Ambegia, S. Ansell, P. Cullis, J. Heyes, L. Palmer and I. MacLachlan, *Biochim. Biophys. Acta*, 2005, **1669**, 155.
50. J. Heyes, K. Hall, V. Tailor, R. Lenz and I. MacLachlan, *J. Controlled Release*, 2006, **112**, 280.
51. T. W. Geisbert, L. E. Hensley, E. Kagan, E. Z. Yu, J. B. Geisbert, K. Daddario-DiCaprio, E. A. Fritz, P. B. Jahling, K. McClintock, J. S. Phelps, A. Lee, A. D. Judge, L. B. Jeffs and I. MacLachlan, *J. Infect. Dis.*, 2006, **193**, 1650.

52. M. J. Hope, M. B. Bally, G. Webb and P. R. Cullis, *Biochim. Biophys. Acta*, 1985, **812**, 55.

53. A. L. Bailey and S. M. Sullivan, *Biochim. Biophys. Acta*, 2000, **1468**, 239.

54. N. Maurer, K. F. Wong, H. Stark, L. Louie, D. McIntosh, T. Wong, P. Scherrer, S. C. Semple and P. R. Cullis, *Biophys. J.*, 2001, **80**, 2310.

55. A. Chonn, P. R. Cullis and D. V. Devine, *J. Immunol*, 1991, **146**, 4234.

56. L. Mayer, M. Bally, P. Cullis, S. Wilson and J. Emerman, *Cancer Lett.*, 1990, **53**, 183.

57. P. Tam, M. Monck, D. Lee, O. Ludkovski, E. Leng, K. Clow, H. Stark, P. Scherrer, R. W. Graham and P. R. Cullis, *Gene Ther.*, 2000, **7**, 1867.

58. Y. Stein, G. Halperin and O. Stein, *FEBS Lett.*, 1980, **111**, 104.

59. M. Bally, L. Mayer, M. Hope and R. Nayar, in *Liposome Technology*, ed. G. Gregoriadis, CRC Press, Boca Raton, 1992, p. 27.

60. J. E. Riviere, T. Martin-Jimenez, S. F. Sundlof and A. L. Craigmill, *J. Vet. Pharmacol. Ther.*, 1997, **20**, 453.

61. R. Z. Yu, R. S. Geary, J. M. Leeds, T. Watanabe, M. Moore, J. Fitchett, J. Matson, T. Burckin, M. V. Templin and A. A. Levin, *J. Pharm. Sci.*, 2001, **90**, 182.

62. D. Lew, S. E. Parker, T. Latimer, A. M. Abai, A. Kuwahara-Rundell, S. G. Doh, Z. Y. Yang, D. LaFace, S. H. Gromkowski, G. J. Nabel, M. Manthorpe and J. Norman, *Hum. Gene Ther.*, 1995, **6**, 553.

63. M. Ogris, S. Brunner, S. Schüller, R. Kircheis and E. Wagner, *Gene Ther.*, 1999, **6**, 595.

64. A. R. Thierry, P. Rabinovich, B. Peng, L. C. Mahan, J. L. Bryant and R. C. Gallo, *Gene Ther.*, 1997, **4**, 226.

65. R. Z. Yu, R. S. Geary, J. M. Leeds, T. Watanabe, J. R. Fitchett, J. E. Matson, R. Mehta, G. R. Hardee, M. V. Templin, K. Huang, M. S. Newman, Y. Quinn, P. Uster, G. Zhu, P. K. Working, M. Horner, J. Nelson and A. A. Levin, *Pharm. Res.*, 1999, **16**, 1309.

66. A. A. Gabizon, *Cancer Invest.*, 2001, **19**, 424.

67. K. A. Gelmon, A. Tolcher, A. R. Diab, M. B. Bally, L. Embree, N. Hudon, C. Dedhar, D. Ayers, A. Eisen, B. Melasky, C. Burge, P. Logan and L. D. Mayer, *J. Clin. Oncol.*, 1999, **17**, 697.

68. P. A. Cossum, H. Sasmor, D. Dellinger, L. Truong, L. Cummins, S. R. Owens, P. M. Markham, J. P. Shea and S. Crooke, *J. Pharmacol. Exp. Ther.*, 1993, **267**, 1181.

69. D. K. Monteith, M. J. Horner, N. A. Gillett, M. Butler, R. Geary, T. Burckin, T. Ushiro-Watanabe and A. A. Levin, *Toxicol. Pathol.*, 1999, **27**, 307.

70. S. P. Henry, D. Monteith and A. A. Levin, *Anticancer Drug Des.*, 1997, **12**, 395.

71. A. D. Judge, K. McClintock, J. R. Shaw and I. MacLachlan, *Mol. Ther.*, 2006, **13**, 328.

72. F. Heil, H. Hemmi, H. Hochrein, F. Ampenberger, C. Kirschning, S. Akira, G. Lipford, H. Wagner and S. Bauer, *Science*, 2004, **303**, 1526.

73. L. Alexopoulou, A. Czopik-Holt, R. Medzhitov and R. Flavell, *Nature*, 2001, **413**, 732.
74. A. Judge, V. Sood, J. Shaw, D. Fang, K. McClintock and I. MacLachlan, *Nat. Biotechnol.*, 2005, **23**, 457.
75. V. Hornung, M. Guenthner-Biller, C. Bourquin, A. Ablasser, M. Schlee, S. Uematsu, A. Noronha, M. Manoharan, S. Akira, A. de Fougerolles, S. Endres and G. Hartmann, *Nat. Med.*, 2005, **11**, 263.
76. M. Sioud, *J. Mol. Biol.*, 2005, **348**, 1079.
77. H. R. Michie, K. R. Manogue, D. R. Spriggs, A. Revhaug, S. Odwyer, C. A. Dinarello, A. Cerami, S. M. Wolff and D. W. Wilmore, *N. Engl. J. Med.*, 1988, **318**, 1481.
78. S. E. Krown, *Semin. Oncol.*, 1986, **13**, 207.
79. G. Suntharalingam, M. R. Perry, S. Ward, S. J. Brett, A. Castello-Cortes, M. D. Brunner and N. Panoskaltsis, *N. Engl. J. Med.*, 2006, **355**, 1018.
80. M. Schlee, V. Hornung and G. Hartmann, *Mol. Ther.*, 2006, **14**, 463.
81. M. Manoharan, *Curr. Opin. Chem. Biol.*, 2004, **8**, 570.
82. C. R. Allerson, N. Sioufi, R. Jarres, T. Prakash, N. Maik, A. Berdeja, L. Wanders, R. H. Griffey, E. E. Swayze and B. Bhat, *J. Med. Chem.*, 2005, **48**, 901.
83. F. Czauderna, M. Fechtner, S. Dames, H. Aygun, A. Klippel, G. J. Pronk, K. Giese and J. Kaufmann, *Nucleic Acids Res.*, 2003, **31**, 2705.
84. T. Prakash, C. R. Allerson, P. Dande, T. Vickers, N. Sioufi, R. Jarres, B. Baker, E. E. Swayze, R. H. Griffey and B. Bhat, *J. Med. Chem.*, 2005, **48**, 4247.
85. Y. Chiu and T. Rana, *RNA*, 2003, **9**, 1034.
86. J. Layzer, A. McCaffrey, A. Tanner, Z. Huang, M. Kay and B. Sullenger, *RNA*, 2004, **10**, 766.
87. J. Elmen, H. Thonberg, K. Ljungberg, M. Frieden, M. Westergaard, Y. Xu, B. Wahren, Z. Liang, H. Orum, T. Koch and C. Wahlestedt, *Nucl. Acids Res.*, 2005, **33**, 439.
88. A. D. Judge, G. Bola, A. C. H. Lee and I. MacLachlan, *Mol. Ther.*, 2006, **13**, 494.
89. A. Allison and G. Gregoriadis, *Nature*, 1974, **252**, 252.
90. W. M. Li, M. B. Bally and M. P. Schutze-Redelmeier, *Vaccine*, 2001, **20**, 148.
91. C. Boeckler, D. Dautel, P. Schelte, B. Frisch, D. Wachsmann, J. P. Klein and F. Schuber, *Eur. J. Immunol.*, 1999, **29**, 2297.
92. S. S. Diebold, T. Kaisho, H. Hemmi, S. Akira and C. Reise Sousa, *Science*, 2004, **303**, 1529.
93. M. A. Behlke, *Mol. Ther.*, 2006, **13**, 644.
94. R. M. Crooke, M. J. Graham, K. M. Lemonidis, C. P. Whipple, S. Koo and R. J. Perera, *J. Lipid Res.*, 2005, **46**, 872.
95. J. Soutschek, A. Akinic, B. Bramlage, K. Charisse, R. Constein, M. Donoghue, S. Elbashir, A. Geik, P. Hadwiger, J. Harborth, J. Matthias, V. Kesavan, G. Lavine, R. K. Pandey, T. Racie, K. G. Rajeev, I. Röhl, I. Toudjarska, G. Wang, S. Wuschko, D. Bumcrot, V. Kotelianski, S. Limmer, M. Manoharan and H.-P. Vornlocher, *Nature*, 2004, **432**, 173.

CHAPTER 12

Vector-Mediated and Viral Delivery of Short Hairpin RNAs

HENRY FECHNER[*,1] AND JENS KURRECK[2,3]

[1] Department of Cardiology and Pneumology, Charité-University Medicine Berlin, Campus Benjamin Franklin, Hindenburgdamm 30, 12200 Berlin, Germany; [2] Institute for Chemistry and Biochemistry, Free University Berlin, Thielallee 63, 14195 Berlin, Germany; [3] Institute of Industrial Genetics, University of Stuttgart, Allmandring 31, 70569 Stuttgart, Germany

12.1 Introduction

RNA interference (RNAi) is an evolutionary-conserved mechanism in eukaryotic cells for post-transcriptional silencing of a target gene in a sequence-specific manner. The phenomenon was originally discovered in the nematode *Caenorhabditis elegans*.[1] Long double-stranded RNA (dsRNA) molecules were found to bind to and promote the degradation of target RNAs, thereby preventing the expression of specific genes. Initially, the application of RNAi was restricted to lower eukaryotes and plants, since long dsRNA molecules cause a strong, unspecific interferon response in mammals.

Elucidation of the RNAi pathway helped to circumvent this problem. The long dsRNA molecule is cleaved by an RNase III-type enzyme termed Dicer into small interfering RNAs (siRNAs) of approximately 20 nucleotides in length. The siRNA is then incorporated into a multimeric protein complex known as the RNA-induced silencing complex (RISC). Activation of RISC is thought to be accompanied by cleavage and removal of one of the two strands of the siRNAs.[2,3] The remaining strand allows RISC to bind to a complementary RNA. The endonuclease activity of RISC then cleaves the target RNA, thereby preventing its translation. These molecular scissors, known as the 'slicer'

RSC Biomolecular Sciences
Therapeutic Oligonucleotides
Edited by Jens Kurreck
© Royal Society of Chemistry 2008

activity of RISC, have been identified to be located in the Argonaute 2 protein, a major component of the complex.[4]

The interferon response in mammalian cells is only induced by dsRNA molecules which are longer than 30 nucleotides, so that there was a chance that siRNAs could be employed to enable RNAi-mediated gene silencing without triggering unspecific cellular processes. In fact, Tuschl and co-workers demonstrated in their milestone publication from 2001 that siRNAs 21 nucleotides in length consisting of a 19 base-pair helix and two nucleotide overhangs at the 3'-ends of both strands were suitable to knockdown gene expression in mammalian cells without eliciting an interferon response.[5] This finding made RNAi one of the most powerful tools in modern molecular biology, which allows functional studies of gene function by generating loss-of-function phenotypes. Furthermore, it opened the road for the development of new therapeutic approaches against those diseases that, to date, cannot satisfactorily be treated.

siRNA molecules are usually synthesized by solid-phase chemistry. A drawback of this approach, however, is the transient nature of siRNA-induced gene silencing. The duration of target-gene knockdown of five to seven days has frequently been observed *in vitro*[6] and *in vivo*.[7] After this period, siRNAs become degraded or they are diluted out through cell division. The development of vector-based systems to synthesize RNAi *in situ* helped to overcome this limitation and is the focus of this chapter. This approach enabled a number of new applications, including the selection of cell lines, which permanently silence a target gene, the development of inducible systems, the generation of transgenic animals and the construction of viral vectors known from gene therapeutic approaches to facilitate efficient delivery.

12.2 Plasmid-derived Short Hairpin RNAs

In 2002, several groups simultaneously introduced the concept of plasmid-induced gene silencing by RNAi (summarized by Shi[8]). The most widely used approach converts the siRNA sequence into a DNA sequence encoding the sense strand of the siRNA, a loop and the antisense strand of the siRNA. Transcription from this DNA template results in a self-complementary RNA molecule referred to as short hairpin RNA (shRNA). These shRNA molecules are recognized as substrates by Dicer and are then processed to give siRNA-type molecules (Figure 12.1).

Usually, shRNA molecules are transcribed intracellularly under the control of RNA polymerase III promoters. These promoters are evolutionary optimized to generate high levels of short, precisely defined RNA molecules. Unlike RNA polymerase II transcripts, the resulting transcripts are neither capped nor polyadenylated. Most commonly used are the promoter of the U6 component of the splicosome[9–11] and the H1 promoter of the RNA component of RNaseP.[12] The U6 or the H1 promoter preferentially initiate transcription at guanine or adenine, respectively, and recognize a series of five to six thymines as a termination signal. It is therefore essential not to use sequences that include longer stretches of consecutive uridines in either of the siRNA strands.

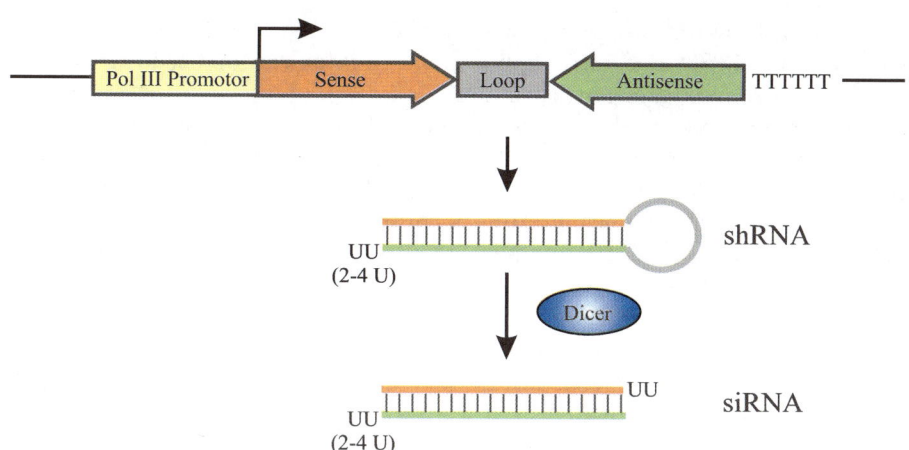

Figure 12.1 Vector expression of short hairpin RNAs (shRNAs). The self-complementary RNA molecules can be expressed under control of polymerase III promoters, like the U6 or H1 promoter. The shRNAs are then processed by the RNase Dicer to yield the siRNA. (Figure adapted with kind permission of the publisher from Jens Kurreck, Oligonucleotide als zellbiologische Werkzeuge, in ed. F. Lottspeich and J. Engels, *Bioanalytik*, Spektrum Verlag, Germany, 2006.)

The exact design of the shRNA determines its efficiency. First, the size and the nucleotide sequence of the loop were shown to play an important role.[12] More recently, Siolas *et al.*[13] demonstrated that the length of the duplex greatly influences the potency of the shRNA as well. Chemically synthesized shRNAs 29 nucleotides in length were found to be much more potent inducers of RNAi than were siRNAs or shRNAs with the standard 19-mer helix. A possible mechanistic explanation for this higher efficiency is that these longer shRNAs are initially processed by Dicer to give 21-mers. According to the current model, Dicer is involved in the loading process of siRNAs into RISC,[14] thus explaining the higher potency of Dicer substrates compared to traditional 21-mer siRNAs.

Transcription with RNA polymerase II yields RNAs that contain a cap at the 5'-end and a poly-A tail at the 3'-end. Only a few reports of work employing RNA polymerase II promoters to generate shRNAs have therefore been published to date. Since the cap and poly-A tail are not compatible with the RNAi machinery, this approach requires a setup with an optimized version of the cytomegalovirus promoter, as well as a minimized poly-A cassette to generate functional shRNA molecules.[15]

Transient transfection of cells with shRNA-expressing plasmids enables the induction of RNAi over an extended period of time compared to the application of chemically synthesized siRNAs. The transfection efficiency of plasmids, however, is usually significantly lower than that of oligonucleotides, resulting in insufficient knockdown. A simple calculation demonstrates this problem: if a plasmid encoding an shRNA that mediates an 80% knockdown of the target gene is transfected into 60% of the cells, the observed decrease in the protein level will be 48%. For most applications, this extent of silencing will be insufficient to investigate gene functions.

To overcome this problem, viral vectors can be used that transfer the shRNA expression cassettes into almost 100% of the targeted cells (*vide infra*) or stable cell lines can be selected, which integrate the shRNA-encoding sequence and its promoter into their genomes. For this purpose, the plasmid usually contains an antibiotic resistance cassette. In one of the first examples, stable cell lines were selected in which the expression of the tumour-suppressor gene p53 was found to be almost completely inhibited, even after two months.[12] It has been observed that the amount of shRNA in stably transfected cells is dramatically reduced compared to the amount of shRNA transiently expressed.[16] The high efficiency, however, means that even low levels of active shRNAs will be sufficient for good knockdown of the target gene.

The plasmid-based RNAi approach also allows the combination of two or more siRNAs. This strategy is desirable for various applications. First, it allows the simultaneous knockdown of two target genes. Second, combination of two or more potent drugs is a common strategy in conventional antiviral therapy to prevent the emergence of escape mutants. It will therefore also be advantageous to combine multiple shRNAs for RNAi-based treatments of viral infections.

Several reports in the literature have shown that extended inhibition of viruses by means of single-gene RNAi will result in the emergence of escape mutants that are no longer susceptible to the used siRNA. This problem is particularly relevant for RNA viruses with an error-prone replication machinery. A single point mutation in the target site was found to be sufficient to render the respective siRNA inactive.[17–19] Interestingly, human immunodeficiency virus 1 (HIV-1) can escape RNAi silencing not only through mutations in the siRNA target site, but also through base substitutions that alter the local RNA structure, rendering the target site inaccessible for the siRNA.[20] In one case, an extended deletion in the HIV-nef gene was observed.[19] This gene is obviously dispensable for virus replication under certain conditions. It is thus advisable to target regions of the viral genome that are essential for the viral life cycle.

As a strategy to prevent viral escape, we developed an siRNA double expression vector (SiDEx), which generates two shRNAs simultaneously.[21] This vector contains two expression cassettes, each consisting of a U6 promoter and the shRNA-encoding complementary DNA (cDNA). At first, we demonstrated that the two U6 promoters do not interfere with one another and both shRNAs inhibit the expression of their target. We then artificially introduced a point mutation into the viral target gene fused to green fluorescent protein (GFP) as a reporter so that the respective mono-expression vector lost its silencing capacity. The second shRNA of the double-expression vector, however, compensated this loss of activity and maintained high inhibitory activity. In subsequent experiments we demonstrated the high efficiency of the double expression cassette in inhibiting replication of infectious coxsackievirus B3.[22]

Berkhout *et al.* used lentiviral vectors to inhibit HIV-1 replication and found a double shRNA-expression vector to delay virus escape.[23] Combination of shRNAs was furthermore found to have an additive inhibitory effect on HIV-1, so that a lentiviral vector was developed that expresses three different shRNAs targeting *gag* and two different sites of *pol* from separate H1 promoters. In an alternative approach, three RNA-based silencing technologies were combined: A triple combination of an shRNA targeting HIV-1 *rev* and *tat* mRNAs, a decoy binding to trans-activation responsive element (TAR) and a ribozyme cleaving the mRNA of the CC chemokine receptor 5 (CCR5) receptor was found to mediate long-term inhibition of HIV-1 in haematopoietic progenitor cells.[24]

12.3 Generation of Transgenic Animals for Knockdown Studies

The development of shRNA expression cassettes was furthermore the basis for the generation of transgenic animals whose cells constitutively or conditionally transcribe shRNAs and silence a target gene. The shRNA-expression vector has successfully been introduced into mouse embryonic stem cells by electroporation.[25] Germline transmission of the shRNA-expression construct was confirmed in numerous F_1 progeny, in which the siRNA was detected and both, the target mRNA and protein level were reduced. Alternatively, transgenic animals were generated by transfer of the shRNA expression cassettes with a lentiviral vector (*vide infra*).[26] A great advantage of the RNAi technology is the possibility to obtain transgenic knockdown in animals other than mice, such as rats, for which knockout of a target gene by homologous recombination in embryonic stem cells is far from being a routine method.[27]

Most of the transgenic knockdown animals described to date were obtained by pronuclear injection and transfection of plasmid-based shRNA constructs or by lentiviral transduction. With these techniques, the transgene is randomly integrated into the genome, which may have a great impact on the efficiency of gene silencing. Depending on the site of transgene integration, the efficiency of gene silencing in those animals was variable because of the inherent problem of position-dependent effects on gene expression. Thus, a thorough characterization of the resulting mouse lines is required to determine the level of RNAi in all relevant organs, which is time-consuming and laborious. To avoid these complications, efforts were made to identify an autosomal locus that allows strong and predictable expression of shRNA transgenes inserted through homologous recombination. The Rosa26 locus is ubiquitously expressed during development and in adult mice, indicating an open chromatin structure, which enables polymerase III-dependent expression of shRNAs. Analysis of mice revealed that this single-copy shRNA configuration indeed mediates body-wide gene silencing with high efficiency, ranging between 80 and 95% knockdown in most organs.[28] Moreover, the insertion of shRNA vectors by recombinase-mediated cassette exchange provided an efficient method for the rapid generation of RNAi transgenic mice.

A technology for the accelerated production of loss-of-function phenotypes is particularly valuable for current research, since sequence analysis of the human and murine genomes has revealed numerous genes whose function is still unknown. Until recently, generation of knockout animals by homologous recombination in murine embryonic stem cells was the method of choice for deciphering gene function *in vivo*. This technology, however, is laborious, and it usually takes one to two years to obtain knockout animals with a homogenous genetic background. RNAi technology now allows generation of knockdown phenotypes much more quickly. Transgenic animals expressing shRNAs are ready for phenotypic analysis and behavioural studies after as little as three months. Two comparative studies reported the phenotypes of knockout and RNAi-knockdown animals to be largely comparable, but some significant differences were observed.[29,30] For example, siRNA-mediated knockdown of peroxisome proliferator activated receptor alpha (Ppara) resulted in animals that developed hypoglycaemia and hypertriglyceridaemia, phenotypes observed in Ppara knockout animals as well. In contrast to the knockout mice, however, fasting was not required to uncover these phenotypes in siRNA-treated animals.[29] The discrepancies are not yet fully understood, but may result from incomplete inhibition of target-gene expression in the transgenic shRNA animals and/or to compensatory mechanisms induced in the knockout animals.

Furthermore, the RNAi technology may be useful in producing transgenic livestock with improved resistance against diseases. It has, for example, been shown that the level of the prion protein (PrP) can be significantly reduced in a transgenic goat fetus and in bovine blastocysts expressing shRNAs against PrP.[31] This approach might help to build up livestock that is protected against transmissible spongiform encephalopathies and may be adapted to generate cattle resistant against the foot-and-mouth disease virus (FMDV) or other pathogens as well. The use of transgenic livestock, however, might not only face technical problems, but will also have to be well thought through with respect to its social and ethical consequences.

12.4   MicroRNA-type shRNAs

Elucidation of the cellular microRNA (miRNA) pathway allowed further improvements of RNAi technology as a research tool. miRNAs are small RNAs endogenously expressed in eukaryotic cells that form a complex network to regulate gene expression (for a review, see Mattick and Makunin[32]). Several hundred miRNAs have been identified and confirmed in the human genome to date. miRNA genes are usually transcribed by RNA polymerase II into long pri-miRNAs, which are cleaved by the RNase Drosha to form pre-miRNAs of approximately 70 nucleotides in length. Pre-miRNAs are then transported to the cytoplasm by binding to exportin-5. In the cytoplasm, Dicer processes the pre-miRNA into the mature ~22 nucleotide-long miRNA that becomes incorporated into RISC. miRNAs usually bind to the 3′-untranslated region of mRNAs in an imperfect manner and are considered to inhibit translation without destroying their target. It has, however, been shown that miRNAs can also induce cleavage of the target mRNA.[33] The extent to which the target

RNA becomes degraded probably depends on the degree of complementarity of the duplex between the miRNA strand and its target.

It has been shown that the natural miRNA sequence can be replaced by an artificial shRNA embedded in the miRNA environment to target a gene of interest.[34] With these constructs RNA polymerase II promoters can be employed to regulate transcription instead of RNA polymerase III promoters, which have been used to express conventional shRNAs. As a first step, RNA polymerase II generates the primary transcripts, which can contain a protein-encoding mRNA upstream of the miRNA sequence for the simultaneous expression of a cDNA (Figure 12.2). Drosha will then cleave the pre-miRNA out of the primary transcript, which will finally be processed by Dicer into small miRNA-type-siRNAs (misiRNAs) that assemble into RISC for gene silencing.

A direct comparison of a conventional shRNA against HIV-1 *tat* and an isosequential shRNA embedded in the backbone of the human miRNA-30 revealed that the latter construct was 80% more effective in reducing HIV-1 p24 antigen production.[35] The authors performed Northern blot analysis to demonstrate that the pre-miRNA construct resulted in a higher level of mature 21-mer siRNA molecules. The high potency of misiRNAs convinced the

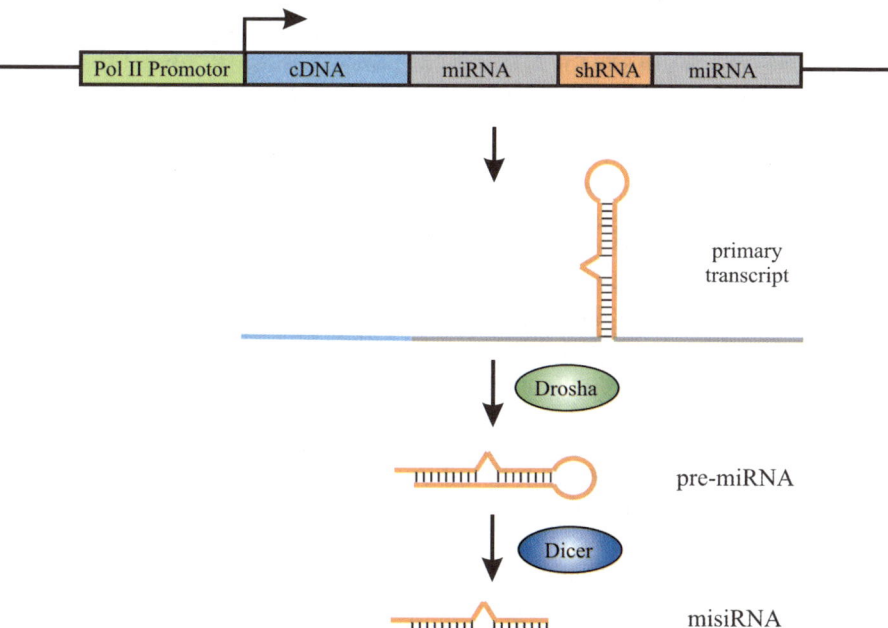

Figure 12.2 Expression of microRNA-type-siRNAs (misiRNAs). To express mis-iRNAs, polymerase II promoters can be employed that generate a transcript with the shRNA embedded in a natural miRNA environment. The transcribed RNA can furthermore encode a protein. Drosha processes the primary transcript into the ∼70 nucleotide-long pre-miRNA, which is exported from the nucleus by exportin-5. In the cytoplasm, Dicer cleaves the pre-miRNA into the mature misiRNA.

Hannon laboratory to construct a second-generation library covering the mouse and human genomes.[36] The library consists of 140 000 shRNAs in the backbone of miRNA-30 targeting approximately 31 000 human and 29 000 mouse genes.

In addition to the higher efficiency, the expression of miRNA-type siRNAs opens up the possibility of two new applications. First, polycistronic expression vectors can be constructed that express multiple misiRNAs from a single transcript.[37] These expression systems can be used either to silence multiple targets simultaneously or to direct several siRNAs against the same target to improve silencing.

Second, a new and not yet fully exploited option is the use of cell-type specific promoters. While RNA polymerase III promoters induce a strong and ubiquitous expression, RNA polymerase II promoters are usually only active in certain cell types. For example, a tissue-specific polymerase II promoter [the proximal promoter (Pp) of the mouse *Rhox5* gene] has been employed to drive the expression of an shRNA targeting the transcription factor Wilms tumour 1. In transgenic mice, the level of the targeted protein was specifically reduced in nurse cells in the testis.[38] Non-specific side effects observed after siRNA treatment mean that it will be advantageous to induce the expression of shRNA only in cells in which the target gene shall be knocked down. Furthermore, it will be interesting to investigate loss-of-function phenotypes in certain organs of the complete organism for research purposes. Given the advantages of the expression cassettes that generate miRNA-type siRNAs, it is likely that these systems will largely displace the conventional shRNA plasmids.

12.5   Inducible RNAi Systems

Pharmacological control of therapeutic gene expression permits the drug-dependent precise titration of gene-product dosage, intermittent or pulsatile treatment and defined termination of therapy.[39] For many biochemical and functional analyses of essential genes, conditional knockout approaches are necessary and conditional RNAi systems may thus be of fundamental impact for use in both experimental and therapeutic applications. For example, knockout of genes essential for embryonic development can lead to the early death of animals. In these cases it is impossible to study gene function at late time points of the development.[40] Inducible gene knockdown overcomes these limitations as it allows the induction of gene silencing at desired points in tissue and organ development. Moreover, reversibly regulated systems enable transient gene silencing and will therefore be useful to study gene function in 'knockout–knockin' models. In such models, target genes are transiently silenced by a knockdown trigger, leading to a loss of their function and the induction of phenotypic changes in cells, tissues or organs. Inhibition of the knockdown trigger would then lead to the restoration of normal gene function in the same target cells. In this way it may be possible to study whether reconstitution of a disrupted gene can lead to the reconstitution of cellular, tissue or organ function.

As conventional techniques employed for conditional models are time- and cost-intensive and often difficult to handle, several platforms based on RNAi technology are currently being developed to generate conditional knockdown models *in vitro* as well *in vivo*.[41,42]

In addition to research approaches, RNAi has a great potential for therapeutic application in the context of molecular and gene therapy. This therapeutic concept aims at correcting gene function to cure diseases by introducing therapeutic nucleic acids into the target cells. While in cancer therapy and antiviral approaches complete inhibition of the expression of tumour and viral genes is desirable, only a specific range or dose of the therapeutic protein will lead to a successful outcome in certain other disorders. This requires the regulation of gene-expression levels within a therapeutic window. Moreover, *in vivo* gene silencing by interfering RNAs can exert serious side effects under certain circumstances. For example, Grimm *et al.* reported such serious side effects following high-level expression of shRNA *in vivo*.[43] 36 out of 49 distinct adeno-associated virus (AAV)–shRNA vectors, unique in length and sequence and directed against six different targets, resulted in dose-dependent liver injury with 23 ultimately causing death. The authors found indications that the toxicity is caused by competition of the artificial shRNAs with the endogenous miRNA pathway. These results clearly demonstrate that for safety reasons regulation of shRNA expression in therapeutic settings is a desirable aim.

Beginning in 2003 several strategies for conditional gene silencing were developed. Two broad types of controllable systems have been established. The first class of systems enables *reversible* gene knockdown while the other class comprises *non-reversible* systems. In reversible systems, the expression of shRNA can be changed from an 'ON' status, in which the shRNA is expressed and the target gene is silenced, to the 'OFF' status, in which the shRNA expression is repressed and the target-gene expression remains unaffected. Of utmost importance in these systems is that the ON–OFF switch is reversible and can be modulated by addition or withdrawal of an inducer. In contrast, the non-reversible systems only enable the induction once, accompanied by disruption or induction of shRNA expression.

The basis of inducible RNAi systems available to date is a wide panel of reversible and non-reversible gene-expression systems that have been developed during the past 15 years. In most reversible systems a drug-dependent regulator protein binds to an inducible response promoter driving transgene expression. Depending on whether the regulator protein is a transactivator or a trans-suppressor, transcription of the transgene is induced or suppressed, respectively.[44] The tetracycline-controlled transcription (Tet) system is currently the most widely used *reversible* gene expression system applied in basic research.[45,46] It combines various desirable properties, including highly inducible expression levels, low background activity, fast kinetics after addition of the well-defined, non-toxic inducer drugs tetracycline (tet) and doxycycline (dox) as well as rapid reversal of induction after its withdrawal. Some additional *reversible* approaches, such as the rapamycin,[39] ecdysone[47] or antiprogestin[48] system, have been employed for *in vitro* and *in vivo* investigations as well. Site-specific recombination of the Cre-lox and FLP-*frt* systems are the most popular approaches for the *non-reversible* induction of gene expression. These systems make use of the ability of the recombinases Cre or FLP to excise a DNA fragment between two lox and frt sites, respectively. These techniques have been developed as indispensable tools for the precise *in vivo* manipulation of the mouse genome, enabling control, in space and time, of the onset of gene knockouts in almost any tissue of the mouse. These features greatly facilitate the creation of sophisticated animal models for human diseases and drug development.[49] Moreover, fusion of Cre recombinase to a mutated ligand-binding domain of the human or mouse oestrogen receptor results in drug-dependent inducible knockout systems.[50,51]

Conditional gene-expression systems were initially developed to deliver cDNA-transgenes and antisense RNAs. In these configurations, however, they were inapplicable for shRNA expression. One of the most important obstacles to applying the systems to RNAi approaches is that shRNAs are usually expressed from polymerase III promoters, while cDNA-transgenes and antisense RNAs are expressed from polymerase II promoters. As outlined above, the RNA polymerase III-type H1 and U6 promoters are widely employed to express shRNAs. These promoters consist of three essential sequence components, the TATA box, a core promoter (PSE) and the distal promoter element (DSE). It has been shown that all three elements are required for the cooperative binding of proteins and the DSE is critical for the stability of the DNA–protein complex and the activation of both proximal elements.[52] However, sequences between the essential elements and between the TATA box and the transcription start site are of minor importance, whereas the distance between the promoter elements seems to be critical for their function. Based on these observations, several modifications of the polymerase III promoters concerning both the essential regulatory promoter elements and the sequences between them were carried out to achieve an inducible regulation of shRNA expression.[53–60] The second important component of systems for conditional gene expression are the regulator proteins. Many of these consist of a drug-dependent DNA-binding domain fused either to a transactivator or to a transrepressor domain. The replacement of a transactivator domain specific for polymerase II promoters by the corresponding one for engineered polymerase III promoters allowed the inducible expression of shRNAs with type III RNA polymerases.[54,61] The third component of an inducible shRNA-expression system is the shRNA itself. Modifications in its loop structure by the insertion of loxP site flanked stuffer (spacer) sequences[62] or its embedding into miRNA sequences[63–66] enable their conditional transcription.

One of the first and still most extensively employed methods for the regulation of shRNA expression is the Tet system.[67] A strong advantage of this system is that all the required components can be expressed from one vector genome.[68–70] Different variants of the Tet system have been developed for the regulation of shRNA expression from polymerase III promoters. They are based on epigenetic suppression, steric hindrance and the use of new transactivators (Figures 12.3A–C).

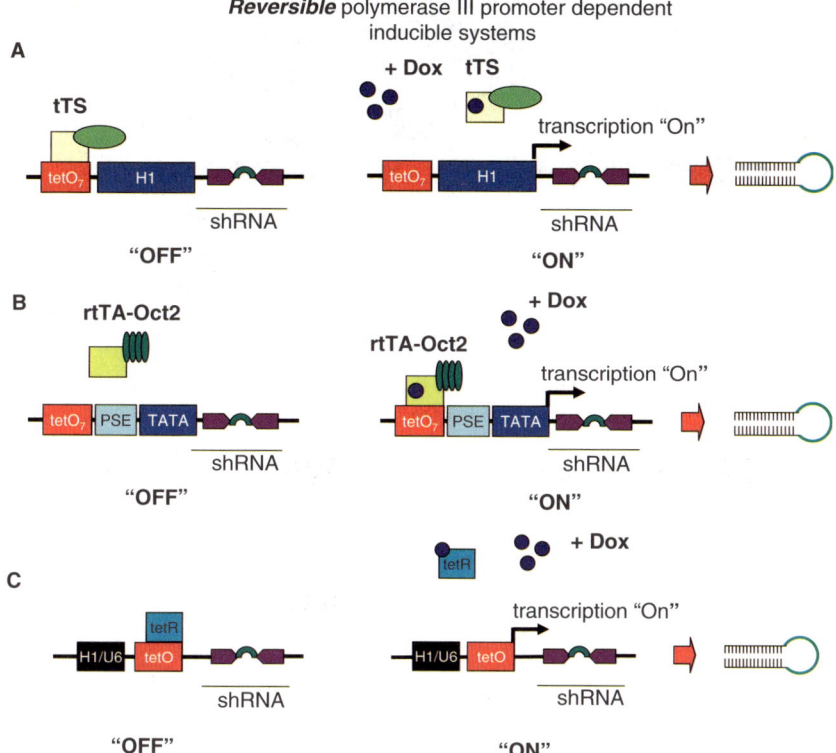

Figure 12.3 Reversible and non-reversible inducible systems for shRNA-mediated gene silencing. (A) Dox-dependent epigenetic suppression of polymerase III promoter. In the absence of Dox, a Dox-controlled transcriptional silencer (tTS) binds to a *tetO$_7$* located upstream of the polymerase III promoter, resulting in suppression of shRNA transcription. In the presence of Dox, tTS is released and generation of shRNA is switched on. (B) Dox-dependent transactivator system. A Dox-controlled transcriptional transactivator (rtTA-Oct2) binds to a *tetO$_7$* in the presence of Dox and transactivates a minimal polymerase III promoter, leading to shRNA transcription. In the absence of Dox, rtTA-Oct2 is unable to bind to *tetO$_7$* and the shRNA is not transcribed. (C) Dox-dependent steric hindrance of polymerase III promoters. The tetR binds to a *tetO* in the absence of Dox. The *tetO* is located downstream of the polymerase III promoter and binding of tetR prevents the expression of the shRNA. In the presence of Dox, the tetR does not interact with the *tetO* and the shRNA is expressed. (D) Dox-dependent system based on a polymerase II promoter. The rtTA binds to the *tetO$_7$* in the presence of Dox and transactivates a minimal CMV promoter, leading to the expression of both a marker GFP gene and a misiRNA from the same transcript. In the absence of Dox, the rtTA is unable to bind to the *tetO$_7$* and expression of GFP and the misiRNA is prevented. (E) Non-reversible polymerase III promoter-dependent inducible system. The shRNA sense and antisense sequences are separated by a stuffer sequence, which is flanked by two loxP sites, resulting in the transcription of a non-functional shRNA transcript. In the presence of Cre recombinase the stuffer sequence is excised and the residual single loxP site forms the loop of a functional shRNA.

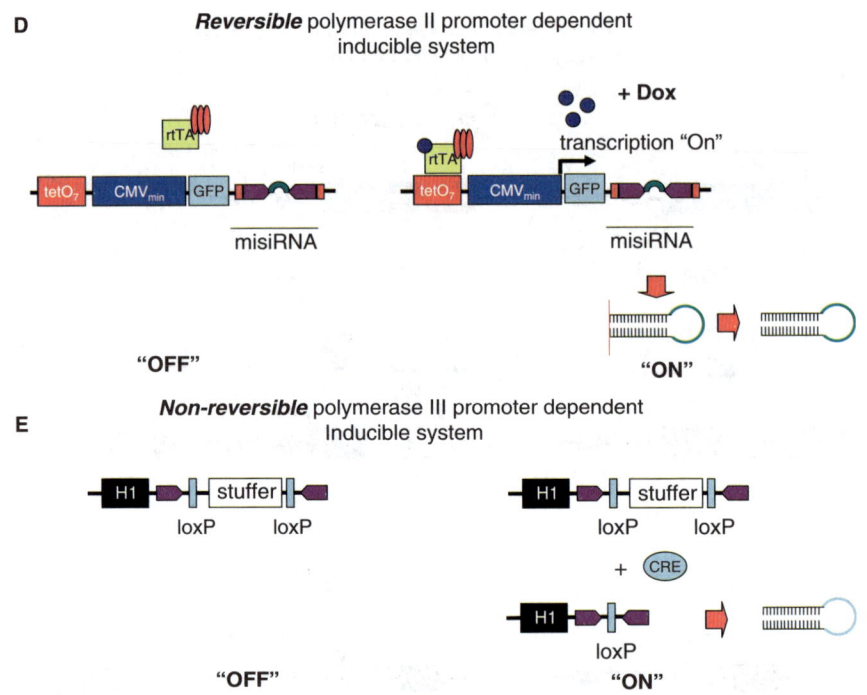

Figure 12.3 (Continued)

  The epigenetic suppression system mediates suppression of the H1, U6 and/ or tRNA^{val} promoter activity by the use of a Dox-controlled transcriptional silencer (tTS; Figure 12.3A). A Krüppel-associated box (KRAB) domain found in many zinc-finger proteins is the key structure of this silencer, as it can suppress both polymerase II and polymerase III promoters in an orientation-independent manner within up to three kilobases from its binding site.[71,72] In tTS (also referred to as tTR-KRAB) the KRAB domain is fused to a tetR DNA-binding domain. In the absence of Dox the tTS binds to the *tetO₇* sequence, which is inserted upstream of a polymerase III promoter, resulting in the blockade of shRNA synthesis. In the presence of Dox, tTS is released from the *tetO₇* sequence, leading to the expression of shRNA.[55,73–76] A reverse set-up is based on the fusion of the KRAB domain to a mutated tetR DNA-binding domain (rtTR), which binds to the *tetO₇* sequence in the presence of Dox.[75] Szulc *et al.*[75] directly compared the tTS dependent Tet-ON system and rtTR-KRAB dependent Tet-OFF system using a single lentiviral vector system. Both variants exhibited tight and efficient control of shRNA expression and enabled time- and dose-dependent as well as reversible knockdown of marker gene expression *in vitro*. Furthermore, Dox-controlled RNAi could be demonstrated in embryonic and CD34 + haematopoietic stem and progenitor cells, as well as for the efficient and reversible regulation of TP53 expression in human tumours xenotransplanted into nude mice. In contrast to these results,

Chen *et al.*[55] did not achieve shRNA-mediated gene knockdown by use of the tTS Tet system. The major difference between both approaches was the use of lentiviral and plasmid vector constructs, respectively. Therefore, it may be assumed that the fidelity of the rtTA-KRAB and tTS-controlled Tet system strongly depends on stable integration of the *tetO*-linked transcription unit into the genomic DNA. This assumption is supported by the finding that the primary mechanism by which the KRAB domain suppresses promoter activity is triggering of heterochromatin formation.[74] Others have shown, however, that tTS is capable of suppressing *tetO*-linked promoters by using adenoviral vectors, which do not integrate into the genomic DNA.[77] Therefore, size, number and/or configuration of the vector genome may be an additional critical component for the silencing activity of tTS.

Another attempt to control shRNA expression comprises Dox-dependent transactivator systems (Figure 12.3B). In this system, a Dox-dependent trans-activator (rtTA) binds to a *tetO_7* motif linked to a minimal polymerase III promoter unit in the presence of Dox and induces shRNA expression. In the absence of Dox, the transactivator is unable to bind and thus does not induce transcription from the promoter unit. As the available rtTAs were incapable of transactivating polymerase III promoters, Amar *et al.*[54] constructed a new Dox-dependent transactivator, rtTA-Oct2^Q, containing a polymerase III specific transactivation domain and inserted it, together with the response elements, into one lentiviral vector genome. With this approach RNAi could be induced in a dose- and time-dependent manner by the administration of Dox, resulting in downregulation of target gene expression by 90%. Moreover, silencing was reversible after withdrawal of Dox. However, the concentration of Dox required for the full induction of shRNA expression was approximately 10-fold higher than that needed in a similar polymerase II promoter transac-tivator system. Although the application of such a high Dox doses may be acceptable for *in vitro* investigations, it will significantly reduce the suitability of this system for *in vivo* application, as high Dox doses may induce side effects.

Another approach for inducible RNAi-mediated gene silencing is based on the steric hindrance of polymerase III promoter based shRNA expression (Figure 12.3C). To this end a *tetO* sequence is inserted upstream of the tran-scription start site to interfere with the function of the TATA box through recruitment of the tetracycline repressor (tetR).[56–59,71,76] In the absence of Dox, tetR binds to the *tetO* and inhibits production of shRNA, whereas in the presence of Dox the repression is relieved by sequestration of the tetR.[56–59,78] In one study, T-REx^{TM} Henrietta Lacks (HeLa) cells expressing TetR were stably transfected with an shRNA expression cassette consisting of an H1-tetO pro-moter for the generation of shRNAs directed against human polo-like kinase 1 (Plk1), which is overexpressed in a broad spectrum of human tumours. While expression of Plk1 was unaffected in the absence of Dox, exposure to Dox led to the repression of Plk1 mRNA to 3% of normal levels and Plk1 protein to 14%. As a result of Plk1 depletion, cell proliferation strongly decreased. These *in vitro* results were confirmed *in vivo* in a xenotransplantation mouse model. After administration of Dox, Plk1 expression was downregulated and tumour

growth significant inhibited.[56] Recently, site-directed integration of the transgene into the mouse genome by recombinase-mediated cassette exchange at the
rosa26 locus was employed to generate transgenic mice, allowing temporal
control of shRNA expression upon induction by Dox.[79] The feature to reversibly switch genes on and off by inducible RNAi approaches is an extra
advantage of this technique compared to conventional knockout applications.
Further studies evaluated the applicability of adenovirus vectors (AdVs) for
tetR dependent gene silencing.[53,58] In one of these studies, one AdV expressed
tetR while a second AdV contained the H1-*tetO*-promoter for the expression
of an shRNA against p300. This study demonstrated Dox-dependent reduction
of p300 mRNA and protein levels in the presence of both vectors and Dox, but
no effect in the absence of the inducer drug.[58] However, the steric hindrance
system has not yet been analysed in detail in one vector transient transduction
system.

As a result of the difficulties of developing new polymerase III dependent
inducible RNAi systems, several groups have focussed on the development of
a polymerase II promoter dependent inducible shRNA expression system
(Figure 12.3D). A major advantage of this strategy is that inducible systems
developed for cDNA-transgene expression can be easily adapted for shRNA
expression, as no modifications in the regulator proteins or response promoters
are necessary. The key step in this development was the generation of shRNAs
designed as primary miRNA transcripts, allowing their expression from polymerase II promoters (see above). To render this system inducible, Stegmeier
et al.[65] constructed lentiviral vectors delivering a Dox-responsive cytomegalovirus (CMV) promoter that controls the generation of a misiRNA transcript
which consists of an artificial shRNA embedded in the environment of the
naturally occurring miR30. The vector was transferred into cell lines, expressing either stable Dox-dependent transactivator or suppressor proteins. In this
way, three variants of the Tet system for inducible polymerase II dependent
RNAi could be compared: the TRex, the Tet-ON and the Tet-OFF system. All
three variants showed tight Dox-dependent regulation of misiRNA expression
and conditional silencing of the targeted RB gene. Moreover, the systems
showed a single-copy proviral integrant provided efficient knockdown in a
polyclonal cell population.[65] Meanwhile, other groups have developed non-viral
and lentiviral vector systems to deliver the Dox-dependent regulatory and response elements in a single vector genome.[64,66] Both vector types enable tight
regulation of misiRNA expression and gene silencing. Moreover, in these inducible systems, a marker gene can be co-expressed with the misiRNA, which
allows direct monitoring of misiRNA expression status. The major advantage of
these vectors, however, is the possibility of inducing gene silencing in a one-step
reaction. This feature will reduce the cost and time required, and guarantees that
regulator proteins and response units enter the same cell, thereby increasing the
efficacy of gene silencing and reducing the risk of side effects.

The Tet system, however, is not the only reversible system that has been
employed to regulate shRNA expression for RNAi approaches. Another
promising example is the ecdysone system, which consists of a nuclear receptor

and a transcription factor that induce the expression of a chimeric transactivator in the presence of the inducer, muristerone A. The transactivator then initiates transcription of the shRNA. Using this system, Gupta *et al.*[61] demonstrated stringent dosage- and time-dependent kinetics of induction with undetectable background expression in the absence of the inducer. Moreover, inducible suppression of human p53 in glioblastoma cells was demonstrated and it was shown that induction is reversible after withdrawal of the inducer, as observed by reappearance of the protein and restoration of the original cell phenotype. A major drawback of the ecdysone system in its present form is the need to apply three independent vectors to induce shRNA expression. While this may still be useful for *in vitro* investigations, where cells can easily be co-transfected or co-transduced with the respective plasmids or viral vectors, it limits its usage *in vivo*.

Since it is widely used for regulated creation of conventional knockouts, the Cre-lox system (Figure 12.3E) has been extensively evaluated as a *non-reversible* approach for conditional RNAi.[42,60,62,80,81] The basic principle of the Cre-lox system is the inactivation of shRNA expression through the separation of interacting sequences by DNA stuffer sequences. Since the stuffer sequences are flanked by loxP sites, they can be eliminated by Cre recombinase, which leads to reconstitution of functional shRNA expression cassettes. As an example, Fritsch *et al.*[62] positioned a neomycin cassette that was flanked by two loxP sites between the sense and the antisense strand of an shRNA. In the absence of Cre, the transcription of the shRNA was prevented by a stretch of four T's in the neomycin cassette, which terminates polymerase III transcription, whereas in the presence of Cre, recombination eliminates the neomycin cassette, allowing read-through of the entire shRNA. In this system, the remaining intercalated lox site represents an essential component of the shRNA, as it forms the loop structure of the shRNA. This Cre-lox strategy allowed repression of gene activity in a time-dependent manner in cell culture, and regulated gene silencing in a time- and tissue-dependent manner in animals. Moreover, the system proved to be non-leaky, *i.e.* no silencing was observed prior to the intended activation. The silencing efficiency of the Cre-lox shRNA system, however, is lower than that of a constitutively expressed shRNA, most likely because of incomplete CRE-induced recombination. Others have designed a Cre-lox system carrying a mouse U6 promoter that was separated from the shRNA by stuffers flanked by modified loxP sites that contained a functional TATA box in its internal stuffer region.[42,80] In a third approach the lox-stuffer-lox fragment was positioned downstream of a H1 promoter.[81] This system was able to achieve conditional and tissue-specific RNAi in Cre-expressing transgenic mice.[80] Both tamoxifen and Dox-inducible Cre expression have been shown to be suitable to generate drug-inducible shRNA transcription based on the Cre-lox system.[60,82] Alternatively to the Cre-lox system, the FLP recombinase-mediated DNA recombination has also been used to generate inducible RNAi in embryonic stem cells.[83]

In summary, a wide panel of *reversible* and *non-reversible* conditional shRNA expression systems have been developed and their feasibility for *in vitro* and *in vivo* application has been convincingly demonstrated. Their specificity, efficiency and relative simplicity mean that these systems may have the potential to replace other techniques traditionally used to generate conditional knockouts in basic research on a large scale. Conditional RNAi may, furthermore, reduce the risk of deleterious side effects following shRNA expression at high level.

12.6 Viral Vectors for Delivery of shRNAs

In recent years, post-transcriptional gene silencing by RNAi has quickly emerged as a powerful tool for genetic analysis in mammalian cells. The long-term success of these approaches, however, will largely depend on their ability to achieve efficient gene silencing *in vivo* and on the applicability of the technology for the treatment of human diseases. Currently, the transfection efficiency of siRNAs for *in vivo* applications is rather low and preferentially leads to enrichment of the siRNAs in the liver, while other organs and tissues are rather refractory to taking up siRNAs.[84] The rapid high-volume injection procedure for systemic siRNA application initially used in mice cannot be adapted for humans.[84,85] Therefore, further improvement of the transfection efficiency constitutes one of the most critical challenges to the use of siRNAs *in vivo*. Treatment of cancer by the knockdown of oncogenes, for example, would require that each malign cell within a tumour is transfected with an siRNA, which is not readily achievable with the currently available siRNA transfection technologies. Similar obstacles prevent the use of plasmids or any other form of non-viral shRNA delivery vectors *in vivo*.

It is thus obvious that the translation of the RNAi approach from *in vitro* to *in vivo* will require tools for efficient delivery or expression of siRNAs. Recent progress made in the development of liposomal agents for the *in vivo* delivery of siRNAs are summarized in Chapters 10 and 11. The present overview focuses on the development of virus-derived shRNA expression vectors that are considered to provide significantly higher efficiency for systemic shRNA delivery than their non-viral counterparts.[86] One of the major prerequisites for the successful applications of this strategy is the choice of an appropriate vector type that will allow efficient delivery of the genetic material to the target tissue without causing deleterious side effects. Lentiviral, adeno-associated virus (AAV) and AdV are currently the most commonly used viral vector types for shRNAs delivery. Each vector type, however, has its specific advantages and limitations as summarized in Table 12.1. Therefore, an 'ideal' vector type for all approaches in research and gene therapy does not exist to date and is unlikely to be developed in the near future. The choice of an appropriate vector depends, rather, on the specific requirements of a given application.

12.6.1 Retrovirus Vectors

The family *Retroviridae* comprises a large group of viruses that infect primarily vertebrates, although infection of other animals has been described in a few

Table 12.1 Main features of viral vectors for the delivery of short hairpin RNAs.

	Retrovirus vectors		Adenovirus vectors			Adeno-associated virus vectors	
	Onco-retrovirus	*Lentivirus*	*First/second generation*	*Guless*	*Oncolytic*	*Monomeric*	*Self complementary*
Target cells	proliferating cells	quiescent and proliferating cells	quiescent and proliferating cells	quiescent and proliferating cells	tumor cells	quiescent and proliferating cells	quiescent and proliferating cells
Immune response	very low	very low	strong	low	strong	low	low
Integrate into the host cell chromosome	yes	yes	no	no	no	rare	rare
Onset of transgene expression	rapid	rapid	rapid	rapid	rapid	slow	rapid
Replication	no	no	no	no	yes	no	no
Transgene expression levels	high	high	very high	very high	very high	high	high
Duration of transgene expression	long term	long term	short/prolong	long term	short	long term	long term
Transgene packaging capacity	$\approx 7\,kb$	$\approx 7\,kb$	$\approx 3\text{--}9\,kb$	up to $\approx 35\,kb$	$< 4\,kb$	$\approx 4\,kb$	$\approx 2\,kb$
Virus titers to produce vector particles/ml	$\approx 10^9$	$\approx 10^9$	$10^{12}\text{--}10^{13}$	$10^{12}\text{--}10^{13}$	$10^{12}\text{--}10^{13}$	$10^{12}\text{--}10^{13}$	$10^{12}\text{--}10^{13}$
Potential risk	Insertional mutagenesis	Insertional mutagenesis	cytotoxicity	cytotoxicity	cytotoxicity	cytotoxicity	cytotoxicity

cases. A common feature of these viruses is that replication involves the process of conversion of the viral RNA genome into double-stranded DNA (dsDNA), which stably integrates into the genome of the target cells as proviral DNA. *Oncovirinae* are a sub-group of *Retroviridae* consisting of oncogenic retroviruses and closely related non-oncogenic viruses. Viruses of a second subgroup, the *lentivirinae*, are associated with slow, progressive diseases affecting the immune system. A prominent representative of this class of viruses is the human immunodeficiency virus (HIV). *Oncovirinae* and *lentivirinae* are the origin of the development of onco-retroviral and lentiviral vectors, respectively, for the delivery of transgenes.[87] Both vector types share several features. Most importantly, they integrate into the host cell genome, thereby transmitting their genome from initially transduced cells to the following generations of cells. As the immune response to the vectors is very low, this feature enables long-term transgene expression. For gene therapeutic applications, onco-retroviral and lentiviral vectors are preferred for transduction cells that divide rapidly and have only a short life span, *e.g.* haematopoietic stem cells.

The supposed advantage of retroviral vectors, *i.e.* the long-term expression of the transgene through stable integration of their proviral DNA into the host genome, however, bears the serious risk of insertional mutagenesis. In fact, in a gene therapy trial to treat X-linked severe combined immunodeficiency (SCID), the development of leukaemia was reported in three children who received this treatment.[88] This study was carried out with an onco-retrovirus to transfer the cDNA of the gamma chain of the cytokine receptor into CD34 + haematopoietic stem cells *ex vivo*. The onco-retrovirus vector was found to be integrated in the proximity of the LMO2 proto-oncogene promoter, which leads to aberrant transcription and expression of LMO2.

Lentiviruses – unlike onco-retroviruses – tend to integrate distally from promoters in introns, potentially limiting their overall oncogenicity.[89] Moreover, they are capable of transducing dividing as well as nondividing cells, thereby broadening the spectrum of potential applications.[90] Lentiviral vectors derived from HIV-1, HIV-2, feline immunodeficiency virus (FIV), simian immunodeficiency virus (SIV) and the equine infectious anaemia virus (EIAV) have been shown to cross-package each other.[91] Furthermore, enveloped pseudotypes can be produced with the G glycoprotein of the vesicular stomatitis virus (VSV-G).[92] These options to combine envelope proteins mean there are virtually no limits with respect to the host range of retroviral vectors.

In 2002, the first investigations were carried out to explore the potential of retroviral and lentiviral vector-mediated RNAi.[93,94] For example, Abbas-Terki *et al.*[94] constructed a lentivirus vector for delivery of shRNAs directed against the enhanced green fluorescent protein (EGFP). Using this vector, expression of the EGFP marker gene could be successfully inhibited. The effect was dose-dependent and silencing persisted for at least 25 days post-infection. Based on these initial reports, onco-retroviral and lentiviral vectors have been intensively employed in basic research, as well as in the development of gene therapeutic applications.

Intensive efforts have been made to develop RNAi as a new approach to treat viral diseases. As an example, vectors expressing shRNAs have successfully been employed to contain the spread of HIV-1 or hepatitis C virus (HCV) by inhibiting viral gene expression or by silencing of the surface receptors on the host cells.[23,24,95–97] As RNA viruses tend to develop escape mutants following RNAi treatment (see above), more recently lentiviral vectors have been developed that express multiple shRNAs directed against both virus RNA and virus receptors. In an attempt to inhibit HCV, Henry *et al.*[95] found that double and triple lentiviral shRNA vectors were independently effective in simultaneously reducing HCV replication, CD81 receptor expression, and HCV surface protein E2 binding, resulting in an increased potency of the antiviral effect. Likewise, multiple shRNAs targeting HIV-1[23] and a lentivirus vector delivering an shRNA against the HIV-1 RNA, a decoy oligonucleotide binding to TAR, and a ribozyme intended to cleave the mRNA of the CCR5 receptor[24] were found to prevent the spread of HIV-1. Both approaches are currently being developed for *ex vivo* therapy of human patients with HIV-1 infections as outlined in Chapters 13 and 14.

In further studies delivery of shRNA expression cassettes by lentiviral vectors has been employed for the generation of transgenic mice. For example, Pfeifer *et al.* constructed a lentiviral vector expressing shRNAs that target the mRNA of the cellular prion protein PrP(C).[98] *In vitro* investigations with this vector demonstrated efficient and stable suppression of the accumulation of the pathogenic prion protein PrP(Sc). Moreover, chimeric mice derived from lentiviral vector-transduced embryonic stem cells showed significantly extended survival after scrapie infection. In this context it is notable that lentiviral vectors have been developed which enable tight and efficient regulation of shRNA and misiRNA expression (for details, see Section 12.5).

12.6.2 Adenovirus Vectors

The adenovirus is a non-enveloped, icosahedral virus of 60–90 nm in diameter with a linear, dsDNA genome of 30–40 kb. Human adenoviruses cause acute respiratory infections, pharyngitis and conjunctivitis; some types are responsible for epidemic keratoconjunctivitis. Adenoviruses have also been associated with gastroenteritis and pneumonia in young children as well as with myocarditis. More than 50 different human serotypes have been identified to date.

The concept of deriving gene-delivery vectors from adenovirus (usually serotypes 2 or 5) exploits the broad tissue tropism and the low pathogenicity of this virus.[99] Major advantages of these vectors include the easy production of high titres of up to 10^{12} to 10^{13} vector particles per millilitre, their ability to transduce quiescent as well as proliferating cells and a very fast and strong transgene expression after transduction of the target cell. Several types of AdVs have been developed. First-generation AdVs are characterized by the substitution of the E1 region of the vector genome by the transgene expression cassette. As the E1A gene is essential for adenovirus replication, the resulting vector lacking the E1 region is replication deficient. These vectors, however, induce a cytotoxic T lymphocyte (CTL)-mediated immune response, which leads to rapid clearance of vector-transduced cells and thus restricts transgene expression to a short duration of only a few weeks.[100] In second-generation AdVs, the E2 and/or E4 regions are deleted as well. In some studies improved persistence of the transgene as well as a reduced inflammatory response could be shown.

In an attempt to make the vectors safer, highly attenuated, so called 'gutless', AdVs have been developed. In these vectors, all adenoviral coding sequences have been deleted and only the 5′- and 3′-inverted terminal repeats (ITRs) and the packaging signal (Ψ) have been kept from the wild-type adenovirus. This vector type allows insertion of up to 36 kb of foreign DNA. Moreover, the lack of viral gene expression from these vectors has been shown to reduce considerably their toxicity and immunogenicity *in vivo*. As a consequence, expression of the transgene was found to persist for up to one year.[101,102]

Oncolytic adenoviruses (oAdV) are a special class of AdVs designed to replicate selectively in tumour cells. This specificity is achieved either by the use of tumour-specific promoters driving the expression of the adenoviral E1A, E1B or E4 genes[103–106] or by mutation of the E1A and/or E1B genes in such a way that their ability to interfere with the expression of cell cycle and pro-apoptotic genes is abolished.[107,108] Their ability to destroy tumour cells makes them a promising new tool in cancer therapy.

AdVs belong to the viral vectors with the fastest and strongest transgene expression. This feature makes them one of the most promising vector systems for the treatment of infectious diseases by means of RNAi. As an example, Sanchez *et al.*[109] developed a first-generation AdV for the expression of shRNAs targeting sequences within the viral L polymerase and Z mRNAs of the lymphocytic choriomeningitis virus (LCMV). Application of this vector resulted in efficient silencing of the target genes and in a strong suppression of the production of infectious LCMV. Moreover, application of the recombinant AdV system effectively cured persistently LCMV-infected cells. In a further study, the antiviral potency of shRNA-expressing AdVs was demonstrated *in vivo* in a model for FMDV infections in guinea pigs and in a clinically more relevant setting in pigs.[110] Three out of five guinea pigs inoculated with an AdV that expressed an shRNA against the FMDV polymerase and challenged with homologous FMDV virus 24 hours later were protected from the major clinical manifestations of the disease. Similar results were obtained with the treatment of pigs. Two out of three pigs that were inoculated with a mixture of AdVs expressing shRNAs either against the FMDV polymerase or against the structural protein D1 and challenged with FMDV 24 hours later were protected from the major clinical symptoms of the disease. These results demonstrate the potential for the application of AdV-mediated shRNA expression to treat infectious diseases.

Other studies investigated the potential of AdV-expressed shRNAs in heart[111–113] and cancer diseases.[114–117] Zhang *et al.*,[116] for example, observed additive effects when inhibiting tumour growth through shRNA-mediated

K-ras knockdown and cancer cell lysis by the oncolytic AdV ONYX-411 *in vitro* as well as *in vivo*. Daily intratumoural injections of the vector significantly reduced the growth of subcutaneous pancreatic cancer xenografts in nude mice by 85.5%, including complete suppression of tumour growth in three out of five mice. The parental oncolytic virus ONYX-411 or an ONYX-411-variant expressing an shRNA against GFP were markedly less effective (47.8% and 44.1% reduction of tumour growth, respectively).

12.6.3 Vectors Based on Adeno-associated Viruses

AAVs belong to the family of *Parvoviridae*, which comprises viruses with non-enveloped, icosahedral capsids with a diameter of about 20 nm containing a linear single strand DNA molecule. The viral genome is relatively small (about 4800 base pairs) and simple in its organization, consisting of only two genes (*rep* and *cap*) that encode four non-structural (Rep) and three structural (VP) proteins. The two genes are embedded between two ITRs.[118] A number of AAV serotypes have been discovered in recent years. While AAV serotypes 1 to 8 were isolated between the late 1960s and 2002, Gao *et al.*[119] detected and isolated numerous new AAV variants, 55 of which were found in human tissues and 53 of which were isolated from non-human primates.

AAV has emerged as an attractive vector for gene therapy because of its broad tissue tropism, efficient transduction and lack of human pathogeneity.[86] AAV vectors consist of a transgene expression cassette flanked by the 5'- and 3'-ITR retained from wild-type AAV. As the *rep* and *cap* genes are lacking, AAV vectors are unable to replicate. The originally developed AAV vector prototypes are derived from AAV-2. These vectors have low transduction efficiencies *in vivo*,[120] but this problem can be overcome, since vector genomes with intact AAV-2 ITRs can be easily packaged into a capsid of another AAV serotype. The generation of these pseudotype AAV vectors has several advantages, as it is well-known that AAV serotypes strongly differ in their tissue tropism and transduction efficiency. The AAV-1 serotype, for example, is highly suitable for muscle-directed gene transfer, while the AAV-9 serotype efficiently transduces the heart.[121–123] The AAV-8 serotype appears to preferentially target the liver. It has been reported that hepatocytes transduced with AAV-8 *in vivo* were found to carry more than 10 vector genome copies per cell.[124] Moreover, as patients may develop antibodies directed against one AAV serotype, packaging of the vector genome into the capsid of another AAV serotype easily enables re-administration of AAV vectors.[125]

Another advantage of AAV vectors is their long-term stability *in vivo*. Detection of transgene expression for up to six years has been reported.[126] This feature makes AAV vectors ideal for the treatment of disorders which require long-term transgene expression, such as haemophilia B[123,125] or chronic neurological[127] and heart muscle diseases.[128] Long-term activity of AAV vectors is somewhat surprising, since the majority of the vectors remain episomal.[129] AAV vectors thus reduce the risk of insertional mutagenesis as compared to retroviral vectors.

A drawback of conventional single-stranded AAV vectors, however, is that maximal transgene expression is only achieved several weeks after vector administration. They are therefore not suitable for short-term *in vitro* and *in vivo* applications. Self-complementary (sc) double-stranded AAV vectors represent a new AAV vector type which overcome this disadvantage, as they reach maximal transgene expression within several days.[130,131] A shortcoming of scAAV vectors is their low packaging capacity of only ~2.2 kb of foreign DNA. While this may limit their use for the expression of many transgenes, it seems to be less problematic in the context of shRNA expression cassettes, which are in general only around 0.4 kb in size.

A number of studies have already been performed to explore the delivery of shRNAs by AAV vectors. Tomar *et al.*[132] carried out one of the first successful approaches to knockdown gene expression by the AAV vector-mediated delivery of shRNAs. The authors generated AAV-2 vector plasmids expressing shRNAs directed against p53 and caspase-8, respectively, under control of either the H1 or the U6 promoter. They used the vector particles to infect cultured HeLa cells and demonstrated efficient and dose-dependent knockdown of p53 and caspase-8 proteins by Western blot analyses. In the same year, Hommel *et al.*[133] reported the first successful RNAi-mediated gene silencing *in vivo* using AVV vectors as a delivery vehicle. For their study, they constructed an AAV vector expressing an shRNA to specifically knock-down *Th*. This gene encodes tyrosine hydroxylase, an enzyme required for dopamine synthesis in midbrain neurons of adult mice. The localized gene knockdown resulted in behavioural changes, including a motor-performance deficit, and reduced responses to a psycho stimulant. Meanwhile several additional studies have confirmed the usefulness of shRNA expressing AAV vectors for *in vivo* gene silencing.[134-137] These successful attempts underline that AAV vectors represent a promising tool for the use of RNAi technology in the context of experimental as well as potential gene therapeutic applications in humans.

The disadvantages of single-stranded AAV use in cell culture experiments mean that several groups have used scAAVs to deliver shRNAs.[138,139] In one example, Han *et al.*[138] inserted an expression cassette to generate an shRNA directed against the E1α subunit gene (PDHA1) of the pyruvate dehydrogenase complex into an scAAV genome. Transduction of rat lung fibroblast cultures with this scAAV vector resulted in up to 80% downregulation of PDHA1 gene expression 72 hours after transduction and in up to 70% downregulation of the target gene 10 days after transduction. So the authors could clearly demonstrate the success of scAAV-mediated shRNA delivery in short-term cell culture experiments.

12.7 Summary

RNAi has rapidly become a standard technology to study the relevance of genes by producing loss-of-function phenotypes. Specific inhibition of gene expression in eukaryotic cells can be achieved by the introduction of siRNAs which are complementary to the respective mRNA. Chemically synthesized siRNAs, however, silence the target gene for a limited period of only a few days.

To overcome this limitation, plasmid-based approaches have been developed. The most widely used systems make use of expression cassettes that generate shRNAs under the control of RNA polymerase III promoters. After being exported to the cytoplasm, the shRNAs are processed to give siRNA by the RNase Dicer. These systems allow prolonged gene silencing, particularly when stably transfected cell lines are selected. Furthermore, they open the possibility of generating transgenic animals for knockdown studies *in vivo*.

Another important feature of vector-based RNAi approaches is the opportunity to create inducible systems. While shRNA expression is usually initiated from RNA polymerase III promoters, tissue-specific RNA polymerase II promoters can be employed for the transcription of shRNAs embedded in naturally occurring miRNA environments.

Finally, viral vectors for the efficient transfer of shRNA expression cassettes have been developed. Major classes are retroviral and adenoviral vectors as well as vectors derived from adeno-associated viruses. Since each vector type has its specific advantages and drawbacks, the appropriate vector has to be chosen based on the requirements of a particular project. Vector-based strategies are not only suitable for research purposes, but also have the potential to allow gene therapeutic applications of RNAi.

Acknowledgement

We thank Mariola Dutkiewicz, Harry Kurreck, Diana Rothe and Erik Wade for critical reading of the manuscript. Financial support of our collaboration by the DFG (SFB/TR 19, TP C1 to JK and C5 to HF and FE785/1-1 to HF) is gratefully acknowledged.

References

1. A. Fire, S. Xu, M. K. Montgomery, S. A. Kostas, S. E. Driver and C. C. Mello, *Nature*, 1998, **391**, 806.
2. C. Matranga, Y. Tomari, C. Shin, D. P. Bartel and P. D. Zamore, *Cell*, 2005, **123**, 607.
3. T. A. Rand, S. Petersen, F. Du and X. Wang, *Cell*, 2005, **123**, 621.
4. J. Liu, M. A. Carmell, F. V. Rivas, C. G. Marsden, J. M. Thomson, J. J. Song, S. M. Hammond, L. Joshua-Tor and G. J. Hannon, *Science*, 2004, **305**, 1437.
5. S. M. Elbashir, J. Harborth, W. Lendeckel, A. Yalcin, K. Weber and T. Tuschl, *Nature*, 2001, **411**, 494.
6. A. Watanabe, M. Arai, M. Yamazaki, N. Koitabashi, F. Wuytack and M. Kurabayashi, *J. Mol. Cell. Cardiol.*, 2004, **37**, 691.
7. T. Christoph, A. Grunweller, J. Mika, M. K. Schafer, E. J. Wade, E. Weihe, V. A. Erdmann, R. Frank, C. Gillen and J. Kurreck, *Biochem. Biophys. Res. Commun.*, 2006, **350**, 238.
8. Y. Shi, *Trends Genet.*, 2003, **19**, 9.

9. P. J. Paddison, A. A. Caudy, E. Bernstein, G. J. Hannon and D. S. Conklin, *Genes Dev.*, 2002, **16**, 948.

10. G. Sui, C. Soohoo, E. B. Affar, F. Gay, Y. Shi, W. C. Forrester and Y. Shi, *Proc. Natl. Acad. Sci. U. S. A.*, 2002, **99**, 5515.

11. J. Y. Yu, S. L. DeRuiter and D. L. Turner, *Proc. Natl. Acad. Sci. U. S. A.*, 2002, **99**, 6047.

12. T. R. Brummelkamp, R. Bernards and R. Agami, *Science*, 2002, **296**, 550.

13. D. Siolas, C. Lerner, J. Burchard, W. Ge, P. S. Linsley, P. J. Paddison, G. J. Hannon and M. A. Cleary, *Nat. Biotechnol.*, 2005, **23**, 227.

14. R. I. Gregory, T. P. Chendrimada, N. Cooch and R. Shiekhattar, *Cell*, 2005, **123**, 631.

15. H. Xia, Q. Mao, H. L. Paulson and B. L. Davidson, *Nat. Biotechnol.*, 2002, **20**, 1006.

16. F. Czauderna, A. Santel, M. Hinz, M. Fechtner, B. Durieux, G. Fisch, F. Leenders, W. Arnold, K. Giese, A. Klippel and J. Kaufmann, *Nucleic Acids Res.*, 2003, **31**, e127.

17. L. Gitlin, S. Karelsky and R. Andino, *Nature*, 2002, **418**, 430.

18. D. Boden, O. Pusch, F. Lee, L. Tucker and B. Ramratnam, *J. Virol.*, 2003, **77**, 11531.

19. A. T. Das, T. R. Brummelkamp, E. M. Westerhout, M. Vink, M. Madiredjo, R. Bernards and B. Berkhout, *J. Virol.*, 2004, **78**, 2601.

20. E. M. Westerhout, M. Ooms, M. Vink, A. T. Das and B. Berkhout, *Nucleic Acids Res.*, 2005, **33**, 796.

21. S. Schubert, H. P. Grunert, H. Zeichhardt, D. Werk, V. A. Erdmann and J. Kurreck, *J. Mol. Biol.*, 2005, **346**, 457.

22. H. Fechner, I. Sipo, D. Westermann, S. Pinkert, X. Wang, L. Suckau, J. Kurreck, H. Zeichhardt, O. Muller, R. Vetter, H.-P. Schultheiss, C. Tschöpe and W. Poller, submitted to *J. Mol. Med.*

23. O. ter Brake, P. Konstantinova, M. Ceylan and B. Berkhout, *Mol. Ther.*, 2006, **14**, 883.

24. M. J. Li, J. Kim, S. Li, J. Zaia, J. K. Yee, J. Anderson, R. Akkina and J. J. Rossi, *Mol. Ther.*, 2005, **12**, 900.

25. M. A. Carmell, L. Zhang, D. S. Conklin, G. J. Hannon and T. A. Rosenquist, *Nat. Struct. Biol.*, 2003, **10**, 91.

26. G. Tiscornia, O. Singer, M. Ikawa and I. M. Verma, *Proc. Natl. Acad. Sci. U. S. A.*, 2003, **100**, 1844.

27. H. Hasuwa, K. Kaseda, T. Einarsdottir and M. Okabe, *FEBS Lett.*, 2002, **532**, 227.

28. J. Seibler, B. Kuter-Luks, H. Kern, S. Streu, L. Plum, J. Mauer, R. Kuhn, J. C. Bruning and F. Schwenk, *Nucleic Acids Res.*, 2005, **33**, e67.

29. A. T. De Souza, X. Dai, A. G. Spencer, T. Reppen, A. Menzie, P. L. Roesch, Y. He, M. J. Caguyong, S. Bloomer, H. Herweijer, J. A. Wolff, J. E. Hagstrom, D. L. Lewis, P. S. Linsley and R. G. Ulrich, *Nucleic Acids Res.*, 2006, **34**, 4486.

30. T. Christoph, G. Bahrenberg, J. De Vy, W. Engelberger, V.A. Erdmann, M. Frech, B. Kogel, T. Rohl, K. Schiene, W. Schroder, J. Seibler and J. Kurreck, *Mol. Cell. Neurosci.*, 2007, doi:10.1016/j.MCH.2007.12.006.

31. M. C. Golding, C. R. Long, M. A. Carmell, G. J. Hannon and M. E. Westhusin, *Proc. Natl. Acad. Sci. U. S. A.*, 2006, **103**, 5285.

32. J. S. Mattick and I. V. Makunin, *Hum. Mol. Genet.*, 2005, **14**, R121.

33. L. P. Lim, N. C. Lau, P. Garrett-Engele, A. Grimson, J. M. Schelter, J. Castle, D. P. Bartel, P. S. Linsley and J. M. Johnson, *Nature*, 2005, **433**, 769.

34. Y. Zeng, E. J. Wagner and B. R. Cullen, *Mol. Cell.*, 2002, **9**, 1327.

35. D. Boden, O. Pusch, R. Silbermann, F. Lee, L. Tucker and B. Ramratnam, *Nucleic Acids Res.*, 2004, **32**, 1154.

36. J. M. Silva, M. Z. Li, K. Chang, W. Ge, M. C. Golding, R. J. Rickles, D. Siolas, G. Hu, P. J. Paddison, M. R. Schlabach, N. Sheth, J. Bradshaw, J. Burchard, A. Kulkarni, G. Cavet, R. Sachidanandam, W. R. McCombie, M. A. Cleary, S. J. Elledge and G. J. Hannon, *Nat. Genet.*, 2005, **37**, 1281.

37. K. H. Chung, C. C. Hart, S. Al-Bassam, A. Avery, J. Taylor, P. D. Patel, A. B. Vojtek and D. L. Turner, *Nucleic Acids Res.*, 2006, **34**, e53.

38. M. K. Rao and M. F. Wilkinson, *Nat. Protoc.*, 2006, **1**, 1494.

39. V. M. Rivera, T. Clackson, S. Natesan, R. Pollock, J. F. Amara, T. Keenan, S. R. Magari, T. Phillips, N. L. Courage, F. Cerasoli, Jr., D. A. Holt and M. Gilman, *Nat. Med.*, 1996, **2**, 1028.

40. A. A. Dorner, F. Wegmann, S. Butz, K. Wolburg-Buchholz, H. Wolburg, A. Mack, I. Nasdala, B. August, J. Westermann, F. G. Rathjen and D. Vestweber, *J. Cell Sci.*, 2005, **118**, 3509.

41. J. Yu and A. P. McMahon, *Genesis*, 2006, **44**, 252.

42. A. Ventura, A. Meissner, C. P. Dillon, M. McManus, P. A. Sharp, L. Van Parijs, R. Jaenisch and T. Jacks, *Proc. Natl. Acad. Sci. U. S. A.*, 2004, **101**, 10380.

43. D. Grimm, K. L. Streetz, C. L. Jopling, T. A. Storm, K. Pandey, C. R. Davis, P. Marion, F. Salazar and M. A. Kay, *Nature*, 2006, **441**, 537.

44. C. Toniatti, H. Bujard, R. Cortese and G. Ciliberto, *Gene Ther.*, 2004 **11**, 649.

45. M. Gossen and H. Bujard, *Proc. Natl. Acad. Sci. U. S. A.*, 1992, **89**, 5547.

46. M. Gossen, S. Freundlieb, G. Bender, G. Muller, W. Hillen and H. Bujard, *Science*, 1995, **268**, 1766.

47. K. S. Christopherson, M. R. Mark, V. Bajaj and P. J. Godowski, *Proc. Natl. Acad. Sci. U. S. A.*, 1992, **89**, 6314.

48. Y. Wang, B. W. O'Malley, S. Y. Tsai and B. W. O'Malley, Jr., *Proc. Natl. Acad. Sci. U. S. A.*, 1994, **91**, 8180.

49. R. Feil, *Handb. Exp. Pharmacol.*, 2007, **178**, 3.

50. R. Feil, J. Brocard, B. Mascrez, M. LeMeur, D. Metzger and P. Chambon, *Proc. Natl. Acad. Sci. U. S. A.*, 1996, **93**, 10887.

51. C. Verrou, Y. Zhang, C. Zurn, W. W. Schamel and M. Reth, *Biol. Chem.*, 1999, **380**, 1435.

52. M. R. Paule and R. J. White, *Nucleic Acids Res.*, 2000, **28**, 1283.
53. T. Hosono, H. Mizuguchi, K. Katayama, Z. L. Xu, F. Sakurai, A. Ishii-Watabe, K. Kawabata, T. Yamaguchi, S. Nakagawa, T. Mayumi and T. Hayakawa, *Hum. Gene Ther.*, 2004, **15**, 813.
54. L. Amar, M. Desclaux, N. Faucon-Biguet, J. Mallet and R. Vogel, *Nucleic Acids Res.*, 2006, **34**, e37.
55. Y. Chen, G. Stamatoyannopoulos and C. Z. Song, *Cancer Res.*, 2003, **63**, 4801.
56. S. Kappel, Y. Matthess, B. Zimmer, M. Kaufmann and K. Strebhardt, *Nucleic Acids Res.*, 2006, **34**, 4527.
57. S. Matsukura, P. A. Jones and D. Takai, *Nucleic Acids Res.*, 2003, **31**, e77.
58. D. Kuninger, D. Stauffer, S. Eftekhari, E. Wilson, M. Thayer and P. Rotwein, *Hum. Gene Ther.*, 2004, **15**, 1287.
59. N. Ke, D. Zhou, E. Chatterton, G. Liu, J. Chionis, J. Zhang, L. Tsugawa, R. Lynn, D. Yu, B. Meyhack, F. Wong-Staal and Q. X. Li, *Exp. Cell Res.*, 2006, **312**, 2726.
60. X. Coumoul, W. Li, R. H. Wang and C. Deng, *Nucleic Acids. Res.*, 2004, **32**, e85.
61. S. Gupta, R. A. Schoer, J. E. Egan, G. J. Hannon and V. Mittal, *Proc. Natl. Acad. Sci. U. S. A.*, 2004, **101**, 1927.
62. L. Fritsch, L. A. Martinez, R. Sekhri, I. Naguibneva, M. Gerard, M. Vandromme, L. Schaeffer and A. Harel-Bellan, *EMBO Rep.*, 2004 **5**, 178.
63. R. A. Dickins, M. T. Hemann, J. T. Zilfou, D. R. Simpson, I. Ibarra, G. J. Hannon and S. W. Lowe, *Nat. Genet.*, 2005, **37**, 1289.
64. A. Epanchintsev, P. Jung, A. Menssen and H. Hermeking, *Nucleic Acids Res.*, 2006, **34**, e119.
65. F. Stegmeier, G. Hu, R. J. Rickles, G. J. Hannon and S. J. Elledge, *Proc. Natl. Acad. Sci. U. S. A.*, 2005, **102**, 13212.
66. K. J. Shin, E. A. Wall, J. R. Zavzavadjian, L. A. Santat, J. Liu, J. I. Hwang, R. Rebres, T. Roach, W. Seaman, M. I. Simon and I. D. Fraser, *Proc. Natl. Acad. Sci. U. S. A.*, 2006, **103**, 13759.
67. M. van de Wetering, I. Oving, V. Muncan, M. T. Pon Fong, H. Brantjes, L. D. van, F. C. Holstege, T. R. Brummelkamp, R. Agami and H. Clevers, *EMBO Rep.*, 2003, **4**, 609.
68. A. M. Feldman and D. McNamara, *N. Engl. J. Med.*, 2000, **343**, 1388.
69. K. Verhoef, G. Marzio, W. Hillen, H. Bujard and B. Berkhout, *J. Virol.*, 2001, **75**, 979.
70. H. Fechner, X. Wang, A. Hurtado Pico, J. Wildner, L. Suckau, S. Pinkert, I. Sipo, S. Weger and W. Poller, *J. Biotechnol.*, 2007, **127**, 560.
71. M. Wiznerowicz, J. Szulc and D. Trono, *Nat. Meth.*, 2006, **3**, 682.
72. U. Deuschle, W. K. Meyer and H. J. Thiesen, *Mol. Cell Biol.*, 1995 **15**, 1907.
73. K. Miyake, J. Flygare, T. Kiefer, T. Utsugisawa, J. Richter, Z. Ma, M. Wiznerowicz, D. Trono and S. Karlsson, *Mol. Ther.*, 2005, **11**, 627.
74. M. Wiznerowicz and D. Trono, *J. Virol.*, 2003, **77**, 8957.

75. J. Szulc, M. Wiznerowicz, M. O. Sauvain, D. Trono and P. Aebischer, *Nat. Meth.*, 2006, **3**, 109.
76. M. Hiraoka-Kanie, M. Miyagishi and J. K. Yamashita, *Bioch. Biophys. Res. Commun.*, 2006, **351**, 669.
77. H. Fechner, X. Wang, M. Srour, U. Siemetzki, H. Seltmann, A. P. Sutter, H. Scherubl, C. C. Zouboulis, R. Schwaab, W. Hillen, H. P. Schultheiss and W. Poller, *Gene Ther.*, 2003, **10**, 1680.
78. Y. Matthess, S. Kappel, B. Spankuch, B. Zimmer, M. Kaufmann and K. Strebhardt, *Oncogene*, 2005, **24**, 2973.
79. J. Seibler, A. Kleinridders, B. Kuter-Luks, S. Niehaves, J. C. Bruning and F. Schwenk, *Nucleic Acids Res.*, 2007, **35**, e54.
80. G. Tiscornia, V. Tergaonkar, F. Galimi and I. M. Verma, *Proc. Natl. Acad. Sci. U. S. A.*, 2004, **101**, 7347.
81. J. E. Heinonen, A. J. Mohamed, B. F. Nore and C.I.E. Smith, *Oligonucleotides*, 2005, **15**, 139.
82. H. S. Chang, C. H. Lin, Y. C. Chen and W. C. Yu, *Am. J. Pathol.*, 2004, **165**, 1535.
83. D. Wegmuller, I. Raineri, B. Gross, E. J. Oakeley and C. Moroni, *Stem Cells*, 2007, **25**, 1178.
84. S. Merl, C. Michaelis, B. Jaschke, M. Vorpahl, S. Seidl and R. Wessely, *Circulation*, 2005, **111**, 1583.
85. D. R. Sorensen, M. Leirdal and M. Sioud, *J. Mol. Biol.*, 2003, **327**, 761.
86. D. Grimm, K. Pandey and M. A. Kay, *Methods Enzymol.*, 2005, **392**, 381.
87. E. Klimatcheva, J. D. Rosenblatt and V. Planelles, *Front. Biosci.*, 1999, **4**, D481.
88. S. Hacein-Bey-Abina, K. C. Von, M. Schmidt, M. P. McCormack, N. Wulffraat, P. Leboulch, A. Lim, C. S. Osborne, R. Pawliuk, E. Morillon, R. Sorensen, A. Forster, P. Fraser, J. I. Cohen, B. G. de Saint, I. Alexander, U. Wintergerst, T. Frebourg, A. Aurias, D. Stoppa-Lyonnet, S. Romana, I. Radford-Weiss, F. Gross, F. Valensi, E. Delabesse, E. Macintyre, F. Sigaux, J. Soulier, L. E. Leiva, M. Wissler, C. Prinz, T. H. Rabbitts, D. F. Le, A. Fischer and M. Cavazzana-Calvo, *Science*, 2003, **302**, 415.
89. X. Wu, Y. Li, B. Crise and S. M. Burgess, *Science*, 2003, **300**, 1749.
90. K. V. Morris and J. J. Rossi, *Gene Ther.*, 2006, **13**, 553.
91. M. T. Browning, R. D. Schmidt, K. A. Lew and T. A. Rizvi, *J. Virol.*, 2001, **75**, 5129.
92. P. L. Sinn, S. L. Sauter and P. B. McCray, Jr., *Gene Ther.*, 2005, **12**, 1089.
93. T. R. Brummelkamp, R. Bernards and R. Agami, *Cancer Cell*, 2002, **2**, 243.
94. T. Abbas-Terki, W. Blanco-Bose, N. Deglon, W. Pralong and P. Aebischer, *Hum. Gene Ther.*, 2002, **13**, 2197.
95. S. D. Henry, W. P. van der, H. J. Metselaar, H. W. Tilanus, B. J. Scholte and L. J. van der Laan, *Mol. Ther.*, 2006, **14**, 485.
96. E. Cave, M. S. Weinberg, T. Cilliers, S. Carmona, L. Morris and P. Arbuthnot, *AIDS Res. Hum. Retroviruses*, 2006, **22**, 401.
97. J. Anderson and R. Akkina, *AIDS Res. Ther.*, 2005, **2**, 1.

98. A. Pfeifer, S. Eigenbrod, S. Al-Khadra, A. Hofmann, G. Mitteregger, M. Moser, U. Bertsch and H. Kretzschmar, *J. Clin. Invest.*, 2006, **116**, 3204.
99. C. Volpers and S. Kochanek, *J. Gene Med.*, 2004, **6**, S164.
100. Y. Yang, Q. Li, H. C. Ertl and J. M. Wilson, *J. Virol.*, 1995, **69**, 2004.
101. N. Morral, W. O'Neal, K. Rice, M. Leland, J. Kaplan, P. A. Piedra, H. Zhou, R. J. Parks, R. Velji, E. Aguilar-Cordova, S. Wadsworth, F. L. Graham, S. Kochanek, K. D. Carey and A. L. Beaudet, *Proc. Natl. Acad. Sci. U. S. A.*, 1999, **96**, 12816.
102. G. Schiedner, N. Morral, R. J. Parks, Y. Wu, S. C. Koopmans, C. Langston, F. L. Graham, A. L. Beaudet and S. Kochanek, *Nat. Genet.*, 1998, **18**, 180.
103. A. N. Lukashev, C. Fuerer, M. J. Chen, P. Searle and R. Iggo, *Hum. Gene Ther.*, 2005, **16**, 1473.
104. K. Doronin, M. Kuppuswamy, K. Toth, A. E. Tollefson, P. Krajcsi, V. Krougliak and W. S. Wold, *J. Virol.*, 2001, **75**, 3314.
105. D. C. Yu, G. T. Sakamoto and D. R. Henderson, *Cancer Res.*, 1999, **59**, 1498.
106. M. Kuppuswamy, J. F. Spencer, K. Doronin, A. E. Tollefson, W. S. Wold and K. Toth, *Gene Ther.*, 2005, **12**, 1608.
107. J. R. Bischoff, D. H. Kirn, A. Williams, C. Heise, S. Horn, M. Muna, L. Ng, J. A. Nye, A. Sampson-Johannes, A. Fattaey and F. McCormick, *Science*, 1996, **274**, 373.
108. C. Heise, T. Hermiston, L. Johnson, G. Brooks, A. Sampson-Johannes, A. Williams, L. Hawkins and D. Kirn, *Nat. Med.*, 2000, **6**, 1134.
109. A. B. Sanchez, M. Perez, T. Cornu and J. C. de la Torre, *J. Virol.*, 2005, **79**, 11071.
110. W. Chen, M. Liu, Y. Jiao, W. Yan, X. Wei, J. Chen, L. Fei, Y. Liu, X. Zuo, F. Yang, Y. Lu and Z. Zheng, *J. Virol.*, 2006, **80**, 3559.
111. A. Rinne, C. Littwitz, M. C. Kienitz, A. Gmerek, L. I. Bosche, L. Pott and K. Bender, *J. Muscle Res. Cell Motil.*, 2006, **27**, 413.
112. H. Fechner, S. Pinkert, X. Wang, I. Sipo, L. Suckau, J. Kurreck, A. Dorner, K. Sollerbrant, H. Zeichhardt, H. P. Grunert, R. Vetter, H. P. Schultheiss and W. Poller, *Gene Ther.*, 2007, **14**, 960.
113. H. Fechner, L. Suckau, J. Kurreck, I. Sipo, X. Wang, S. Pinkert, S. Loschen, J. Rekittke, S. Weger, D. Dekkers, R. Vetter, V. A. Erdmann, H. P. Schultheiss, M. Paul, J. Lamers and W. Poller, *Gene Ther.*, 2007 **14**, 211.
114. Y. Li, H. Li, G. Yao, W. Li, F. Wang, Z. Jiang and M. Li, *Cancer Gene Ther.*, 2007, **14**, 748.
115. C. Chetty, P. Bhoopathi, P. Joseph, S. Chittivelu, J. S. Rao and S. Lakka, *Mol. Cancer Ther.*, 2006, **5**, 2289.
116. Y. A. Zhang, J. Nemunaitis, S. K. Samuel, P. Chen, Y. Shen and A. W. Tong, *Cancer Res.*, 2006, **66**, 9736.
117. E. N. Gurzov and M. Izquierdo, *Gene Ther.*, 2006, **13**, 1.
118. D. Grimm and J. A. Kleinschmidt, *Hum. Gene Ther.*, 1999, **10**, 2445.

119. G. Gao, L. H. Vandenberghe, M. R. Alvira, Y. Lu, R. Calcedo, X. Zhou and J. M. Wilson, *J. Virol.*, 2004, **78**, 6381.
120. H. Nakai, C. E. Thomas, T. A. Storm, S. Fuess, S. Powell, J. F. Wright and M. A. Kay, *J. Virol.*, 2002, **76**, 11343.
121. C. A. Pacak, C. S. Mah, B. D. Thattaliyath, T. J. Conlon, M. A. Lewis, D. E. Cloutier, I. Zolotukhin, A. F. Tarantal and B. J. Byrne, *Circ. Res.*, 2006, **99**, e3.
122. K. Inagaki, S. Fuess, T. A. Storm, G. A. Gibson, C. F. McTiernan, M. A. Kay and H. Nakai, *Mol. Ther.*, 2006, **14**, 45.
123. T. VandenDriessche, L. Thorrez, A. Costa-Sanchez, I. Petrus, L. Wang, L. Ma, L. de Waele, Y. Iwasaki, V. Gillijns, J. M. Wilson, D. Collen and M. K. Chuah, *J. Thromb. Haemost.*, 2007, **5**, 16.
124. G. P. Gao, M. R. Alvira, L. Wang, R. Calcedo, J. Johnston and J. M. Wilson, *Proc. Natl. Acad. Sci. U. S A.*, 2002, **99**, 11854.
125. A. C. Nathwani, J. T. Gray, J. McIntosh, C. Y. Ng, J. Zhou, Y. Spence, M. Cochrane, E. Gray, E. G. Tuddenham and A. M. Davidoff, *Blood*, 2007, **109**, 1414.
126. K. S. Bankiewicz, J. Forsayeth, J. L. Eberling, R. Sanchez-Pernaute, P. Pivirotto, J. Bringas, P. Herscovitch, R. E. Carson, W. Eckelman, B. Reutter and J. Cunningham, *Mol. Ther.*, 2006, **14**, 564.
127. H. Fu, L. Kang, J. S. Jennings, S. S. Moy, A. Perez, J. Dirosario, D. M. McCarty and J. Muenzer, *Gene Ther.*, 2007, **14**, 1065.
128. A. S. Pachori, L. G. Melo, L. Zhang, S. D. Solomon and V. J. Dzau, *J. Am. Coll. Cardiol.*, 2006, **47**, 635.
129. D. M. McCarty, S. M. Young and R. J. Samulski, *Annu. Rev. Genet.*, 2004, **38**, 819.
130. Z. Wang, H. I. Ma, J. Li, L. Sun, J. Zhang and X. Xiao, *Gene Ther.*, 2003, **10**, 2105.
131. D. M. McCarty, H. Fu, P. E. Monahan, C. E. Toulson, P. Naik and R. J. Samulski, *Gene Ther.*, 2003, **10**, 2112.
132. R. S. Tomar, H. Matta and P. M. Chaudhary, *Oncogene*, 2003, **22**, 5712.
133. J. D. Hommel, R. M. Sears, D. Georgescu, D. L. Simmons and R. J. DiLeone, *Nat. Med.*, 2003, **9**, 1539.
134. A. M. Babcock, D. Standing, K. Bullshields, E. Schwartz, C. M. Paden and D. J. Poulsen, *Mol. Ther.*, 2005, **11**, 899.
135. M. Gorbatyuk, V. Justilien, J. Liu, W. W. Hauswirth and A. S. Lewin, *Virus Res.*, 2007, **47**, 1202.
136. A. Tessitore, F. Parisi, M. A. Denti, M. Allocca, U. Di Vicino, L. Domenici, I. Bozzoni and A. Auricchio, *Mol. Ther.*, 2006, **14**, 692.
137. X. Wang, L. Skelley, R. Cade and Z. Sun, *Gene Ther.*, 2006, **13**, 1097.
138. Z. Han, M. Gorbatyuk, J. Thomas, Jr., A. S. Lewin, A. Srivastava and P. W. Stacpoole, *Mitochondrion*, 2007, **7**, 253.
139. D. Xu, D. McCarty, A. Fernandes, M. Fisher, R. J. Samulski and R. L. Juliano, *Mol. Ther.*, 2005, **11**, 523.

CHAPTER 13

Development of an RNAi-Based Gene Therapy against HIV-1

OLIVIER TER BRAKE AND BEN BERKHOUT

Laboratory of Experimental Virology, Department of Medical Microbiology, Center for Infection and Immunity Amsterdam (CINIMA), Academic Medical Center of the University of Amsterdam, Meibergdreef 15, 1105 AZ Amsterdam, the Netherlands

13.1 Introduction

The virus that causes acquired immune deficiency syndrome (AIDS) was isolated in 1983.[1] This pathogen, the human immunodeficiency virus type 1 (HIV-1), was classified as a retrovirus belonging to the subfamily of lentiviruses. Currently, HIV-1 infection is a global epidemic with 33 million people infected worldwide (http://www.unaids.org). In 2007, 2.5 million people were infected and over 2.1 million people died of AIDS. In spite of strong efforts to develop a vaccine or cure for this virus infection, none have been successful so far. Multiple antiretroviral drugs have been developed that inhibit HIV-1 replication. Initially, patients were treated with a single drug, which resulted in the emergence of drug-resistant viruses. Only when patients were treated with a combination of antiretroviral drugs, known as highly active antiretroviral therapy (HAART), was the infection effectively controlled.[2–4] However, adherence to the drug regimen is an important prerequisite for the success of HAART, which can be further complicated by severe side effects of the treatment. Suboptimal levels of medication, resulting from irregular intake, can result in the selection of drug-resistant virus variants and treatment failure. Thus, there is a need for new therapeutic approaches to treat HIV-1 infection.

RSC Biomolecular Sciences
Therapeutic Oligonucleotides
Edited by Jens Kurreck
© Royal Society of Chemistry 2008

The recent discovery of RNA interference (RNAi),[5] a potent sequence-specific gene regulatory mechanism, offers a new means to inhibit invading viruses.[6,7] Efficient inhibition of HIV-1 replication can be achieved when an RNAi inhibitor against the virus is stably expressed in T cell lines as a short hairpin RNA (shRNA). However, similar to current antiviral drugs used in the clinic, the application of a single inhibitor against HIV-1 is not sufficient to maintain inhibition. Several *in vitro* studies have demonstrated that HIV-1 can escape from inhibition by mutation of its RNAi target sequence.[8–11] In addition, viral escape may result from mutations outside the target region that stabilize a repressive structure in the viral RNA genome.[12,13]

In this chapter, we focus on the viral escape routes and strategies to counteract these. Three different strategies can be envisioned. The first and most obvious approach, which is similar to HAART, is to combine multiple shRNAs directed against highly conserved HIV-1 sequences. Second, highly conserved target sequences may be identified in the HIV-1 genomes that have severely restricted escape options, and some targets perhaps that do not allow any escape mutations to arise. When a highly effective primary shRNA is combined with second generation shRNAs against these few escape options, viral escape routes may be effectively blocked. Third, in addition to shRNAs against HIV-1, shRNAs directed against cellular factors that are required for HIV-1 replication may be included. Finally, we describe how these strategies can be translated into a clinical application for the treatment of people infected with HIV-1.

13.2 The HIV-1 Life Cycle

HIV-1 infects cells of the immune system, in particular CD4-positive cells, which include T lymphocytes, monocytes and macrophages. HIV-1 infection is associated with a progressive loss of CD4 T cells and a concomitant increase in viral load. This triggers the gradual collapse of the immune system, resulting in opportunistic infections and various cancers, which define the onset of AIDS and ultimately results in death.

A schematic of the HIV-1 life cycle is shown in Figure 13.1A. The virion attaches to the cell *via* the envelope protein that binds to the CD4 receptor. This binding induces a conformational change in the envelope that allows binding to a co-receptor, CXC chemokine receptor 4 (CXCR4) or CC chemokine receptor 5 (CCR5). Upon binding of the co-receptor, the viral and cellular membranes are fused and the virion core is released into the cytoplasm. Within the virion core, the viral RNA genome is converted into DNA by the viral reverse transcriptase enzyme. The core is transported to the nucleus, where the DNA is integrated into the host genome by the viral integrase enzyme to form proviral DNA as shown in Figure 13.1B. Viral transcription from the long terminal repeat (LTR) promoter is initiated by host–cell transcription factors. The early viral transcripts are fully spliced and encode two essential proteins, Tat and Rev. Tat potently activates viral transcription *via* interaction with an RNA

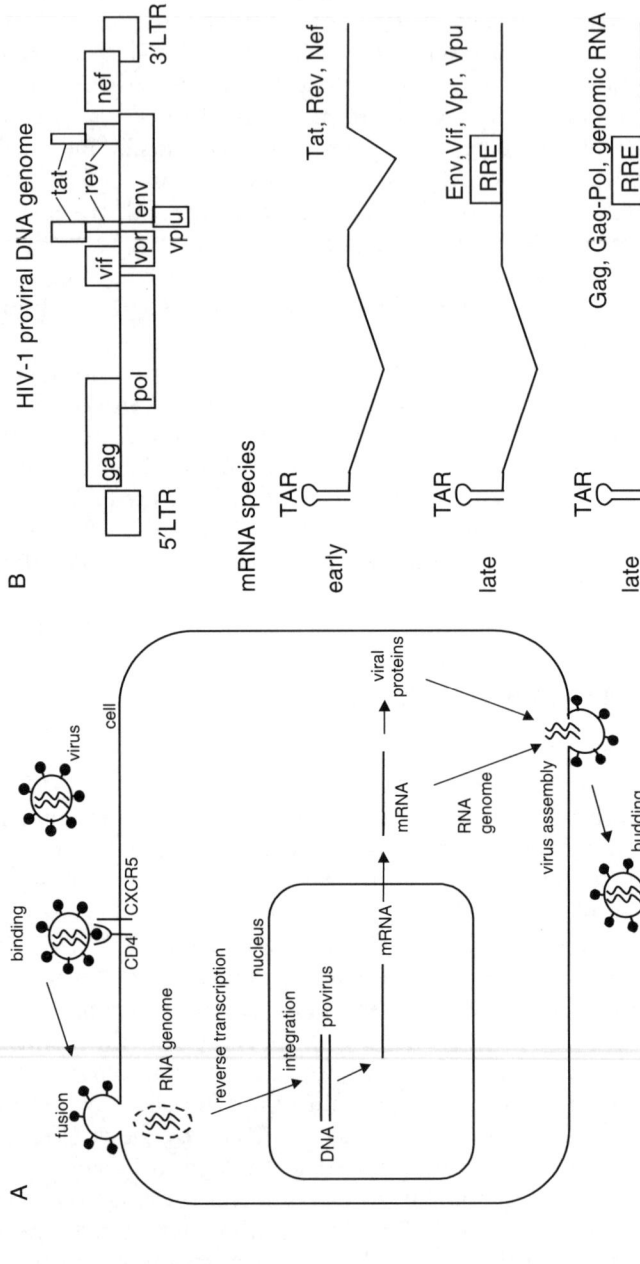

Figure 13.1 Schematic of the HIV-1 life cycle (A) and HIV-1 genome organization with early and late viral gene expression (B).

hairpin at the extreme 5′-end of the mRNA, the *trans*-activating response (TAR) element. Rev interacts with the Rev-responsive element (RRE) in viral mRNAs to facilitate nuclear export of unspliced and singly-spliced mRNAs. These late mRNAs encode Gag, Pol and Env which, together with the viral RNA genome, assemble at the cell membrane to form new virions that are released from the cell by budding. A detailed description of the HIV-1 life cycle is given in Gomez and Hope,[14] and Luciw.[15]

13.3 RNA Interference as an Antiviral Strategy

RNAi is an ancient gene regulation mechanism present in plants, insects and mammals. Its role in mammals and humans is the processing of microRNA (miRNA), small non-coding RNA that regulates cellular gene expression (reviewed in He and Hannon[16]). The RNAi pathway can also be induced by artificial substrates, transiently by small-interfering RNA (siRNA)[17] and stably by intracellularly-expressed shRNAs, which are processed into siRNAs (Figure 13.2).[18] Perfect complementarity of the siRNA with the target sequence results in cleavage of the mRNA.[19] Since its discovery, RNAi has been widely used in gene knockdown studies to analyze gene function. Recently, siRNA

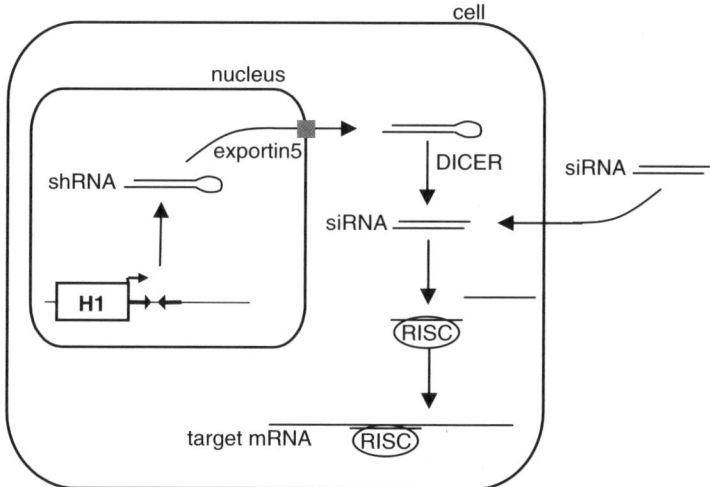

Figure 13.2 Induction of the RNAi mechanism by transfected siRNA or intracellularly-expressed shRNA. shRNA can be expressed from a transfected or stably integrated expression cassette, usually from a polymerase III promoter (for instance H1). The shRNA is exported from the nucleus to the cytoplasm by Exportin-5, and processed by the Dicer enzyme into an siRNA. Alternatively, the siRNA is directly transfected into the cell. The passenger strand is removed from the siRNA and the guide strand is incorporated into the RNA-induced silencing complex (RISC). The siRNA guides the RISC to a complementary target mRNA, which is cleaved.

libraries were developed against the human genome that enabled high-throughput gene knockdown studies.[20–24] It is also attractive to use RNAi technology specifically to inhibit viruses. For acute virus infections like influenza virus A and respiratory syncytial virus (RSV), siRNAs are particular useful, since local delivery to the lungs seems feasible.[25] For chronic infections with HIV-1, hepatitis B and hepatitis C virus, a constant supply of siRNA is required, which makes a gene therapy strategy attractive, providing a constant local supply of antiviral siRNAs.

RNAi has been shown to be very effective in inhibiting HIV-1 replication in stably-transduced T cell lines that express a single antiviral shRNA. However, as with conventional single-drug treatments of HIV-1 infections in patients, RNAi-resistant viruses emerge. Single point mutations and deletions within the target sequence, or mutations outside the target that result in an RNA structure change, have been observed as potential escape routes for the virus to become resistant. These results were obtained with a shRNA directed against the non-essential Nef gene, which allows the virus more freedom in the selection of resistance mutations. For the development of an effective RNAi gene therapy, it is preferred to target highly conserved HIV-1 sequences. Highly conserved targets may allow only a few escape routes, as many RNAi resistance mutations will not be selected because they impose a high fitness cost.

Two forms of HIV-1 RNA can theoretically be targeted by RNAi – the incoming RNA genome, as it is present within the virion particles, and the *de novo* synthesized viral transcripts (Figure 13.1). Although several studies have suggested that the incoming RNA genome is a target for RNAi,[26–30] other studies have presented good evidence against this possibility.[31–34] It is likely that the viral RNA genome is protected from RNAi within the core particle because the RNAi machinery cannot access the RNA.[34] This inability to inactivate the incoming HIV-1 RNA genome is unfortunate, since it would prevent the establishment of the provirus in newly infected cells. While targeting the newly synthesized viral transcripts, it may be beneficial to select well-conserved target sequences that are present in all HIV-1 RNAs, spliced and unspliced (Figure 13.1B). This includes sequences in the untranslated leader or trailer region, or in the early fully spliced transcripts, *e.g.* sequences present in the Tat and Rev coding sequences.

13.3.1 Combinatorial RNAi Approach 1: Highly Conserved Targets

One strategy to counteract viral escape is to use multiple shRNAs targeting different conserved HIV-1 regions (Figure 13.3).[11,35–38] Conserved targets in the HIV-1 RNA genome will not allow for a deletion-based escape route, as occurred with shRNA targeting the non-essential Nef gene sequences.[9] One can try to estimate the chance of escape with one *versus* multiple shRNAs.[11,35] First, we make the assumption that deletion is not an option for the virus; only that point mutations are allowed to occur. In principle, a single point mutation

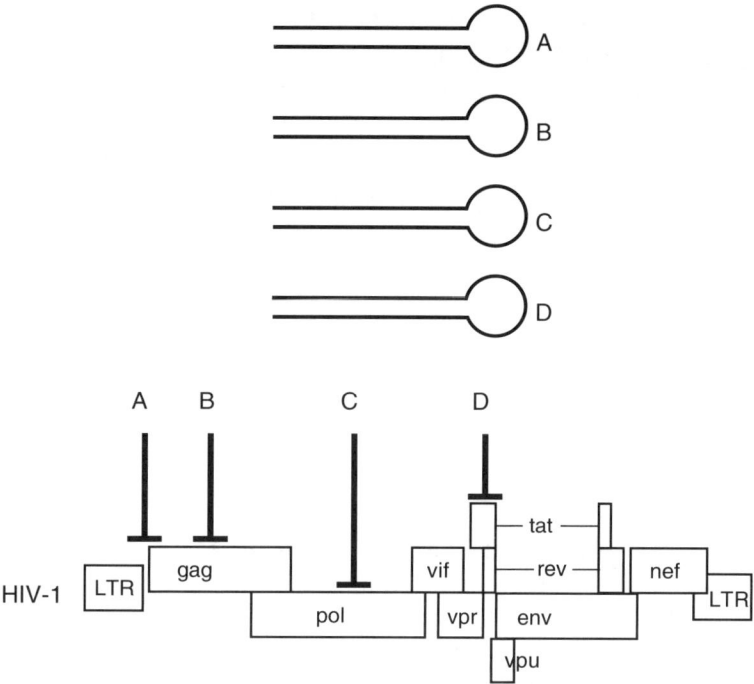

Figure 13.3 Combinatorial RNAi approach 1. Multiple shRNAs are directed against multiple HIV-1 sequences.

can make the virus insensitive to RNAi. The error rate of the reverse transcriptase of HIV-1 is 3×10^{-5},[39] and consequently the chance of viral escape for a 19 nucleotide target in a single infection is $19 \times (3 \times 10^{-5}) = 5.7 \times 10^{-4}$. Studies in the field of drug resistance indicate that an untreated HIV-infected individual has a virus population size of 10^4 to 10^5. This means that for each shRNA, several potential escape variants are already present before the start of therapy. Thus, the emergence of a drug-resistant variant seems inevitable when a single shRNA is used. When multiple shRNAs (N) are used simultaneously, the likelihood of obtaining a drug-resistant variant drops exponentially with the number of shRNAs $(5.7 \times 10^{-4})^N$. If we assume that there is already resistance to at least one of the shRNAs used in a patient, then the chance of a resistant variant emerging is $(5.7 \times 10^{-4})^{N-1}$. For instance, if four shRNAs are used simultaneously, the chance of escape is 1.9×10^{-10}. Although this chance seems remote given the average viral load in a patient, it cannot be excluded that multi-shRNA resistant mutants can evolve through recombination.

Several studies have tested siRNAs against highly conserved targets within the HIV-1 genome. All of these studies used siRNA design criteria to select the inhibitors; in contrast, we have performed a large screen of 86 shRNAs directed at highly conserved HIV-1 sequences, for which we excluded siRNA design

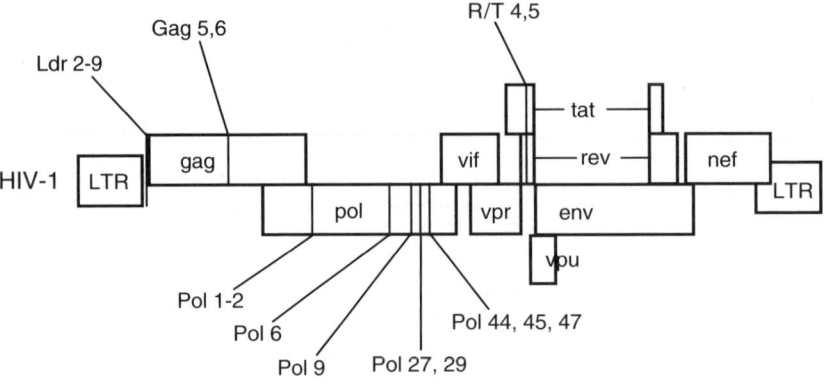

Figure 13.4 Highly conserved HIV-1 sequences that can be effectively targeted with RNAi.

criteria.[40a] In this way, we identified 21 potent shRNAs against highly conserved HIV-1 targets (Figure 13.4). These inhibitors are ideal candidates for a multiple shRNA approach. We showed that the combined expression of multiple shRNAs results in more potent inhibition compared to the expression of only a single shRNA. As a consequence, viral escape was delayed when we combined two shRNAs[40a] and prevented when four shRNAs were expressed.[40b] This finding is important for the development of a multi-shRNA-based gene therapy, since it shows that combinatorial RNAi works through the same principles that make HAART successful.

Several studies have described HIV-1 escape from RNAi.[8,9,30,41,42] Interestingly, the majority of viral escape mutants have a single point mutation within the target sequence. However, there is some evidence that prolonged RNAi inhibition selects for multiple point mutations, indicating that single point mutations may not provide full resistance.[9] These results underscore the high sequence specificity of RNAi, which contrasts with miRNA-induced translational inhibition, for which perfect complementarity is not required. Perhaps one may have expected that a single point mutation would not lead to viral resistance if translational inhibition remained feasible. However, there is an important difference between the siRNA and miRNA routes that provides a simple explanation for this apparent discrepancy. miRNA inhibition *via* translational repression requires multiple target sequences within the 3'-untranslated region of the mRNA. With siRNA-mediated RNAi, only a single target site is required that can be positioned within an open reading frame. In fact, when multiple shRNAs are combined to target different HIV-1 regions, the combined action on multiple targets starts to mimic the miRNA scenario. When, indeed, an miRNA-like mechanism is activated, this means that a single point mutation would not suffice for resistance. In a multiple-shRNA gene therapy, such an additional effect would significantly increase the genetic barrier for HIV-1 to become resistant.

13.3.2 Combinatorial RNAi Approach 2: Second-Generation shRNAs

The multiple shRNA approach should prevent the evolution of RNAi-resistant HIV-1 escape variants. As an alternative to preventing viral escape, we introduced the concept of second generation shRNAs aimed at preferred escape mutants.[11] This approach has the potential to block viral escape, yet target only a single target site (Figure 13.5). Two recent studies also addressed this issue. Sabariegos *et al.* systematically introduced single nucleotide substitutions at all 19 positions of an siRNA target sequence within the HIV-1 integrase gene.[10] Antiviral activity of the siRNA was abolished by the majority of mutations. When they tested a second generation siRNA that compensated for a fully resistant mutation at position 9 of the target sequence, virus inhibition was restored. Nishitsuji *et al.* studied HIV-1 escape in lentiviral vector-transduced cells expressing shRNAs against conserved targets in the HIV-1 U3 region, integrase and tat genes.[30] Resistant viruses emerged against all shRNAs, of which most contained a single point mutation within the target site. HIV-1 molecular clones were constructed of two escape mutants against the integrase shRNA, mutated either at position 4 or 9 within the target sequence. In addition, they constructed lentiviral vectors expressing a second-generation shRNA against these escape mutants. When transduced cells were infected with

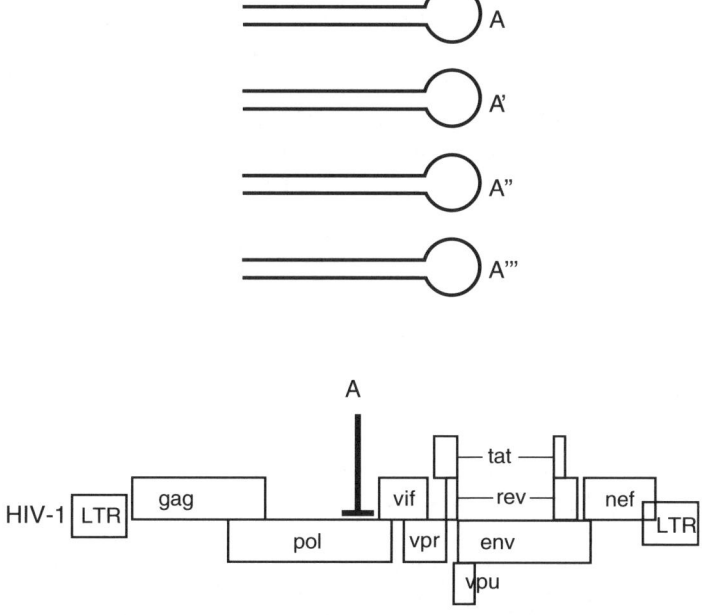

Figure 13.5 Combinatorial RNAi approach 2. A primary shRNA (A) is combined with second-generation shRNAs directed against escape mutants (A′, A″ and A‴).

the matching HIV-1 molecular clone, inhibition was indeed restored. Interestingly, the mutant viruses reverted back to the wild-type sequence. This is surprising since other escape routes are theoretically possible, but apparently not used because only the wild-type virus has an optimal fitness.

These studies confirm that the concept of preventing viral escape with a second generation of siRNAs or shRNAs is a valid approach. Ideally, highly conserved regions of the viral genome are targeted, such that the number of viral escape routes is severely restricted. Only when the number of second generation shRNAs that are required is limited will this concept have therapeutic relevance. Thus, large scale viral escape studies should be performed to evaluate fully the potential of the second generation approach. We recently initiated an extensive analysis of viral escape with four of the 21 potent shRNA inhibitors against highly conserved targets in essential viral genes to address this question.[42b] Preliminary results show that HIV-1 can escape exclusively through point mutations within the target sequence, as no deletions were observed. In some cases, the top two escape routes accounted for at least half of all escapes. Thus, with only two second generation shRNAs, the majority of escape routes may be blocked. Studying viral escape for the remaining 17 shRNAs should reveal whether escape is even more restricted for some of these targets.

13.3.3 Combinatorial Approach 3: Include Cellular Factors as RNAi Targets

We have already discussed that RNAi cannot prevent proviral establishment. In addition, RNAi directed at viral transcripts produced after provirus establishment may not completely silence virus production. Thus, RNAi directed at HIV-1 itself could still allow a very low level of virus spreading. As an alternative to prevent provirus establishment, cellular receptors or co-factors involved in the initial phase of infection can be targeted. In addition, cellular factors involved in infectious virus production may be targeted to prevent virus spread. The genetic barrier for viral escape may be higher when cellular factors are targeted, since HIV-1 needs to adapt to alternative cellular co-factors, which is more complicated or even impossible when no alternative cellular functions are available. Thus, the combination of RNAi directed at HIV-1 and cellular co-factors (Figure 13.6) may result in an even more potent and durable inhibition.

However, the selection of host gene targets requires caution. Knocking down cellular factors that are essential for HIV-1 may not only impair virus replication, but may also be detrimental to the cell and the host. Importantly, since an RNAi-based lentiviral vector gene therapy will, preferably, use a low multiplicity of infection, cellular factors will only be knocked down partially. Several cellular factors involved in the HIV-1 life cycle have already been considered as possible RNAi targets – the CD4 receptor and co-receptors CCR5 and CXCR4,[43–45] integration factors like BAF1, emerin and lens epithelium-derived growth factor (LEDGF)/p75,[46–48] transcription factors

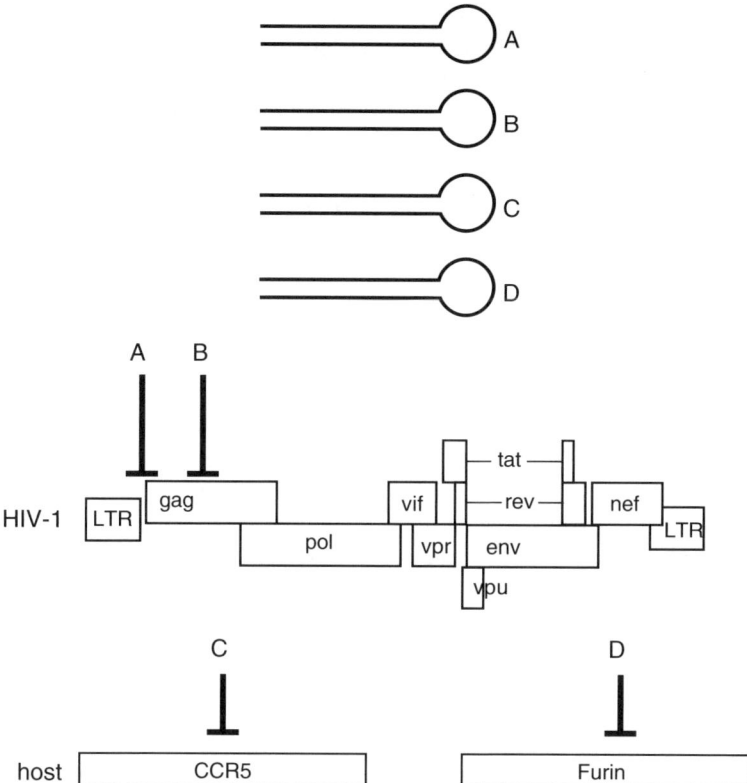

Figure 13.6 Combinatorial RNAi approach 3. Combination of shRNAs directed against HIV-1 (A and B) and cellular factors CCR5 (C) and furin (D).

like nuclear factor kappa beta (NFκB), p21-activated kinase (PAK-1) and cyclin T1,[31,49,50] or furin, which is involved in envelope protein maturation.[49] Not all cellular co-factors are ideal targets. For instance, knocking down CD4 is not desirable in a therapeutic setting because it is essential for the immune system. Also, an exceptionally high lentiviral vector copy number ('intensified RNAi') may be required to achieve complete knockdown to obtain a therapeutic effect, as demonstrated for LEDGF/p75, which facilitates HIV-1 integration.

The CCR5 receptor is an obvious and attractive target in a multiple shRNA approach. HIV-1-infected people who carry a defective CCR5 gene, CCR5-Δ32, show a delay in progression to AIDS and people homozygous for CCR5-Δ32 are largely protected from HIV-1 infection.[51,52] Most importantly, people carrying this defective gene are healthy. Thus, targeting CCR5, resulting in perhaps only a partial knockdown, has a potential therapeutic benefit for HIV-infected patients. These examples highlight that the only targets that can be selected are those that are dispensable for the host (cell), and for which a partial knock-down impairs HIV-1 replication. The recent development of RNAi libraries against the human genome, either using transfected siRNAs or viral vectors, will assist in the identification of novel cellular factors involved in HIV-1 replication.[20–24] All cellular factors that show up as potential therapeutic targets will require extensive validation and, in particular, toxicity screening, similar to what is required for anti-HIV shRNAs.

13.4   Gene Therapy for HIV-1 Infection

Although current HAART is highly effective in suppressing HIV-1 replication, there are downsides to the use of chemotherapy.[3] Medication has to be taken lifelong, is often toxic and a high level of adherence is required to prevent the virus from developing drug resistance. Thus, there is an obvious need for alternative therapeutic approaches. One such strategy would be to use gene therapy to deliver antiviral genes that interfere with viral replication to make cells resistant to HIV-1. An advantage of such an approach would be that the antivirals are produced exclusively in the treated cells, which may have the benefit of reducing toxic side effects associated with the systemic application of chemotherapy. In addition, patients are no longer required to take daily medication, which may reduce the chance of drug resistance. These features of anti-HIV gene therapy have the potential to improve significantly the quality of life for HIV-1-infected patients. The current status of gene therapy strategies to treat HIV and/or AIDS are described in Strayer *et al.*,[53] and Wolkowicz and Nolan.[54]

Recombinant viral vectors, based on viruses, are most often used for gene delivery (reviewed in Goncalves[55]). A typical production scheme for a viral vector is shown in Figure 13.7A. Viral proteins for packaging the viral genome are provided in *trans*. Viral genes associated with pathogenicity are removed from the viral vector genome, and only the *cis*-acting elements required for packaging and gene transfer are retained. Within the viral genome, an expression cassette for a therapeutic gene is inserted. The vector plasmid is co-transfected with the packaging plasmids in a producer cell, resulting in the production of viral vector particles. These vector particles are not replication competent, but can be used for a single round of infection, which results in delivery of the vector genome and its therapeutic cargo to the target cells (transduction).

Many different cell types are infected by HIV-1. However, since T cells constitute the major cell population implicated in HIV infection and progression to AIDS, making these cells resistant to HIV-1 is a key aspect of anti-HIV gene therapy.[53,54] Currently, two optional strategies exist to make T cells resistant to HIV-1 (Figure 13.7B). The most simple scenario is to isolate peripheral blood mononuclear cells from a patient and purify the CD4 + T cells. These cells are subsequently treated with a viral vector expressing an anti-HIV gene to make them resistant to HIV-1. The transduced cells are then engrafted back into the patient, where they will survive, improving the immunity of the patient. However, these T cells will have a limited life span, and thus periodic infusions may be required.

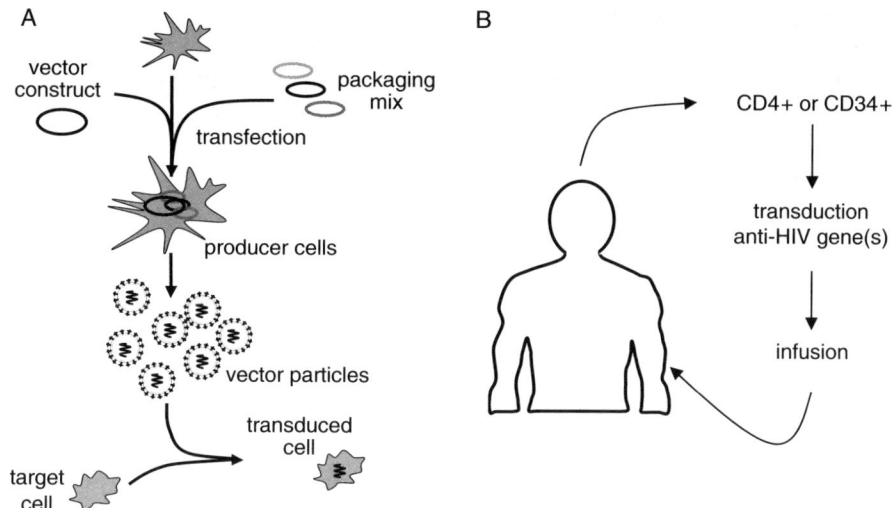

Figure 13.7   Gene therapy of HIV-1 infected patients. (A) Schematic diagram of viral vector production. (B) Viral vector with an anti-HIV gene is used to transduce a patient's CD4+ or CD34+ cells, which are infused back into the patient.

It is perhaps more desirable to transduce haematopoietic blood stem cells. Blood stem cells continuously give rise to all myeloid and lymphoid lineages, and engraftment of autologous transduced blood stem cells will result in a steady production of T cells, but, for example, also monocytes and macrophages that resist HIV-1 infection. Thus, targeting CD34+ stem cells with only a single gene therapy treatment may suffice for a sustained therapeutic effect. HIV-1-resistant cells will preferentially survive over unprotected cells, which are either killed by the virus or removed upon virus infection by the immune system. This may lead to partial reconstitution of the immune system. Therefore, CD4+ T cells and CD34+ blood stem cells are the prime targets for gene therapy approaches to treat HIV-1 infection. Several clinical trials have been performed, predominantly with retroviral vectors delivering anti-HIV genes into the T cells or blood stem cells of patients.[53] Retroviral vectors were used because they can transduce both cell types and stably integrate into the genome of the transduced cells. The antiviral genes used in these studies interfere with early viral gene expression (Figure 13.1B). The majority of studies used transdominant negative Rev proteins that inhibit the wild-type Rev protein in *trans*.[56–61] Alternatively, an RRE decoy was expressed that sequesters the Rev protein or cellular co-factors.[62] Other antivirals (*e.g.* ribozymes and antisense molecules), whose antiviral mechanism rely on base-pairing with complementary HIV-1 sequences, have also been tested.[58,63–65] All these studies represent small scale clinical trials with only a few patients, and the patients actually continued on regular HAART treatment. Thus, it is difficult to assess the therapeutic effect of the gene therapy. Nevertheless, the procedures were shown to be safe, and transduced cells could be detected long term. Preferential survival of transduced cells was occasionally reported, confirming that anti-HIV gene therapy has the potential to improve the patient's immunity.[57,58,60] The recent discovery of RNAi and its proven efficacy against HIV-1 validates efforts to translate RNAi-based gene therapies into a clinical application.

13.4.1   Lentiviral Vector Development

Of crucial importance for the clinical development of gene therapy strategies is that the vector can be produced to high titres and that the vector is genetically stable. The lentiviral vector is very efficient in transducing CD34+ blood stem cells or CD4+ T cells. However, this vector system is based on HIV-1, and expression of anti-HIV shRNAs from the lentiviral vector may cause unwanted complications. Indeed, it has been reported that targeting of the Gag-Pol and Rev mRNAs of the packaging system can reduce the transduction titer.[66,67] We systematically addressed this issue with a selected set of anti-HIV shRNAs.[68] All possible routes by which anti-HIV shRNA expression can interfere with lentiviral vector production were investigated. With shRNAs against Gag-Pol mRNA, both the capsid and transduction titres were reduced, which was resolved with a human codon-optimized Gag-Pol version. Targeting of Rev was not problematic. RNAi attack on the vector genome is a serious possibility, which should be avoided by selecting shRNAs for which the target is absent in the lentiviral vector or by modification of the target sequence. Strikingly, the shRNA encoding sequence within the vector genome was not self-targeted by the shRNA because of stable RNA folding and occlusion of the target sequence.[12,13]

We proposed adapting the lentiviral vector to remove perfect complementarity between shRNAs and potential target sequences. However, introducing RNAi-resistant mutations within the lentiviral vector backbone adds a potential risk of transfer of RNAi resistance to the wild-type HIV-1 genome. This danger may be limited because the lentiviral vector is self-inactivating – it has a deletion in the HIV-1 LTR promoter that severely cripples the promoter activity, thus reducing the risk of mobilization by wild-type HIV-1.[69,70] The term self-inactivating implies that the promoter is not active after lentiviral transduction, but full length lentiviral vector genome transcripts can be detected, albeit at a low level.[71,72] Thus, if a modified lentiviral vector with RNAi-resistant mutations is used, transfer of RNAi resistance to HIV-1 is a serious concern. Most effective shRNAs target the essential open reading frames, Gag and Pol. An interesting solution would be to use RNAi-resistance mutations that, at the same time, insert stop-codons within these open reading frames. A putative recombination event will result in an HIV-1 variant that is RNAi resistant, yet replication defective. Potential mobilization of the vector genome may also be restricted by improving the lentiviral vector design. An interesting approach in this respect is to incorporate target sequences for endogenous miRNAs within the lentiviral genome.[73,74] Full length lentiviral vector genomes will be degraded by RNAi, thus severely reducing the likelihood of mobilization and recombination.

It is well known that HIV-1 is recombination-prone, and that the presence of multiple shRNA expression cassettes with repeated promoter sequences within the lentiviral vector may pose serious problems. The lentiviral vector can recombine at the repeat sequences during the transduction process, resulting in deletion of one or multiple expression cassettes.[75] Thus, to avoid recombination of the lentiviral vector, multiple shRNAs should be expressed from different promoters. Indeed, when we expressed four shRNAs from four different promoters, the vector genome stability was improved and no deletions were observed.[40b]

The fact that the shRNA sequence within the lentiviral vector genome is not a target for RNAi should greatly facilitate the incorporation of multiple shRNA expression cassettes without seriously reducing the transduction titre. We have observed that the titre of multiple shRNA lentiviral vectors was reduced, as compared with the control empty vector. These titre reductions were in accordance with previous reports, indicating that an increase in vector genome size has a negative impact on transduction titer.[76] This effect is attributed to a defect in RNA encapsidation or the reduced nuclear export of vector genomes. Perhaps the transduction titre may be improved by increasing vector genome availability and packaging into vector particles by overexpression of chromosome region maintenance 1 (CRM1) and/or DEAD (Asp-Glu-Ala-Asp)-box RNA helicase (DDX3) during vector production. Both co-factors can enhance the transport of vector genomes from the nucleus to the cytoplasm.[77–79] In addition, different orientations and insertion sites of the shRNA cassettes should be compared to find the optimal setting for vector expression and genome stability.

13.4.2 Preclinical Evaluation: Safety and Efficacy

We have discussed several promising anti-HIV combinatorial RNAi strategies. However, important issues in relation to safety and efficacy need to be addressed before clinical trials can be considered. Several potential side effects have been reported for siRNAs and shRNAs. First, it was recently shown that *in vivo* expression of shRNAs led to fatalities in mice through oversaturation of the miRNA–siRNA pathways.[80] Mice were treated in these experiments with high doses of a viral vector, resulting in extremely high copy numbers per cell. This result clearly demonstrates that shRNA overdosing can be lethal. Second, shRNA expression could lead to induction of the interferon pathway. The initial paradigm was that only double-stranded RNA (dsRNA) larger than 30 base pairs could induce this pathway; recently small dsRNA was also shown to evoke this reaction.[81–85] This effect is dose dependent, and some sequence motifs were implicated in the activation of this response.[83,85] Finally, RNAi could also induce off-target effects, in which siRNAs silence partially complementarily transcripts through an miRNA-like mechanism. Such an effect requires complementarity between the siRNA seed region and the 3'-untranslated region of a target gene.[86–88] As such, any effective siRNA will, in theory, have numerous potential off-target genes. When multiple shRNAs are combined, the number of potential off-target genes will also go up, thus increasing the chance of a negative impact on the treated cells.

These potential side effects are a genuine concern for the development of a multiple shRNA approach against HIV-1, and the potential risks should be properly assessed in pre-clinical evaluations. It is important to note that our experimental design has always been with safety in mind. For instance, we have always used a low multiplicity of infection to obtain cells with a single vector copy to avoid high expression levels that may induce unwanted side effects. Under these conditions, the RNAi effect remained stable for at least 100 days, and growth rates of transduced cell lines and primary T cells were similar to those of untransduced cells. We also did not observe induction of interferon-β or protein kinase R (PKR) phosphorylation in transient transfection experiments with the multiple shRNA lentiviral vector constructs. Combined, these data indicate that the proposed gene therapy approach is likely to be safe.

Nevertheless, a more thorough evaluation is required to assess properly the safety and efficacy of multiple shRNA gene therapy. In our experimental setup, we exclusively used HIV-1 infectious molecular clone LAI as the source of the virus, which contrasts with the heterogeneous virus population (quasispecies) found in patients. Also, the transduced T cell lines and differentiated primary human T cells are obviously quite different from the haematopoietic CD34+ stem cells that will be transduced in an eventual therapeutic setting. These stem cells will develop into different lineages, and expression of multiple shRNAs may influence this development. For instance, miRNAs are involved in gene regulation to control haematopoiesis in mice.[89,90] Saturation of the miRNA pathway may result in a disturbance of haematopoiesis. In addition, human immune cells have been shown to induce inflammatory cytokines and an interferon response upon transfection of siRNAs or shRNAs. Similar responses may be activated in immune cells derived from transduced CD34+ cells. However, when CD34+ cells were transduced with a lentiviral vector expressing anti-HIV shRNAs and stimulated to differentiate *in vitro*, such immunostimulatory effects were not induced, indicating that stable expression may avoid these effects.[91] In addition, cellular miRNAs are often expressed from human polymerase III promoters, which indicates that an expression strategy for shRNAs using these promoters may not be that different from their natural counterpart.[92]

Recently, a humanized mouse model was developed that is ideally suited for the preclinical assessment of safety and efficacy. This humanized $Rag^{2-/-}\gamma c^{-/-}$ mouse sustains long term multi-lineage human haematopoiesis and is capable of mounting immune responses.[93] In this model, human CD34+ cells have been engrafted that were transduced with a lentiviral vector expressing a shRNA against the p53 gene, which resulted in stable shRNA expression and down-modulation of the p53 gene.[94] p53 knockdown did not affect the development of CD34+ cells into various mature leukocyte subsets, including T cells, but it conferred resistance to p53-dependent apoptotic stimuli on the T cells.

This indicates that, at least with single shRNA expression against a cellular target, the procedure should, in principle, be safe. These mice can also be infected with HIV-1, resulting in viraemia and CD4+ T cell depletion, which resembles the main features of human HIV-1 infection.[95-97] Thus, this model is ideally suited to studying the safety and efficacy of multi-shRNA lentiviral vector gene therapy. The development of the immune system can be closely monitored, and these mice can be infected with HIV-1 to study the therapeutic effects. When the multiple shRNA lentiviral vector is shown to be safe and effective, a Phase I clinical trial can be initiated.

Acknowledgements

RNAi research in the Berkhout laboratory is sponsored by NWO-CW (Top grant) and ZonMw (Vici grant and Translational Gene Therapy grant).

References

1. F. Barre-Sinoussi, J. C. Chermann, F. Rey, M. T. Nugeyre, S. Chamaret, J. Gruest, C. Dauguet, C. Axler-Blin, F. Vezinet-Brun, C. Rouzioux, W. Rozenbaum and L. Montagnier, *Science*, 1983, **220**, 868.
2. S. G. Deeks, *BMJ*, 2006, **332**, 1489.
3. D. D. Richman, *Nature*, 2001, **410**, 995.
4. B. R. Schackman, K. A. Gebo, R. P. Walensky, E. Losina, T. Muccio, P. E. Sax, M. C. Weinstein, G. R. Seage III, R. D. Moore and K. A. Freedberg, *Med. Care*, 2006, **44**, 990.
5. A. Fire, S. Xu, M. K. Montgomery, S. A. Kostas, S. E. Driver and C. C. Mello, *Nature*, 1998, **391**, 806.
6. G. G. Carmichael, *Nature*, 2002, **418**, 379.
7. J. Haasnoot, E. M. Westerhout and B. Berkhout, *Nat. Biotechnol.*, 2007, **25**, 1435.
8. D. Boden, O. Pusch, F. Lee, L. Tucker and B. Ramratnam, *J. Virol.*, 2003, **77**, 11531.
9. A. T. Das, T. R. Brummelkamp, E. M. Westerhout, M. Vink, M. Madiredjo, R. Bernards and B. Berkhout, *J. Virol.*, 2004, **78**, 2601.
10. R. Sabariegos, M. Gimenez-Barcons, N. Tapia, B. Clotet and M. A. Martinez, *J. Virol.*, 2006, **80**, 571.
11. O. ter Brake and B. Berkhout, *J. RNAi Gene Sil.*, 2005, **1**, 56.
12. E. M. Westerhout, M. Ooms, M. Vink, A. T. Das and B. Berkhout, *Nucleic Acids Res.*, 2005, **33**, 796.
13. E. M. Westerhout and B. Berkhout, *Nucleic Acids Res.*, 2007, **35**, 4322.
14. C. Gomez and T. J. Hope, *Cell Microbiol.*, 2005, **7**, 621.
15. P. Luciw, in *Virology*, ed. B. N. Fields, D. M. Knipe and P. M. Howley, Lippincott-Raven Publishers, New York, 3rd edn, 1996.
16. L. He and G. J. Hannon, *Nat. Rev. Genet.*, 2004, **5**, 522.
17. S. M. Elbashir, J. Harborth, W. Lendeckel, A. Yalcin, K. Weber and T. Tuschl, *Nature*, 2001, **411**, 494.

18. T. R. Brummelkamp, R. Bernards and R. Agami, *Science*, 2002, **296**, 550.
19. S. M. Hammond, E. Bernstein, D. Beach and G. J. Hannon, *Nature*, 2000, **404**, 293.
20. M. Miyagishi, S. Matsumoto and K. Taira, *Virus Res.*, 2004, **102**, 117.
21. P. J. Paddison, J. M. Silva, D. S. Conklin, M. Schlabach, M. Li, S. Aruleba, V. Balija, A. O'Shaughnessy, L. Gnoj, K. Scobie, K. Chang, T. Westbrook, M. Cleary, R. Sachidanandam, W. R. McCombie, S. J. Elledge and G. J. Hannon, *Nature*, 2004, **428**, 427.
22. D. Huesken, J. Lange, C. Mickanin, J. Weiler, F. Asselbergs, J. Warner, B. Meloon, S. Engel, A. Rosenberg, D. Cohen, M. Labow, M. Reinhardt, F. Natt and J. Hall, *Nat. Biotechnol.*, 2005, **23**, 995.
23. D. E. Root, N. Hacohen, W. C. Hahn, E. S. Lander and D. M. Sabatini, *Nat. Methods*, 2006, **3**, 715.
24. K. Berns, E. M. Hijmans, J. Mullenders, T. R. Brummelkamp, A. Velds, M. Heimerikx, R. M. Kerkhoven, M. Madiredjo, W. Nijkamp, B. Weigelt, R. Agami, W. Ge, G. Cavet, P. S. Linsley, R. L. Beijersbergen and R. Bernards, *Nature*, 2004, **428**, 431.
25. V. Bitko, A. Musiyenko, O. Shulyayeva and S. Barik, *Nat. Med.*, 2005, **11**, 50.
26. G. A. Coburn and B. R. Cullen, *J. Virol.*, 2002, **76**, 9225.
27. J. M. Jacque, K. Triques and M. Stevenson, *Nature*, 2002, **418**, 435.
28. J. Capodici, K. Kariko and D. Weissman, *J. Immunol.*, 2002, **169**, 5196.
29. P. J. Joshi, T. W. North and V. R. Prasad, *Mol. Ther.*, 2005, **11**, 677.
30. H. Nishitsuji, M. Kohara, M. Kannagi and T. Masuda, *J. Virol.*, 2006, **80**, 7658.
31. R. M. Surabhi and R. B. Gaynor, *J. Virol.*, 2002, **76**, 12963.
32. W. Y. Hu, C. P. Myers, J. M. Kilzer, S. L. Pfaff and F. D. Bushman, *Curr. Biol.*, 2002, **12**, 1301.
33. H. Nishitsuji, T. Ikeda, H. Miyoshi, T. Ohashi, M. Kannagi and T. Masuda, *Microbes Infect.*, 2004, **6**, 76.
34. E. M. Westerhout, O. ter Brake and B. Berkhout, *Retrovirology*, 2006, **3**, 57.
35. B. Berkhout, *Curr. Opin. Mol. Ther.*, 2004, **6**, 141.
36. G. J. Hannon and J. J. Rossi, *Nature*, 2004, **431**, 371.
37. P. Shankar, N. Manjunath and J. Lieberman, *JAMA*, 2005, **293**, 1367.
38. M. Stevenson, *Nat. Rev. Immunol.*, 2003, **3**, 851.
39. L. M. Mansky, *Virology*, 1996, **222**, 391.
40. (a) O. ter Brake, P. Konstantinova, M. Ceylan and B. Berkhout, *Mol. Ther.*, 2006, **14**, 883.
    (b) O. ter Brake, K. 't. Hooft, Y. P. Liu, M. Centlivre, K. J. von Eije and B. Berkhout, *Mol. Ther.*, 2008, in press.
41. S. K. Lee, D. M. Dykxhoorn, P. Kumar, S. Ranjbar, E. Song, L. E. Maliszewski, V. Francois-Bongarcon, A. Goldfeld, M. N. Swamy, J. Lieberman and P. Shankar, *Blood*, 2005, **106**, 818.
42. (a) H. J. Unwalla, H. T. Li, I. Bahner, M. J. Li, D. Kohn and J. J. Rossi, *J. Virol.*, 2006, **80**, 1863.

(b) K. J. von Eije, O. ter Brake and B. Berkhout, *J. Virol.*, 2007, in press.
43. J. Anderson and R. Akkina, *AIDS Res. Ther.*, 2005, **2**, 1.
44. J. Anderson, A. Banerjea and R. Akkina, *Oligonucleotides*, 2003, **13**, 303.
45. E. Song, S. K. Lee, D. M. Dykxhoorn, C. Novina, D. Zhang, K. Crawford, J. Cerny, P. A. Sharp, J. Lieberman, N. Manjunath and P. Shankar, *J. Virol.*, 2003, **77**, 7174.
46. M. Llano, D. T. Saenz, A. Meehan, P. Wongthida, M. Peretz, W. H. Walker, W. Teo and E. M. Poeschla, *Science*, 2006, **314**, 461.
47. G. Maertens, P. Cherepanov, W. Pluymers, K. Busschots, E. De Clercq, Z. Debyser and Y. Engelborghs, *J. Biol. Chem.*, 2003, **278**, 33528.
48. J. M. Jacque and M. Stevenson, *Nature*, 2006, **441**, 641.
49. D. G. Nguyen, K. C. Wolff, H. Yin, J. S. Caldwell and K. L. Kuhen, *J. Virol.*, 2006, **80**, 130.
50. Y. L. Chiu, H. Cao, J. M. Jacque, M. Stevenson and T. M. Rana, *J. Virol.*, 2004, **78**, 2517.
51. R. Liu, W. A. Paxton, S. Choe, D. Ceradini, S. R. Martin, R. Horuk, M. E. MacDonald, H. Stuhlmann, R. A. Koup and N. R. Landau, *Cell*, 1996, **86**, 367.
52. A. M. Roda Husman, H. Blaak, M. Brouwer and H. Schuitemaker, *J. Immunol.*, 1999, **163**, 4597.
53. D. S. Strayer, R. Akkina, B. A. Bunnell, B. Dropulic, V. Planelles, R. J. Pomerantz, J. J. Rossi and J. A. Zaia, *Mol. Ther.*, 2005, **11**, 823.
54. R. Wolkowicz and G. P. Nolan, *Gene Ther.*, 2005, **12**, 467.
55. M. A. Goncalves, *Bioessays*, 2005, **27**, 506.
56. U. Ranga, C. Woffendin, S. Verma, L. Xu, C. H. June, D. K. Bishop and G. J. Nabel, *Proc. Natl. Acad. Sci. U. S. A.*, 1998, **95**, 1201.
57. C. Woffendin, U. Ranga, Z. Yang, L. Xu and G. J. Nabel, *Proc. Natl. Acad. Sci. U. S. A.*, 1996, **93**, 2889.
58. R. A. Morgan, R. Walker, C. S. Carter, V. Natarajan, J. A. Tavel, C. Bechtel, B. Herpin, L. Muul, Z. Zheng, S. Jagannatha, B. A. Bunnell, V. Fellowes, J. A. Metcalf, R. Stevens, M. Baseler, S. F. Leitman, E. J. Read, R. M. Blaese and H. C. Lane, *Hum. Gene Ther.*, 2005, **16**, 1065.
59. E. M. Kang, M. De Witte, H. Malech, R. A. Morgan, S. Phang, C. Carter, S. F. Leitman, R. Childs, A. J. Barrett, R. Little and J. F. Tisdale, *Blood*, 2002, **99**, 698.
60. G. M. Podsakoff, B. C. Engel, D. A. Carbonaro, C. Choi, E. M. Smogorzewska, G. Bauer, D. Selander, S. Csik, K. Wilson, M. R. Betts, R. A. Koup, G. J. Nabel, K. Bishop, S. King, M. Schmidt, C. von Kalle, J. A. Church and D. B. Kohn, *Mol. Ther.*, 2005, **12**, 77.
61. M. H. Malim, W. W. Freimuth, J. Liu, T. J. Boyle, H. K. Lyerly, B. R. Cullen and G. J. Nabel, *J. Exp. Med.*, 1992, **176**, 1197.
62. D. B. Kohn, G. Bauer, C. R. Rice, J. C. Rothschild, D. A. Carbonaro, P. Valdez, Q. Hao, C. Zhou, I. Bahner, K. Kearns, K. Brody, S. Fox, E. Haden, K. Wilson, C. Salata, C. Dolan, C. Wetter, E. Aguilar-Cordova and J. Church, *Blood*, 1999, **94**, 368.

63. B. L. Levine, L. M. Humeau, J. Boyer, R. R. MacGregor, T. Rebello, X. Lu, G. K. Binder, V. Slepushkin, F. Lemiale, J. R. Mascola, F. D. Bushman, B. Dropulic and C. H. June, *Proc. Natl. Acad. Sci. U. S. A.*, 2006, **103**, 17372.

64. R. G. Amado, R. T. Mitsuyasu, J. D. Rosenblatt, F. K. Ngok, A. Bakker, S. Cole, N. Chorn, L. S. Lin, G. Bristol, M. P. Boyd, J. L. MacPherson, G. C. Fanning, A. V. Todd, J. A. Ely, J. A. Zack and G. P. Symonds, *Hum. Gene Ther.*, 2004, **15**, 251.

65. J. L. MacPherson, M. P. Boyd, A. J. Arndt, A. V. Todd, G. C. Fanning, J. A. Ely, F. Elliott, A. Knop, M. Raponi, J. Murray, W. Gerlach, L. Q. Sun, R. Penny, G. P. Symonds, A. Carr and D. A. Cooper, *J. Gene Med.*, 2005, **7**, 552.

66. A. Banerjea, M. J. Li, G. Bauer, L. Remling, N. S. Lee, J. Rossi and R. Akkina, *Mol. Ther.*, 2003, **8**, 62.

67. L. J. Chang, X. Liu and J. He, *Gene Ther.*, 2005, **12**, 1133.

68. O. ter Brake and B. Berkhout, *J. Gene Med.*, 2007, **9**, 743.

69. H. Miyoshi, U. Blomer, M. Takahashi, F. H. Gage and I. M. Verma, *J. Virol.*, 1998, **72**, 8150.

70. R. Zufferey, T. Dull, R. J. Mandel, A. Bukovsky, D. Quiroz, L. Naldini and D. Trono, *J. Virol.*, 1998, **72**, 9873.

71. A. C. Logan, D. L. Haas, T. Kafri and D. B. Kohn, *J. Virol.*, 2004, **78**, 8421.

72. H. Hanawa, D. A. Persons and A. W. Nienhuis, *J. Virol.*, 2005, **79**, 8410.

73. B. D. Brown, M. A. Venneri, A. Zingale, S. L. Sergi and L. Naldini, *Nat. Med.*, 2006, **12**, 585.

74. C. Simon-Mateo and J. A. Garcia, *J. Virol.*, 2006, **80**, 2429.

75. W. An and A. Telesnitsky, *Virology*, 2001, **286**, 475.

76. M. Kumar, B. Keller, N. Makalou and R. E. Sutton, *Hum. Gene Ther.*, 2001, **12**, 1893.

77. I. Popa, M. E. Harris, J. E. Donello and T. J. Hope, *Mol. Cell Biol.*, 2002, **22**, 2057.

78. M. Fornerod, M. Ohno, M. Yoshida and I. W. Mattaj, *Cell*, 1997, **90**, 1051.

79. V. S. Yedavalli, C. Neuveut, Y. H. Chi, L. Kleiman and K. T. Jeang, *Cell*, 2004, **119**, 381.

80. D. Grimm, K. L. Streetz, C. L. Jopling, T. A. Storm, K. Pandey, C. R. Davis, P. Marion, F. Salazar and M. A. Kay, *Nature*, 2006, **441**, 537.

81. A. J. Bridge, S. Pebernard, A. Ducraux, A. L. Nicoulaz and R. Iggo, *Nat. Genet.*, 2003, **34**, 263.

82. V. Hornung, M. Guenthner-Biller, C. Bourquin, A. Ablasser, M. Schlee, S. Uematsu, A. Noronha, M. Manoharan, S. Akira, A. de Fougerolles, S. Endres and G. Hartmann, *Nat. Med.*, 2005, **11**, 263.

83. S. Pebernard and R. D. Iggo, *Differentiation*, 2004, **72**, 103.

84. M. A. Robbins and J. J. Rossi, *Nat. Med.*, 2005, **11**, 250.

85. M. Sioud, *J. Mol. Biol.*, 2005, **348**, 1079.

86. A. Birmingham, E. M. Anderson, A. Reynolds, D. Ilsley-Tyree, D. Leake, Y. Fedorov, S. Baskerville, E. Maksimova, K. Robinson, J. Karpilow, W. S. Marshall and A. Khvorova, *Nat. Methods*, 2006, **3**, 199.
87. A. L. Jackson, J. Burchard, J. Schelter, B. N. Chau, M. Cleary, L. Lim and P. S. Linsley, *RNA*, 2006.
88. X. Lin, X. Ruan, M. G. Anderson, J. A. McDowell, P. E. Kroeger, S. W. Fesik and Y. Shen, *Nucleic Acids Res.*, 2005, **33**, 4527.
89. C. Z. Chen, L. Li, H. F. Lodish and D. P. Bartel, *Science*, 2004, **303**, 83.
90. C. Z. Chen and H. F. Lodish, *Semin. Immunol.*, 2005, **17**, 155.
91. M. A. Robbins, M. Li, I. Leung, H. Li, D. V. Boyer, Y. Song, M. A. Behlke and J. J. Rossi, *Nat. Biotechnol.*, 2006, **24**, 566.
92. G. M. Borchert, W. Lanier and B. L. Davidson, *Nat. Struct. Mol. Biol.*, 2006, **13**, 1097.
93. N. Legrand, K. Weijer and H. Spits, *J. Immunol.*, 2006, **176**, 2053.
94. R. Gimeno, K. Weijer, A. Voordouw, C. H. Uittenbogaart, N. Legrand, N. L. Alves, E. Wijnands, B. Blom and H. Spits, *Blood*, 2004, **104**, 3886.
95. S. Baenziger, R. Tussiwand, E. Schlaepfer, L. Mazzucchelli, M. Heikenwalder, M. O. Kurrer, S. Behnke, J. Frey, A. Oxenius, H. Joller, A. Aguzzi, M. G. Manz and R. F. Speck, *Proc. Natl. Acad. Sci. U. S. A.*, 2006, **103**, 15951.
96. B. K. Berges, W. H. Wheat, B. E. Palmer, E. Connick and R. Akkina, *Retrovirology*, 2006, **3**, 76.
97. L. Zhang, G. I. Kovalev and L. Su, *Blood*, 2007, **109**, 2978.

CHAPTER 14
RNA Based Therapies for Treatment of HIV Infection

LISA SCHERER,[1] MARC S. WEINBERG[2]
AND JOHN J. ROSSI[*,1]

[1] Division of Molecular Biology, City of Hope Beckman Research Institute, Duarte, CA; [2] Department of Molecular Medicine and Hematology, University of the Witwatersrand Medical School, Wits, South Africa

14.1 Introduction: Challenges to Conventional Therapies

Controlling human immunodeficiency virus (HIV) infection continues to be a major challenge, both in underdeveloped and developed nations, despite the fact that the currently employed drug cocktails have markedly changed the profile of progression to acquired immune deficiency syndrome (AIDS) in HIV-infected individuals. Despite the successes, highly active antiretroviral therapy (HAART) treatment is not without significant problems and drawbacks. Because of pharmacokinetic differences between individuals, there are multiple drug-related toxicities leading to non-adherence problems. There is a need for personalized dosing regimens and combinations and continued therapeutic monitoring of the drugs themselves. There continue to be drug failures for those on HAART as a consequence of viral resistance and other complications caused by a lifelong regimen of chemotherapy. The importance of new anti-retroviral drug development cannot be overstated, but given that HAART therapy is lifelong, and there will always be related toxicities, new approaches for treating HIV infection are desirable. The increase in side effects with long-term conventional treatment is caused, in part, by the improved lifespan

RSC Biomolecular Sciences
Therapeutic Oligonucleotides
Edited by Jens Kurreck
© Royal Society of Chemistry 2008

brought on by the very success of antiretroviral therapies. In addition, treatment guidelines traditionally have not recommended initiating therapy in the early stages of infection, despite the risks associated with loss of immunological function, increased likelihood of transmission, and the development of a larger pool of viral sub-species that serve as a reservoir for potential resistance. However, there has also been a recent shift towards starting retroviral therapy earlier, before CD4 counts drop below $200 \times 10^6/L$ (OARAC, http://AIDSinfo. nih.gov).[1,2] As a possible means of circumventing some of the problems associated with HAART, a number of investigators are focusing their attention on gene therapy either as a stand-alone approach or as an adjuvant to pharmacological drug regimens.

In this review, we focus on the progress in developing RNA-based anti-HIV gene therapeutics for long-term applications, with particular attention to molecular targets and their mechanisms of action within the context of the special challenges posed by HIV. Gene-based approaches present conundrums and trade offs analogous to those of conventional drugs. One consideration is the issue of viral *versus* cellular targets. RNA antivirals can be designed with high specificity and HIV-1 products are the preferred target; however, many viral RNAs are highly abundant and viral escape is a major problem that can only be partially ameliorated by targeting highly conserved sequences. Cellular targets are far less prone to mutational escape, and are often in lower abundance, but the side effects of downregulating cellular targets for the long term are unknown. For the sake of discussion, we also divide RNA inhibitors into two classes. The first comprises autonomous RNA inhibitors, such as ribozymes and aptamers, which interact with their respective targets in a highly selective fashion. However, co-localization of these RNA molecules with their targets is achieved largely *via* diffusion, making them behave unpredictably in the cellular milieu. The second class comprises RNA triggers, which include small interfering RNAs (siRNAs). This class of inhibitors utilizes endogenous cellular proteins to find the target sequences and, because of this, exhibit highly efficient target inhibition *via* messenger RNA (mRNA) degradation. Their efficiency, however, can exacerbate related-sequence off-target effects and saturation of the cellular pathway can adversely compromise cellular metabolism and development. Finally, we evaluate RNA gene therapies in the light of the emerging consensus that combinatorial gene therapeutics have the greatest likelihood of success, analogous to HAART therapy.

14.2 Potential HIV Targets for RNA-based Therapeutics

Viral proteins and cellular partners involved in the early events in the emergence of HIV from latency to active replication have been a focus of drug development, since the problems associated with both emergence of viral resistance and perturbation of cellular metabolism, as well as viral knockdown, vastly increase after the onset of active HIV replication. The *trans*-activating response (TAR)–Tat interaction has received special attention because of its central role in the HIV transcriptional *trans*-activation, but cellular co-factors such as nuclear factor kappaB (NF-κB) are also targets of drug development.[3] Not surprisingly, drugs against cellular co-factors have high toxicity, but are still being pursued as options in the event of failure of traditional chemotherapy. Other recently identified co-factors in the activation of HIV transcription include the Werner's syndrome helicase[4] and p90 ribosomal S6 kinase 2 (RSK2).[5]

There has also been a great deal of interest regarding the TAR-binding protein (TRBP) and its effects on HIV replication. Initially identified as an activator of HIV transcription, its subsequent recognition as a partner of Dicer led to the suggestion that TAR subverts the RNA interference (RNAi) pathway during HIV infection by sequestering TRBP.[6,7] However, HIV infection does not cause general downregulation of endogenous microRNAs; in fact, some are upregulated.[8] The effects of siRNA knockdown of TRBP indicates that the activation activities of TRBP predominate in HIV infection at several points during replication, consistent with observations that low endogenous expression of TRBP is directly linked to low HIV replication, as in astrocytes.[9–11]

Additional cellular factors involved in HIV replication continue to be identified. These include 11 cellular microRNAs that are upregulated and the miR-17/92 cluster that is downregulated as a result of HIV infection.[8] p300/CBP-associated factor (PCAF) is a likely target of two microRNAs (miRNAs) derived from the polycistronic pri-miR-17/92, miR-17-5p and miR-17-20a. PCAF is a co-factor for recruiting Tat to TAR and relieving the transcriptional elongation block in transactivation of integrated HIV-1. HIV-induced downregulation of miR-17-5p and miR-17-20a is therefore expected to lead to an increase in PCAF levels in HIV-infected cells and higher rates of HIV replication. Consistent with this model, locked nucleic acids (LNAs) against miR-17-5p and miR-17-20a as well as nuclear-transfected siRNAs targeting the primary and pre-microRNAs for this cluster enhanced HIV replication in Jurkat cells.[8]

14.2.1 Ribozymes, External Guide Sequences and Aptamers as Therapeutics

Ribozymes are antisense RNAs that are also capable of enzymatically cleaving targeted mRNAs. While the majority of attention in RNA-based therapeutics has turned toward triggers of the RNAi mechanism, ribozymes targeting both viral and cellular mRNAs continue to be tested. A pair of ribozymes derived from the minus and plus strands of tobacco ringspot virus [(−)sTRSV and (+)sTRSV)], respectively, were designed to target separate sequences in the HIV-LTR. These combined ribozymes inhibited HIV replication when tethered downstream of a sequence which was antisense to TAR.[12] A different approach combined the HIV packaging signal with a ribozyme or a short HIV antisense strand in a Tat-inducible lentiviral vector. Both the ribozyme and antisense constructs resulted in the production of defective HIV virions when cells expressing these constructs were challenged with HIV.[13] A novel chimeric anti-Tat ribozyme–TAR decoy chimera in which the TAR decoy sequence was an extension of the ribozyme catalytic core gave greater inhibition than either separate component.[14] A combination of a ribozyme and a DNAzyme, each targeting different sequences in HIV, showed good antiviral efficacy when transfected into macrophages.[15] Dual expression of ribozymes targeting CXC chemokine receptor 4 (CXCR4) and CC chemokine receptor 5 (CCR5) HIV co-receptors reduced HIV replication of both T- and M-tropic strains in PBMCs.[16] Two separate clinical trials have employed ribozymes targeting the *tat* and *rev* mRNAs. Both of these trials utilized retroviral vectors to introduce the ribozyme genes into hematopoietic stem cells. Although no efficacies have been reported from these trials, they both clearly show that it is safe to mobilize stem cells in HIV patients, to genetically modify these cells with the retroviral ribozyme vectors and to re-infuse the cells into the patients.[17,18] Ribozymes should continue to be valuable in multiplexing strategies against HIV where multiple mechanisms of knockdown may be advantageous.

Transfer RNAs (tRNAs) represent 2% of the total cellular RNA, and undergo several processing steps during maturation including removal of the 5′-leader sequence by the RNAse P complex and trimming of the 3′-trailer sequence by tRNAse Z^L in human cells.[19] Although there are some sequence constraints, specificity is largely conferred by structure, so external guide sequences (EGSs) that form sufficiently tRNA-like structures upon annealing with their target sequence can induce target cleavage. Expression can be regulated and has been used to inhibit cytomegalovirus (CMV) and HIV replication.[20–23]

Specificity is a concern because of the short length of the targeting sequences; 12–14 nucleotides for EGS elements. Mutations within the target sequence are likely to abrogate the efficacy, nevertheless the dependence on the cellular RNA processing machinery for cleaving the EGS-bound target adds to the robustness. The EGS approach may be a viable option for inhibiting HIV, particularly in combination with other methods, when the only available highly conserved target sequences are 10–14 nucleotides long.

14.2.2 RNAi Approaches to Inhibit HIV Replication

RNAi is a regulatory mechanism of most eukaryotic cells that uses small double-stranded RNA (dsRNA) molecules as triggers to direct homology-dependent control of gene activity.[24] Known as small interfering RNAs (siRNA) these dsRNA molecules are ~21–22 base pairs long and have characteristic two-nucleotide 3′-overhangs that allow them to be recognized by the enzymatic machinery of RNAi that eventually leads to homology-dependent degradation of the target mRNA. In mammalian cells siRNAs are produced from cleavage of longer dsRNA precursors by the RNase III endonuclease Dicer.[25] Dicer is complexed with two RNA binding proteins, the TAR-RNA binding protein (TRBP) and PACT, which are involved in the hand-over of siRNAs to the RNA-induced silencing complex (RISC).[26] The core components of RISC are the Argonaute (Ago) family members. In humans there are eight members of this family but only Ago-2 possesses an active catalytic domain for cleavage activity.[27,28] While siRNAs loaded into RISC are double-stranded, Ago-2 cleaves and releases the 'passenger' strand leading to an activated form of RISC with a single-stranded 'guide' RNA molecule that directs the specificity of the target recognition by intermolecular base pairing.[29] Rules that govern selectivity of strand loading into RISC are based upon the differential thermodynamic stabilities of the ends of the siRNAs.[30,31] The less thermodynamically stable end is favored for binding to the PIWI domain of Ago-2.

Control of disease-associated genes makes RNAi an attractive choice for future therapeutics. Basically every human disease caused by activity from one or a few genes should be amenable for RNAi-based intervention. This list includes cancer, autoimmune diseases, dominant genetic disorders and viral infections. RNAi can be triggered by two different pathways:

1. An RNA-based approach in which synthetic effector siRNAs are delivered by various carriers to target cells as preformed 21 base duplexes.
2. *Via* DNA-based strategies in which the siRNA effectors are produced by intracellular processing of longer RNA hairpin transcripts (see Chapter 12).[24,32]

The latter approach is primarily based on nuclear synthesis of short hairpin RNAs (shRNAs) that are transported to the cytoplasm *via* the miRNA export pathway and are processed into siRNAs by Dicer. While direct use of synthetic siRNA effectors is simple and usually results in potent gene silencing, the effect is transient. In a clinical setting this means repeated treatments would usually have to be administered, and these are large and costly drugs. In the case of HIV infection, this would be a lifelong treatment. DNA-based RNAi drugs, however, have the potential of being stably introduced when used in a gene therapy setting, allowing, in principle, a single treatment of shRNA genes delivered by viral vectors.

HIV was the first infectious agent targeted by RNAi, perhaps because the life cycle and pattern of gene expression of HIV is well understood. Synthetic siRNAs and expressed shRNAs have been used to target virtually all of the HIV-encoded RNAs in cell lines, including *tat, rev, gag, pol, nef, vif, env, vpr* and the LTR.[33–37] Subsequent work showed a host of other viruses, including hepatitis B virus (HBV), hepatitis C virus (HCV), poliovirus, respiratory syncytial virus (RSV) and others were targetable by RNAi (recently reviewed in Leonard and Schaffer[38]).

Despite the early successes of RNAi-mediated inhibition of HIV-encoded RNAs in cell lines, targeting the virus directly represents a substantial challenge for clinical applications because the high viral mutation rate will lead to mutants that can escape being targeted.[39–42] An alternative approach to avoid this problem is to target cellular transcripts that encode functions required for

IIIV-1 entry and replication. To this end cellular co-factors such as NF-κB, the HIV receptor CD4, and the co-receptors CCR5 and CXCR4 have all been downregulated with the result of blocking viral replication or entry.[35,36,43–45] The macrophage-tropic CCR5 co-receptor holds particular promise as a target. This receptor is not essential for normal immune function, and individuals homozygous for a 32 base pair deletion in this gene are resistant to HIV infection whereas individuals who are heterozygous for this deletion have delayed progression to AIDS.[46–48] CXCR4 is essential for hematopoietic stem cell homing to marrow and subsequent T-cell differentiation.[49–51] Targeting this receptor is therefore not a good choice for anti-HIV therapy, nor is targeting the essential CD4 receptor. A possible exception is in dendritic cells where the DC-SIGN receptor can be targeted by siRNAs to prevent infection.[52] Targeting only the CCR5 co-receptor may also present problems since HIV-1 switches to CXCR4 tropism during the course of AIDS, sometimes creating a more virulent infection.[53] Thus, there are drawbacks in solely targeting cellular HIV co-factors and viral targets will need to be included in any successful strategy using RNAi. It may someday be possible to use RNAi to prevent viral transmission by employing siRNAs as a microbicides.[54]

Viral targets should be sequences that are highly conserved throughout the various clades to ensure efficacy against all viral strains and to minimize the emergence of viral mutants resistant to RNAi. Multiplexing siRNAs and/or shRNAs by simultaneously targeting several sites in the virus is an option that should be fully explored and carefully examined for efficacy, inhibition of viral mutants and potential toxicity. Since the shRNA pathway impinges on the endogenous miRNA pathway, there is ample opportunity for off-target effects and competition with miRNAs for loading into RISC.[55,56] An additional potential concern is the putative inhibition of RNAi *via* HIV Tat and TAR. HIV-1 Tat has been demonstrated to bind and inhibit Dicer.[57] TAR also binds TRBP which is a Dicer co-factor and is a component of RISC.[58] Moreover, unlike other components of RISC, TRBP is made in limited amounts in the cell and hence binding to the TAR RNA could sequester TRBP from interacting with RISC and perhaps limit the effectiveness of an RNAi-based therapy. Binding of TRBP by TAR may also be a factor in the observed changes in miRNA profiles in HIV-infected cells.[59]

An alternative approach to relying solely upon RNAi as an anti-HIV approach is mixing a single shRNA with other antiviral genes to provide a potent combinatorial approach. This has been successfully accomplished by co-expressing an anti *tat/rev* shRNA, a nucleolar localizing TAR decoy and an anti-CCR5 ribozyme in a single vector backbone (Figure 14.1).[60] A somewhat different combination used an shRNA with a dominant negative Rev M10 protein in a co-expression system.[61] Perhaps other, more potent, combinations of shRNAs with mixtures of non-shRNA antivirals will be developed in the near future for testing in pre-clinical settings.

A few early reports showed that both siRNAs and shRNAs induced type I interferons and interferon-regulated gene expression, suggesting that small RNAs could activate proteins such as protein kinase R (PKR) and

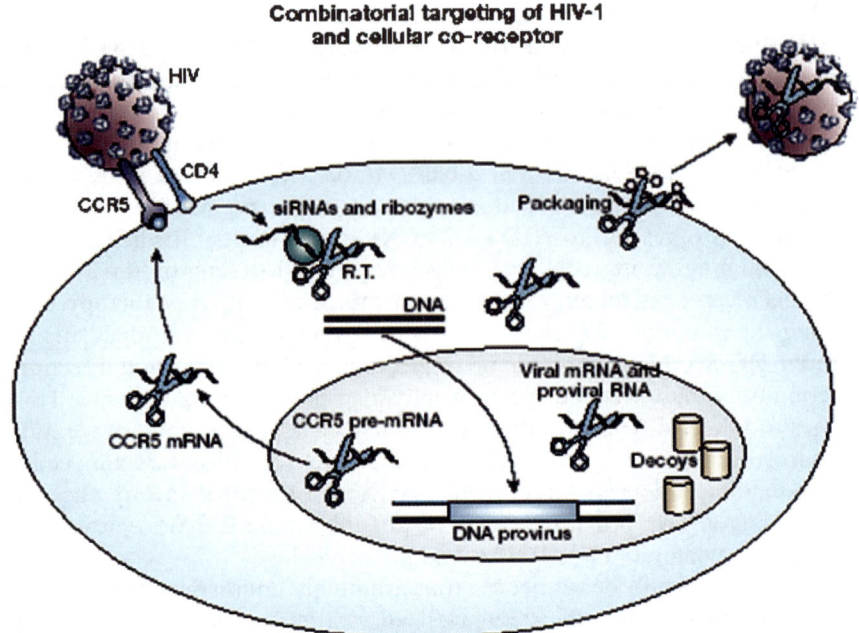

Figure 14.1   Combinatorial targeting of CCR5 and HIV targets with small RNAs. Depicted is HIV binding to the CD4 and CCR5 receptors. The scissors indicate messages (both cellular and HIV encoded) which can be cleaved (downregulated) by either a ribozyme or an siRNA. The cylinders in the nucleus represent decoys which can bind HIV Tat or Rev proteins.

2′-5′-OAS.[62,63] Other potential toxicity issues reside around the ability of some siRNAs to activate the Toll-like receptors in immune cells. This is a sequence-specific effect and clearly a problem when siRNAs are delivered by lipid vehicles, but has not yet been shown to be a problem with expressed shRNAs.[64–66] Thus, in applying RNAi-based approaches for the treatment of HIV infection, approaches which are effective against HIV but mitigate or avoid toxicities need to be developed.

Using multiple shRNAs to target separate conserved sites in HIV, akin to the HAART approach, is more promising, as it prevents cross-resistance between different RNAi effectors or between RNAi effectors and conventional pharmaceuticals. Multiple RNAi effectors would thus have the advantage of limiting escape and targeting a range of sequences, as is found in different viral genotypes or quasispecies (see Chapter 13).[67,68] Viruses which escape the antiviral effects of RNAi can be re-inhibited by targeting different sequences and thus a multiple inhibitory approach should aim to target distinct genomic regions of HIV-1 or, alternatively, target host-derived factors which contribute to viral replication.

A variety of approaches are being explored to express multiple siRNAs.[69] The target sequence of a typical siRNA is 21 nucleotides. Long hairpin RNAs (lhRNAs) greater than 50 base pairs in length can be expressed in cells and create multiple siRNAs *via* Dicer-mediated processing.[70–73] Expressed lhRNAs have been shown to be effective in cell culture against targets for HCV and HIV and *in vivo* for targets against HBV.[71,73–76] However, processing and knockdown efficacy of these substrates is asymmetric, being greatest at the base of the hairpin and tapering off across the full length of the duplex.[71,73] 50 base pair U6 lhRNAs against a conserved HIV-1 *int* or *tat/rev* region suppressed HIV replication in a variant resistant to a shorter shRNA in the same target.[73,74]

Another approach is the use of polycistronic shRNAs, in which several hairpins are expressed as a single transcript. Unlike the lhRNAs, which typically target adjacent sequences, different short hairpins (usually 19 to 29 nucleotides in length) in the primary transcript can be directed against widely separated targets. The individual units can be either simple hairpins or modeled on microRNAs, and potentially processed individually for better control of their individual activity.

Finally, multiple shRNAs may be expressed from individual promoters, typically the U6 and H1 polymerase III promoters. Recent studies show that it is possible to saturate the RNAi pathway by over-expressing shRNAs, resulting in cellular toxicity, particularly with the U6 promoter.[55,56] We recently described the use of tRNA^{Lys3}–shRNA chimeric cassettes that mediate graded shRNA knockdown which may be valuable in multiplexing strategies, particularly if the principles can be extended to other tRNA isotypes.[77] Use of lower multiplicities of infection when introducing multiplexed shRNAs in lentiviral backbones can alleviate toxicity; in one study three separate shRNAs targeted to regions within *pol* and *gag* were capable of inhibiting HIV-1 with the addition of each shRNA-expressing cassette providing an additive inhibitory effect.[78]

14.3 RNAi-induced Transcriptional Gene Silencing

Transcriptional gene silencing (TGS) is a phenomenon first described in plant and fungal cells where siRNAs and the RNAi machinery mediate repression of gene expression through chromatin changes. Exogenous siRNAs targeting specific promoters in mammalian cells have been shown to affect expression of the corresponding gene.[71,79–86] Although the mechanism is not fully understood, the possibility of using TGS to intervene earlier in the viral replication cycle by repressing transcription is very attractive, especially since it avoids the opportunities for viral escape that active replication provides. Also, strategies targeting HIV cellular co-factors that by themselves mediate only moderate suppression of HIV, but avoid toxicity, may be more effective in combination with TGS.

14.4   Delivery of Anti-HIV RNAs to Hematopoietic Cells

Delivery of siRNAs or shRNAs to HIV-1 infected cells is also a challenging problem. The target cells are primarily T-lymphocytes, monocytes, dendritic cells and macrophages. Since synthetic siRNAs will not persist for long periods in cells, they would have to be delivered repetitively for years to treat the infection effectively. Systemic delivery of siRNAs to T-lymphocytes is probably not feasible. Therefore a potential method is to isolate T-cells from patients, followed by transduction, expansion of the transduced cells and re-infusion. In an ongoing clinical trial, T-lymphocytes from HIV-infected individuals are transduced *ex vivo* with a lentiviral vector encoding an anti-HIV antisense RNA.[87,88] The transduced cells are subsequently expanded and re-infused into patients. This type of therapeutic approach could also be applicable to vectors harboring genes that encode siRNAs. A different approach is to transduce isolated hematopoietic progenitor, or stem cells with vectors harboring the therapeutic genes. This approach has the advantage that following maturation and differentiation of these cells, all the hematopoietic lineages capable of being infected by the virus are transduced and protected. Hematopoietic stem cells are mobilized from the patients and transduced *ex vivo* prior to re-infusion. Two clinical trials in which retroviral vectors expressing ribozymes were transduced into mobilized, autologous hematopoietic stem cells have demonstrated the feasibility of this approach.[17,87]

14.5   Future Prospects

The successful use of RNA mediators of anti-HIV activity in human hematopoietic cells has now been validated by many different investigators. With the development of genetically modified viral vectors that are capable of transducing hematopoietic cells with therapeutic RNA encoding constructs, there will be more proof-of-concept studies in animal models within the next couple of years. Ongoing clinical trials using anti-HIV ribozymes and antisense RNAs have demonstrated the safety of hematopoietic-based gene therapies. The next step is to prove the efficacy of these RNA-based inhibitors. Once efficacious use of antiviral RNAs in a gene therapy setting has been demonstrated, this will open the door for expanded clinical applications. We have previously described a triple combination lentiviral construct comprised of a U6-driven TAR RNA decoy appended upon a U16 small nucleolar RNA (snoRNA) for nucleolar localization, a U6-promoted shRNA targeted to both *tat* and *rev* open reading frames, and a VA1-promoted, chimeric anti-CCR5 *trans*-cleaving hammerhead ribozyme. The triple construct efficiently transduced human progenitor CD34+ cells and demonstrated improved suppression of HIV-1 over 42 days when compared to a single anti-*tat/rev* shRNA or double combinations of shRNA–ribozyme or decoy.[88] This triple vector is currently being used in human clinical trials involving lentiviral vector delivery to hematopoietic progenitor cells (trial being conducted at the City of Hope National Medical Center, Duarte, CA, USA).

References

1. S. G. Deeks, *BMJ*, 2006, **332**, 1489.
2. A. N. Phillips, B. G. Gazzard, N. Clumeck, M. H. Losso and J. D. Lundgren, *BMJ*, 2007, **334**, 76.
3. M. Stevens, E. De Clercq and J. Balzarini, *Med. Res. Rev.*, 2006, **26**, 595.
4. A. Sharma, S. Awasthi, C. K. Harrod, E. F. Matlock, S. Khan, L. Xu, S. Chan, H. Yang, C. K. Thammavaram, R. A. Rasor, D. K. Burns, D. J. Skiest, C. Van Lint, A. M. Girard, M. McGee, R. J. Monnat Jr and R. Harrod, *J. Biol. Chem.*, 2007, **282**, 12048.
5. C. Hetzer, D. Bisgrove, M. S. Cohen, A. Pedal, K. Kaehlcke, A. Speyerer, K. Bartscherer, J. Taunton and M. Ott, *PLoS ONE*, 2007, **2**, e151.
6. A. D. Haase, L. Jaskiewicz, H. Zhang, S. Laine, R. Sack, A. Gatignol and W. Filipowicz, *EMBO Rep.*, 2005, **6**, 961.
7. Y. Bennasser, M. L. Yeung and K. T. Jeang, *J. Biol. Chem.*, 2006, **281**, 27674.
8. R. Triboulet, B. Mari, Y. L. Lin, C. Chable-Bessia, Y. Bennasser, K. Lebrigand, B. Cardinaud, T. Maurin, P. Barbry, V. Baillat, J. Reynes, P. Corbeau, K. T. Jeang and M. Benkirane, *Science*, 2007, **315**, 1579.
9. H. S. Christensen, A. Daher, K. J. Soye, L. B. Frankel, M. R. Alexander, S. Laine, S. Bannwarth, C. L. Ong, S. W. Chung, S. M. Campbell, D. F. Purcell and A. Gatignol, *J. Virol.*, 2007, **81**, 5121.
10. S. Bannwarth, S. Laine, A. Daher, N. Grandvaux, G. Clerzius, A. C. Leblanc, J. Hiscott and A. Gatignol, *J. Mol. Biol.*, 2006, **355**, 898.
11. C. L. Ong, J. C. Thorpe, P. R. Gorry, S. Bannwarth, A. Jaworowski, J. L. Howard, S. Chung, S. Campbell, H. S. Christensen, G. Clerzius, A. J. Mouland, A. Gatignol and D. F. Purcell, *J. Virol.*, 2005, **79**, 12763.
12. M. S. Weinberg and J. J. Rossi, *FEBS Lett.*, 2005, **579**, 1619.
13. S. Gu, J. Ji, J. D. Kim, J. K. Yee and J. J. Rossi, *Oligonucleotides*, 2006, **16**, 287.
14. A. Barroso-DelJesus, E. Puerta-Fernandez, N. Tapia, C. Romero-Lopez, F. J. Sanchez-Luque, M. A. Martinez and A. Berzal-Herranz, *RNA Biol.*, 2005, **2**, 75.
15. V. Sood, H. Unwalla, N. Gupta, S. Chakraborti and A. C. Banerjea, *AIDS*, 2007, **21**, 31.
16. A. Qureshi, R. Zheng, T. Parlett, X. Shi, P. Balaraman, S. Cheloufi, B. Murphy, C. Guntermann and P. Eagles, *Biochem. J.*, 2006, **394**, 511.
17. A. Michienzi, D. Castanotto, N. Lee, S. Li, J. A. Zaia and J. J. Rossi, *Ann. N. Y. Acad. Sci.*, 2003, **1002**, 63.
18. F. K. Ngok, R. T. Mitsuyasu, J. L. Macpherson, M. P. Boyd, G. P. Symonds and R. G. Amado, *Methods Mol. Biol.*, 2004, **252**, 581.
19. H. Yan, N. Zareen and L. Levinger, *J. Biol. Chem.*, 2006, **281**, 3926.
20. Y. Habu, N. Miyano-Kurosaki, M. Kitano, Y. Endo, M. Yukita, S. Ohira, H. Takaku, M. Nashimoto and H. Takaku, *Nucleic Acids Res.*, 2005, **33**, 235.

21. M. Ikeda, Y. Habu, N. Miyano-Kurosaki and H. Takaku, *Nucleosides, Nucleotides, Nucleic Acids*, 2006, **25**, 427.
22. E. Kovrigina, L. Yang, E. Pfund and S. Altman, *RNA*, 2005, **11**, 1588.
23. H. Li, P. Trang, K. Kim, T. Zhou, S. Umamoto and F. Liu, *RNA*, 2006, **12**, 63.
24. G. J. Hannon and J. J. Rossi, *Nature*, 2004, **431**, 371.
25. J. Bai, S. Gorantla, N. Banda, L. Cagnon, J. Rossi and R. Akkina, *Mol. Ther.*, 2000, **1**, 244.
26. Y. Lee, J. Han, K. H. Yeom, H. Jin and V. N. Kim, *Cold Spring Harb. Symp. Quant. Biol.*, 2006, **71**, 51.
27. J. Liu, M. A. Carmell, F. V. Rivas, C. G. Marsden, J. M. Thomson, J. J. Song, S. M. Hammond, L. Joshua-Tor and G. J. Hannon, *Science*, 2004.
28. G. Meister, M. Landthaler, A. Patkaniowska, Y. Dorsett, G. Teng and T. Tuschl, *Mol. Cell*, 2004, **15**, 185.
29. G. Tang, *Trends Biochem. Sci.*, 2005, **30**, 106.
30. A. Khvorova, A. Reynolds and S. D. Jayasena, *Cell*, 2003, **115**, 209.
31. D. S. Schwarz, G. Hutvagner, T. Du, Z. Xu, N. Aronin and P. D. Zamore, *Cell*, 2003, **115**, 199.
32. L. J. Scherer and J. J. Rossi, *Nat. Biotechnol.*, 2003, **21**, 1457.
33. J. M. Jacque, K. Triques and M. Stevenson, *Nature*, 2002, **418**, 435.
34. N. S. Lee, T. Dohjima, G. Bauer, H. Li, M. J. Li, A. Ehsani, P. Salvaterra and J. Rossi, *Nat. Biotechnol.*, 2002, **20**, 500.
35. M. A. Martinez, B. Clotet and J. A. Este, *Trends Immunol.*, 2002, **23**, 559.
36. C. D. Novina, M. F. Murray, D. M. Dykxhoorn, P. J. Beresford, J. Riess, S. K. Lee, R. G. Collman, J. Lieberman, P. Shankar and P. A. Sharp, *Nat. Med.*, 2002, **8**, 681.
37. G. A. Coburn and B. R. Cullen, *J. Virol.*, 2002, **76**, 9225.
38. J. N. Leonard and D. V. Schaffer, *Gene Ther.*, 2005, **79**, 1654.
39. D. Boden, O. Pusch, F. Lee, L. Tucker and B. Ramratnam, *J. Virol.*, 2003, **77**, 11531.
40. E. M. Westerhout, M. Ooms, M. Vink, A. T. Das and B. Berkhout, *Nucleic Acids Res.*, 2005, **33**, 796.
41. A. T. Das, T. R. Brummelkamp, E. M. Westerhout, M. Vink, M. Madiredjo, R. Bernards and B. Berkhout, *J. Virol.*, 2004, **78**, 2601.
42. R. Sabariegos, M. Gimenez-Barcons, N. Tapia, B. Clotet and M. A. Martinez, *J. Virol.*, 2006, **80**, 571.
43. J. Anderson and R. Akkina, *Retrovirology*, 2005, **2**, 53.
44. P. Cordelier, B. Morse and D. S. Strayer, *Oligonucleotides*, 2003, **13**, 281.
45. R. M. Surabhi and R. B. Gaynor, *J. Virol.*, 2002, **76**, 12963.
46. J. Eugen-Olsen, A. K. Iversen, P. Garred, U. Koppelhus, C. Pedersen, T. L. Benfield, A. M. Sorensen, T. Katzenstein, E. Dickmeiss, J. Gerstoft, P. Skinhoj, A. Svejgaard, J. O. Nielsen and B. Hofmann, *AIDS*, 1997, **11**, 305.
47. P. Garred, J. Eugen-Olsen, A. K. Iversen, T. L. Benfield, A. Svejgaard and B. Hofmann, *Lancet*, 1997, **349**, 1884.

48. M. Samson, F. Libert, B. J. Doranz, J. Rucker, C. Liesnard, C. M. Farber, S. Saragosti, C. Lapoumeroulie, J. Cognaux, C. Forceille, G. Muyldermans, C. Verhofstede, G. Burtonboy, M. Georges, T. Imai, S. Rana, Y. Yi, R. J. Smyth, R. G. Collman, R. W. Doms, G. Vassart and M. Parmentier, *Nature*, 1996, **382**, 722.
49. T. Lapidot, *Ann. N. Y. Acad. Sci.*, 2001, **938**, 83.
50. T. Lapidot and O. Kollet, *Leukemia*, 2002, **16**, 1992.
51. J. Kahn, T. Byk, L. Jansson-Sjostrand, I. Petit, S. Shivtiel, A. Nagler, I. Hardan, V. Deutsch, Z. Gazit, D. Gazit, S. Karlsson and T. Lapidot, *Blood*, 2004, **103**, 2942.
52. M. P. Nair, J. L. Reynolds, S. D. Mahajan, S. A. Schwartz, R. Aalinkeel, B. Bindukumar and D. Sykes, *AAPS J.*, 2005, **7**, E572.
53. K. K. Arien, Y. Gali, A. El-Abdellati, L. Heyndrickx, W. Janssens and G. Vanham, *Virology*, 2005.
54. D. Palliser, D. Chowdhury, Q. Y. Wang, S. J. Lee, R. T. Bronson, D. M. Knipe and J. Lieberman, *Nature*, 2006, **439**, 89.
55. D. S. An, F. X. Qin, V. C. Auyeung, S. H. Mao, S. K. Kung, D. Baltimore and I. S. Chen, *Mol. Ther.*, 2006, **14**, 494.
56. D. Grimm, K. L. Streetz, C. L. Jopling, T. A. Storm, K. Pandey, C. R. Davis, P. Marion, F. Salazar and M. A. Kay, *Nature*, 2006, **441**, 537.
57. Y. Bennasser, S. Y. Le, M. Benkirane and K. T. Jeang, *Immunity*, 2005, **22**, 607.
58. A. Gatignol, S. Laine and G. Clerzius, *Retrovirology*, 2005, **2**, 65.
59. M. L. Yeung, Y. Bennasser, T. G. Myers, G. Jiang, M. Benkirane and K. T. Jeang, *Retrovirology*, 2005, **2**, 81.
60. M. J. Li, J. Kim, S. Li, J. Zaia, J. K. Yee, J. Anderson, R. Akkina and J. J. Rossi, *Mol. Ther.*, 2005, **12**, 900.
61. H. J. Unwalla, H. T. Li, I. Bahner, M. J. Li, D. Kohn and J. J. Rossi, *J. Virol.*, 2006, **80**, 1863.
62. A. J. Bridge, S. Pebernard, A. Ducraux, A. L. Nicoulaz and R. Iggo, *Nat. Genet.*, 2003, **34**, 263.
63. C. A. Sledz, M. Holko, M. J. de Veer, R. H. Silverman and B. R. Williams, *Nat. Cell Biol.*, 2003, **5**, 834.
64. V. Hornung, M. Guenthner-Biller, C. Bourquin, A. Ablasser, M. Schlee, S. Uematsu, A. Noronha, M. Manoharan, S. Akira, A. de Fougerolles, S. Endres and G. Hartmann, *Nat. Med.*, 2005, **11**, 263.
65. A. D. Judge, V. Sood, J. R. Shaw, D. Fang, K. McClintock and I. MacLachlan, *Nat. Biotechnol.*, 2005, **23**, 457.
66. M. A. Robbins, M. Li, I. Leung, H. Li, D. V. Boyer, Y. Song, M. A. Behlke and J. J. Rossi, *Nat. Biotechnol.*, 2006, **24**, 566.
67. L. J. Chang, X. Liu and J. He, *Gene Ther.*, 2005, **12**, 1133.
68. B. Berkhout and J. Haasnoot, *FEBS Lett.*, 2006.
69. D. H. Kim and J. J. Rossi, *Nat. Rev. Genet.*, 2007, **8**, 173.
70. H. Akashi, M. Miyagishi, T. Yokota, T. Watanabe, T. Hino, K. Nishina, M. Kohara and K. Taira, *Mol. Biosyst.*, 2005, **1**, 382.

71. M. S. Weinberg, A. Ely, S. Barichievy, C. Crowther, S. Mufamadi, S. Carmona and P. Arbuthnot, *Mol. Ther.*, 2007, **15**, 534.
72. A. Strat, L. Gao, T. Utsuki, B. Cheng, S. Nuthalapaty, J. M. Mathis, Y. Odaka and T. Giordano, *Nucleic Acids Res.*, 2006, **34**, 3803.
73. M. Sano, H. Li, M. Nakanishi and J. J. Rossi, *Mol. Ther.*, 2007.
74. H. Nishitsuji, M. Kohara, M. Kannagi and T. Masuda, *J. Virol.*, 2006, **80**, 7658.
75. P. Konstantinova, O. ter Brake, J. Haasnoot, P. de Haan and B. Berkhout, *Retrovirology*, 2007, **4**, 15.
76. P. Konstantinova, W. de Vries, J. Haasnoot, O. ter Brake, P. de Haan and B. Berkhout, *Gene Ther.*, 2006, **13**, 1403.
77. L. J. Scherer, R. Frank and J. J. Rossi, *Nucleic Acids Res.*, 2007, **35**, 2620.
78. O. ter Brake, P. Konstantinova, M. Ceylan and B. Berkhout, *Mol. Ther.*, 2006, **14**, 883.
79. D. H. Kim, L. M. Villeneuve, K. V. Morris and J. J. Rossi, *Nat. Struct. Mol. Biol.*, 2006, **13**, 793.
80. B. A. Janowski and D. R. Corey, *Nucleic Acids Symp. Ser.*, 2005, **49**, 367.
81. B. A. Janowski, J. Hu and D. R. Corey, *Nat. Protoc.*, 2006, **1**, 436.
82. B. A. Janowski, K. E. Huffman, J. C. Schwartz, R. Ram, D. Hardy, D. S. Shames, J. D. Minna and D. R. Corey, *Nat. Chem. Biol.*, 2005, **1**, 216.
83. B. A. Janowski, K. E. Huffman, J. C. Schwartz, R. Ram, R. Nordsell, D. S. Shames, J. D. Minna and D. R. Corey, *Nat. Struct. Mol. Biol.*, 2006, **13**, 787.
84. B. A. Janowski, K. Kaihatsu, K. E. Huffman, J. C. Schwartz, R. Ram, D. Hardy, C. R. Mendelson and D. R. Corey, *Nat. Chem. Biol.*, 2005, **1**, 210.
85. B. A. Janowski, S. T. Younger, D. B. Hardy, R. Ram, K. E. Huffman and D. R. Corey, *Nat. Chem. Biol.*, 2007, **3**, 166.
86. I. Martianov, A. Ramadass, A. Serra Barros, N. Chow and A. Akoulitchev, *Nature*, 2007, **445**, 666.
87. R. G. Amado, R. T. Mitsuyasu, J. D. Rosenblatt, F. K. Ngok, A. Bakker, S. Cole, N. Chorn, L. S. Lin, G. Bristol, M. P. Boyd, J. L. MacPherson, G. C. Fanning, A. V. Todd, J. A. Ely, J. A. Zack and G. P. Symonds, *Hum. Gene Ther.*, 2004, **15**, 251.
88. M. Li and J. J. Rossi, *Methods Mol. Biol.*, 2005, **309**, 261.

Subject Index